Advances in Materials Science for Environmental and Nuclear Technology

Advances in Materials Science for Environmental and Nuclear Technology

Ceramic Transactions, Volume 222

Edited by
Kevin Fox
Elizabeth Hoffman
Navin Manjooran
Gary Pickrell

WILEY
A John Wiley & Sons, Inc., Publication

Published by John Wiley & Sons, Inc., Hoboken, New Jersey.
Published simultaneously in Canada.

Library of Congress Cataloging-in-Publication Data is available.

ISBN 978-0-470-92729-8

10 9 8 7 6 5 4 3 2 1

Contents

NANOTECHNOLOGY FOR ENERGY

Preface

The Materials Science and Technology 2009 Conference and Exhibition (MS&T'09) was held October 25–29, 2009, in Pittsburgh, Pennsylvania. A major theme of the conference was Environmental and Energy Issues. Papers from three of the symposia held under that theme are included in this volume. These symposia include Materials Solutions for the Nuclear Renaissance; Green Engineering and Environmental Stewardship; and Nanotechnology for Energy. These symposia included a variety of presentations with sessions focused on sustainable energy, photovoltaics, nanowires and composites, energy harvesting, catalysts, thin films, corrosion, nuclear fuels, materials in aggressive environments, glass and ceramics for waste disposition, modeling and thermal properties, and education. Also included was a series of invited presentations and an international panel discussion on cement waste forms.

The Green Engineering and Environmental Stewardship symposium was sponsored by the Nuclear and Environmental Technology Division (NETD) of The American Ceramic Society while the Materials Solutions for the Nuclear Renaissance symposium was sponsored by NETD and ASM International.

The success of these symposia and the publication of the proceedings could not have been possible without the support of Greg Geiger of The American Ceramic Society and the other organizers of the program. The program organizers included Allen Apblett, Melody Carter, Alex Cozzi, James Earthman, Rod Ewing, Zachary Grasley, Carol Jantzen, Bill Lee, Josef Matyas, Arianna Morales, Tatsuki Ohji, Ramana Reddy, Mrityunjay Singh, and Srinath Viswanathan. Their assistance, along with that of the session chairs, was invaluable in ensuring the creation of this volume.

KEVIN FOX, Savannah River National Laboratory
ELIZABETH HOFFMAN, Savannah River National Laboratory
NAVIN MANJOORAN, Siemens AG
GARY PICKRELL, Virginia Polytechnic Institute and State University

Materials Solutions for the Nuclear Renaissance

IRRADIATION EFFECTS IN CERAMICS FOR PLUTONIUM DISPOSITION

C. Davoisne[1], W. E. Lee[1], M. C. Stennett[2], N. C. Hyatt[2], N. Peng[3] and C. Jeynes[3]
[1] Department of Materials, Imperial College London, London , United Kingdom.
[2]Immobilisation Science Laboratory, Department of Engineering Materials, University of Sheffield, Sheffield, United Kingdom.
[3]Ion Beam Centre, University of Surrey, Guildford, United Kingdom.

ABSTRACT

Understanding the effect of radiation damage and noble gas accommodation in potential ceramic hosts for plutonium disposition is important for determining long-term behaviour during geological disposal. Polycrystalline samples of zirconolite and Nd-doped zirconolite were irradiated ex-situ with 2 MeV Kr^+ ions to simulate plutonium nuclei recoil which occurs during alpha decay. The microstructural and chemical modifications induced by the irradiation were investigated by Transmission Electron Microscopy (TEM) on thin sections prepared by Focused Ion Beam (FIB). During this study, the feasibility of TEM section preparation from pristine and irradiated ceramics by FIB was demonstrated. After Kr^+ irradiation, the samples were observed to be amorphous at the surface and an interface between the pristine and irradiated parts of the specimen was identified. No significant chemical modification was observed.

INTRODUCTION

Civil nuclear and military nuclear programs generate high level radioactive waste (HLW). Glasses are widely used for the immobilisation of HLW but in the last few decades ceramics have been developed for plutonium disposition and for use as Inert Matrix Fuels (IMF). Candidate ceramics must fulfil some key requirements which include chemical flexibility, to facilitate the incorporation of plutonium and actinides, and good resistance to aqueous dissolution[1-3]. Based on these criteria, zirconolite is a strong candidate phase[4-7] for plutonium disposition.

An understanding of the effect of radiation damage and noble gas accommodation during storage is key to developing a robust safety case. During storage, alpha decay induces the formation of helium and the recoil of the heavy plutonium nuclei induces a high concentration of defects. These result in the formation of gas bubbles, defects and voids which can bring about swelling and cracking of ceramic wasteforms. Different approaches have been used to determine the impact of alpha decay; these include the study of natural analogues[5;8;9] and in-situ ion irradiation of thin sections in a TEM[7;10-12]. Ex-situ studies, such as reported here, allow simultaneous analysis of pristine and irradiated material by TEM.

In this study, we examined the effect of alpha recoil in zirconolite. The irradiation was performed ex-situ using krypton ions.The microstructural and chemical evolution was investigated by TEM and X-ray Energy Dispersive Spectroscopy (EDS) on thin sections prepared by FIB.

EXPERIMENTAL PROCEDURE

Irradiations

The samples studied were zirconolite, $CaZrTi_2O_7$, and zirconolite doped with neodymium as a surrogate for plutonium, $(Ca_{0.8}Nd_{0.2})Zr(Ti_{1.8}Al_{0.2})O_7$. The samples were prepared at the

Immobilisation Science Laboratory in Sheffield as described in Stennett et al.[13]. They were implanted with Krypton ions (2 MeV) at room temperature at the Surrey Ion Beam Center using a 2MV Van de Graaf Implanter. The dose rate was 0.25 $\mu A\ cm^{-2}$ and the dose reached was 5×10^{15} ions cm^{-2}. The ion range and the number of displacements per atom (dpa) were determined using the Monte Carlo code SRIM[14-16]. The displacement energies were fixed for all the elements at 50 eV[17-19]. The calculated results are reported in Table I.

Table I. Average ion range and number of dpa obtained by TRIM calculation

Material	Density (g/cm^3)	Average ion depth (nm)	Straggling (nm)	Number of dpa
Zirconolite	4.498	771.6	182	5.58
Nd-zirconolite	4.498	789.3	190	5.80

Sample characterisation

The microstructural and chemical characterisation was carried out on a JEOL 2000FX operating at 200keV and a FEI Titan 80/300 fitted with a monochromator. Both microscopes were fitted with EDS systems and the sample composition was determined using the method developed by Van Cappellen and Doukhan[20] based on the electroneutrality of ionic crystals.

The TEM sections from the two sample-types were prepared by FIB using a FEI FIB200-SIMS with a Gallium beam operating at 30 keV. We decide to use FIB to prepare the samples because the irradiated area was located less than 1μm from the surface and the TEM section could be selected from the region of interest (i.e. inside a grain or on a grain boundary). The samples were initially coated with a gold layer (~ 50 nm thick) to avoid charge effects and to protect the surface during imaging. The sample was then placed into the FIB chamber and a platinum layer of approximately 1μm, Figure 1.a), was deposited on the area of interest to protect the surface from damage during milling and imaging. The sample was milled until an electron transparent section was obtained, Figure 1.b). The lift-out of the section was performed ex-situ using a micromanipulator and a thin glass needle. The section was attracted and adhered to the tip by the electrostatic effect of the glass needle and transferred on a copper grid with carbon film or holey carbon for imaging in the TEM.

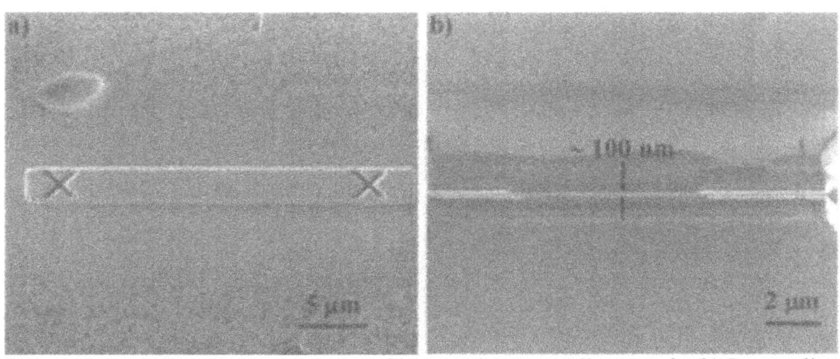

Figure 1. a) Platinum deposition on the area of interest in zirconolite sample; b) Perpendicular view of the TEM section after milling (the central part is thinner than the edges).

RESULTS AND DISCUSSION

Prepared sample

The zirconolite was characterised by TEM after FIB preparation in order to estimate the damage induced by the gallium beam. A typical section prepared across a grain boundary is shown in Figure 2. The sections contain three distinct areas:

- Region "a" consists of protective gold and platinum layers and has a depth of ≤ 1μm. In some sections, the protective layers had been completely removed during milling, exposing the surface of the sample to the gallium ion beam. In these areas a well defined amorphous zone approximately 40-50 nm in depth could be observed.
- Region "c" is the part of the sample damaged by the gallium beam during the edges cutting procedure. This ranges in depth between 650 and 850 nm. The EDS analysis, Figure 3.a), revealed the presence of gallium in high concentration.
- Region "b" is the part of the sample unaltered during the milling process. Although a trace of implanted gallium was observed in the EDS analysis, Figure 3.b), no damage induced by the gallium beam was observed in this part of the sample.

In some sections, black precipitates enriched with gallium are observed, probably arising from precipitation of gallium during milling. The experimental compositions are in good agreement with the theoretical one and are reported in Table II. The little differences observed are probably due to the sample processing.

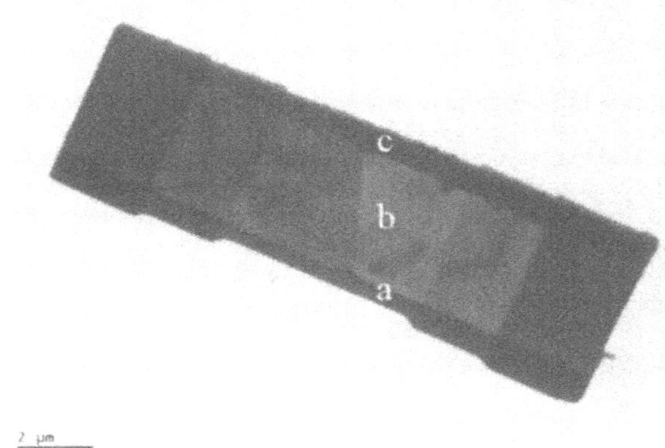

Figure 2. Global view of a TEM section prepared by FIB in zirconolite. It contains 3 distinct areas: a) the platinum and gold protective layers (~1 μm), b) ceramic unaltered by the gallium beam (3-5 μm) and c) the area damaged by the gallium beam during the edge cutting (650-850 nm).

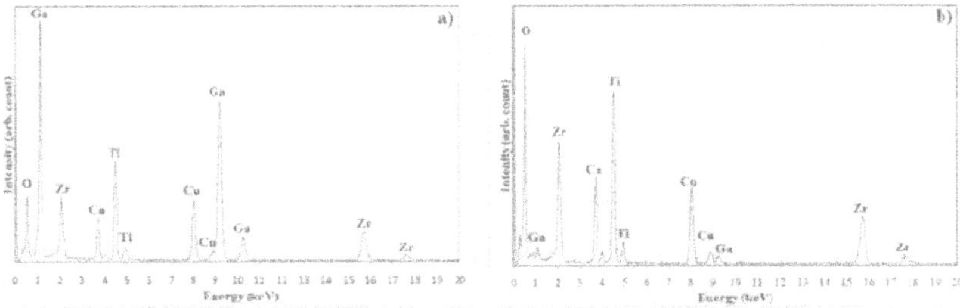

Figure 3. a) EDS spectrum acquired from region "c" in Figure 2; b) EDS spectrum acquired from region "b".

Table II: Experimental (exp) and theoretical (th) sample composition on pristine and irradiated samples

Sample		O (at%)	Ca (at%)	Ti (at%)	Zr (at%)	Al (at%)	Nd (at%)
Zirconolite	exp	63.97 ± 0.09	8.12 ± 0.29	16.37 ± 0.33	11.55 ± 0.28	/	/
	th	63.64	9.09	18.18	9.09	/	/
Nd-zirconolite	exp	63.61 ± 0.12	6.37 ± 0.27	14.38 ± 0.57	11.53 ± 0.57	1.81 ± 0.29	2.00 ± 0.19
	th	63.64	7.27	16.36	9.09	1.82	1.82
Nd-zirconolite irradiated	exp	63.96 ± 0.06	6.33 ± 0.11	14.49 ± 0.43	11.62 ± 0.52	1.87 ± 0.26	1.75 ± 0.27

Kr irradiated sample

Figure 4. shows a TEM section, from within a single grain, prepared in the Kr^+ irradiated Nd-doped zirconolite. Compared to unirradiated samples, a new well defined region is observed from the surface of the sample, Figure 4.d). TRIM calculations indicated that the damage repartition in the sample should extend to approximately 1 μm in depth which is consistent with the region "d". It extends to 1.1 μm into the surface of the sample and is amorphous according to diffraction patterns acquired from the area, Figure 5.a). For zirconolite, the amorphisation dose using krypton ions has been estimated at ~ 4×10^{14} ions.cm^{-2} [7;9;11]. The presence of an amorphous layer in our sample which was irradiated at a dose of 5×10^{15} ions.cm^{-2} is therefore consistent with the literature. EDS measurements performed in the irradiated area are reported in Table II. The average composition shows no significant deviation from the pristine sample.

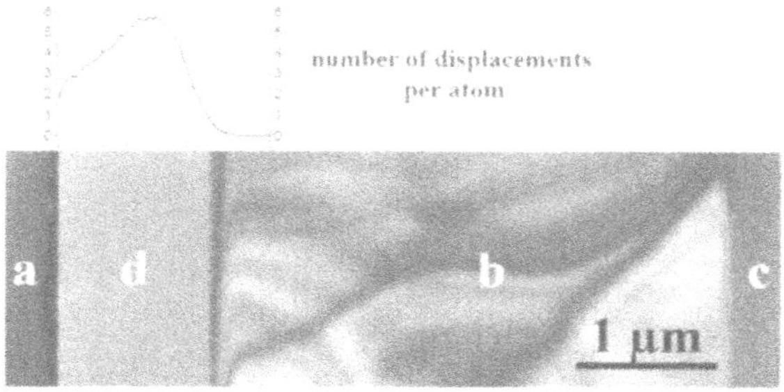

Figure 4. TEM section prepared by FIB in Kr irradiated Nd-zirconolite. 4 distinct areas are present: a) the platinum and gold layer (~1 μm), b) pristine undamaged sample (2-4μm), c) the area damaged by the gallium ions during the edge cutting (650-850 nm) and d) the area damaged by the Kr irradiation (~ 1.1 μm). The calculated damage distribution is shown above the image.

Figure 5. is an enlarge view at the interface between the irradiated and pristine regions of the sample. An interfacial region of 40-60 nm in thickness could be observed. Study of the region by microdiffraction, Figure 5.b), indicated an evolution in the crystal structure in the interface. The change in the relative intensities, the disappearance of the zirconolite superlattice maxima ($h = k = 4n$ and $l = 2m + 1$; $h = k = 4n + 2$ and $l = 2m$) and lines where $h = k = 2n + 1$ indicated loss of the zirconolite superlattice although the fluorite subcell was retained. Such modifications have been reported previously[5;7-12] and indicate progressive cation disordering induced by the irradiation.

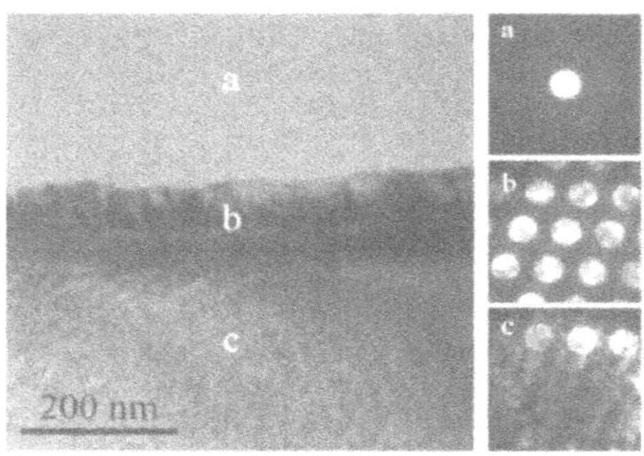

Figure 5. Enlarge view of the interface between the irradiated and pristine parts of the Nd doped zirconolite. Three areas are visible: a) irradiated region with amorphous diffraction pattern; b) the interfacial region which remains crystalline but has a different structure from the pristine sample; c) the pristine area which has the zirconolite structure, [1-10] zone axis shown.

Figure 6. shows HRTEM images and associated FFT from areas on both sides of the interface in a single grain at a zone axis in the undamaged part . Figure 6.a) contains information from both the undamaged and partly irradiated areas: the microstructural evolution begins with the appearance of disordered regions with the fluorite structure type within the zirconolite. Figure 6.b) contains information from both the partly and fully irradiated areas: the zirconolite disappears and is replaced by disordered crystalline and amorphous regions. Considering the HRTEM images were taken from a single grain, the information provided by the FTT at the bottom of Figure 6.b) is indicative of an amorphous area and not an off-zone crystal. This microstructure is similar to that observed in natural zirconolite samples studied by Ewing and Headley[8], which have received doses of ~ 3×10^{25} alpha m^{-3}, or during in-situ TEM studies of synthetic zirconolite[11;12]

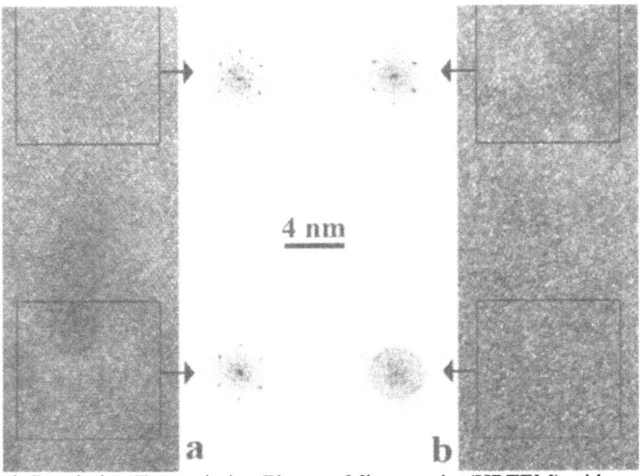

Figure 6. High Resolution Transmission Electron Micrographs (HRTEM) with associated Fast Fourier Transforms (FFT). a) shows images and FFT from pristine (top) and partly irradiated regions of the sample (bottom) and b) shows images and FFT from the partly irradiated (top) and fully irradiated amorphous regions (bottom).

CONCLUSION

We have demonstrated that the preparation of TEM sections in pristine and irradiated ceramics by FIB is feasible to study specific areas of bulk samples. While artefacts such as gallium precipitates and thin amorphous regions are produced by the ion beam milling process they can be easily identified.

The Kr^{+} dose used in this study was shown to induce amorphisation of Nd-doped zirconolite and an interfacial region between the amorphous and pristine parts of the sample was observed. No real difference in the chemistry between the irradiated and unirradiated areas was observed by EDS.

The effect of radiation damage and noble gas accommodation (krypton and helium irradiations) on other ceramic candidates, for plutonium disposition and for use as inert matrix fuels, is currently being studied.

REFERENCES

[1]W. J. Weber, R. C. Ewing, C. R. A. Catlow, T. D. de la Rubia, L. W. Hobbs, C. Kinoshita, H. Matzke, A. T. Motta, M. Nastasi, E. K. H. Salje, E. Vance and S. Zinkle, Radiation effects in crystalline ceramics for the immobilization of high-level nuclear waste and plutonium, J. Mater. Res., 13, 1434-84 (1998).
[2]W. E. Lee, M. I. Ojovan, M. C. Stennett and N. C. Hyatt, Immobilisation of radioactive waste in glasses, glass composite materials and ceramics, Adv. Appl. Ceram., 105, 3-12 (2006).

[3]R. C. Ewing, Ceramic matrices for plutonium disposition, Prog. Nucl. Energy, 49, 635-43 (2007).
[4]A. E. Ringwood, S. E. Kesson, N. G. Ware, W. Hibberson and A. Major, Immobilisation of high level nuclear reactor wastes in SYNROC, Nature, 278, 219-23 (1979).
[5]W. Sinclair and A. E. Ringwood, Alpha-recoil damage in natural zirconolite and perovskite, Geochem. J., 15, 229-43 (1981).
[6]V. M. Oversby, R. A. van Konynenburg, W. E. Glassley and P. G. Curtis, Immobilization in ceramic waste forms of the residues from treatment of mixed wastes, Mater. Res. Soc. Symp. Proc., 333, 285-92 (1994).
[7]K. L. Smith, N. J. Zaluzec and G. R. Lumpkin, In situ studies of ion irradiated zirconolite, pyrochlore and perovskite, J. Nucl. Mater., 250, 36-52 (1997).
[8]R. C. Ewing and T. J. Headley, Alpha-recoil damage in natural zirconolite (CaZrTi2O7), J. Nucl. Mater., 119, 102-9 (1983).
[9]R. C. Ewing and L. M. Wang, Amorphization of zirconolite: alpha-decay event damage versus krypton ion irradiation, Nucl. Instrum. Methods Phys. Res. Sect. B, 65, 319-23 (1992).
[10]T. J. White, Temperature dependence of amorphization for zirconolite and perovskite irradiated with 1 MeV krypton ions, Mater. Res. Soc. Symp. Proc., 353, 1413-20 (1995).
[11]S. X. Wang, L. M. Wang, R. C. Ewing, G. S. Was and G. R. Lumpkin, Ion irradiation-induced phase transformation of pyrochlore and zirconolite, Nucl. Instrum. Methods Phys. Res. Sect. B, 148, 704-9 (1999).
[12]S. X. Wang, G. R. Lumpkin, L. M. Wang and R. C. Ewing, Ion irradiation-induced amorphization of six zirconolite compositions, Nucl. Instrum. Methods Phys. Res. Sect. B, 166-167, 293-8 (2000).
[13]M. C. Stennett, N. C. Hyatt, D. Reid, E. Maddrell, N. Peng, C. Jeynes, K. Kirkby and J. Woicik, Characterisation of Ion Beam Irradiated Zirconolite for Pu Disposition, Mater. Res. Soc. Symp. Proc., 1124, Q10-09 (2009).
[14]J. F. Ziegler, J. P. Biersack and U. Littmark. The Stopping and Range of Ions in Matter. Pergamon Press, New york (1985).
[15]J. F. Ziegler, SRIM-2003, Nucl. Instrum. Methods Phys. Res. Sect. B, 219-220, 1027-36 (2004).
[16]J. F. Ziegler, J. P. Biersack and M. D. Ziegler. SRIM the Stopping and Range of Ions in Matter. SRIM Co. (2008).
[17]S. J. Zinkle and C. Kinoshita, Defect production in ceramics, J. Nucl. Mater., 251, 200-17 (1997).
[18]B. D. Begg, N. J. Hess, W. J. Weber, R. Devanathan, J. P. Icenhower, S. Thevuthasan and B. P. McGrail, Heavy-ion irradiation effects on structures and acid dissolution of pyrochlores, J. Nucl. Mater., 288, 208-16 (2001).
[19]K. L. Smith, M. Colella, R. Cooper and E. R. Vance, Measured displacement energies of oxygen ions in titanates and zirconates, J. Nucl. Mater., 321, 19-28 (2003).
[20]E. Van Cappellen and J. C. Doukhan, Quantitative transmission X-ray microanalysis of ionic compounds, Ultramicroscopy, 53, 343-9 (1994).

SYTHESIS AND STRUCTURES OF $Gd_2(Zr_{2-x}Ce_x)O_7$: A MODEL CERAMIC SYSTEM FOR PLUTONIUM DISPOSITION

D.P. Reid, M.C. Stennett, N.C. Hyatt.
Immobilisation Science Laboratory, Department of Engineering Materials, The University of Sheffield, Mappin Street, Sheffield, S1 3JD, UK.

ABSTRACT

We present an initial overview of the $Gd_2(Zr_{2-x}Ce_x)O_7$ phase diagram, of interest as a model system for ceramic disposition of Pu (with Ce as a Pu surrogate). X-ray diffraction (XRD) and transmission electron microscopy (TEM) were used to determine the cation and oxygen vacancy ordering mechanisms in this solid solution. The pyrochlore structure was shown to remain stable with Ce^{4+} substitutions of up to 12.5 mol%. The defect fluorite phase stability range is characterised by the absence of oxygen vacancy ordering, but retention of weak cation ordering. Ce^{4+} rich compositions exhibit ordering consistent with the C-type structure. X-ray absorption near edge structures (XANES) confirmed the presence of Ce^{4+} as the dominant species in compositions across this system, remaining analogous to Pu^{4+}.

INTRODUCTION

A wide range of ceramic matrices have been proposed as potential candidates for the disposition of actinides[1,2]. In particular, zirconate ceramics with the pyrochlore structure are promising matrices for the immobilisation of plutonium, due to their chemical durability and radiation tolerance (resistance to amorphisation)[3]. This study focussed on characterisation of the $Gd_2(Zr_{2-x}Ce_x)O_7$ solid solution, where cerium was used as an inactive surrogate for plutonium. Tetravalent cerium has a very similar ionic radius to tetravalent plutonium (for a given coordination number) making it a good surrogate from a crystallographic viewpoint[4].

EXPERIMENTAL METHODS

$Gd_2(Zr_{2-x}Ce_x)O_7$ samples (with $0 \le x \le 2$ in 0.25 intervals) were prepared by solid state reaction between Gd_2O_3, ZrO_2 and CeO_2 in stoichiometric ratios. Reagents were intimately mixed for 16 hours using a rotary ball mill with ZrO_2 media and isopropanol as a carrier fluid. The mixed powders were reacted in alumina crucibles at 1500 °C for 4 hours in air with a ramp rate of 5 °C min^{-1}. Powder cakes were reground using a planetary ball mill for 2 minutes, to break up the agglomerates formed during the reaction process, and sieved through a 250 µm mesh. 10 mm diameter compacts were formed using a cold uniaxial press at 200 MPa, for 2 mins and then sintered at 1650 °C for 8 hours in air.

X-ray diffraction patterns were acquired using a STOE STADI-P diffractometer (Darmstadt, Germany) operating in transmission mode at 40 kV and 30 mA, with Mo $K\alpha_1$ radiation (λ = 0.7093 Å) and a linear position sensitive detector (PSD). Secondary and back scattered electron images of ceramic microstructures were acquired using a JEOL JSM 6400 scanning electron microscope operating at a voltage of 10 kV and a working distance of 18 mm. Quantitative elemental analyses from individual grains of sintered pellets, was undertaken by energy dispersive X-ray spectroscopy (EDS) with acquisition times of 240 s.

Ce L_{III} edge X-ray absorption near edge structures (XANES) were acquired on beamline X23A2 of the National Synchrotron Light Source (NSLS), Brookhaven National Laboratory (BNL), USA. The beamline utilises an upwards reflecting, fixed exit, Si(311) double crystal monochromator of a Golovchenko-Cowan design. The energy of the Ce L_{III} edge, which was calibrated using the K-edge of a V foil with E_0 = 5465 eV. XANES data were acquired in transmission mode using a finely ground specimen dispersed in BN to achieve a thickness of one absorption length. Background removal and normalisation of the data was performed using the program Athena[5].

For TEM investigation, pellets were sectioned using a diamond saw, ground to ~ 100 μm using SiC paper and thinned to electron transparency using a Gatan dual ion beam mill (Pleasanton, USA) operating at 6 kV with a total specimen current of 0.6 mA. Electron diffraction patterns of principal zone axes were acquired using a Philips EM420T microscope (Eindhoven, Holland) operating at 120 kV.

X-RAY DIFFRACTION

X-ray diffraction patterns for compositions in the $Gd_2(Zr_{2-x}Ce_x)O_7$ system are shown in Figure 1a. The principal reflections were indexed based on the fluorite subcell, as denoted by the subscript F. Figure 2 shows the calculated fluorite subcell lattice parameter (a_F) as a function of composition. A linear increase in lattice parameter with increasing cerium content was observed, consistent with substitution of the larger cerium ion (Ce^{4+} ionic radius = 0.87 Å) for the smaller zirconium ion (Zr^{4+} ionic radius = 0.72 Å), in six fold coordination[6]. Close inspection of the X-ray diffraction pattern of the x = 0 composition revealed a weak reflection at 2θ = 6.68°, in Figure 1b, that was indexed as a ½$(111)_F$ superlattice reflection, equivalent to the $(111)_P$ reflection from the pyrochlore structure with $a_P = 2a_F$. This superlattice reflection was also present in the diffraction pattern of the x = 0.25 composition, but was observed to be of very weak intensity and diffuse in nature. However, the ½$(111)_F$ reflection was absent from the diffraction patterns of compositions with $x \geq 0.50$.

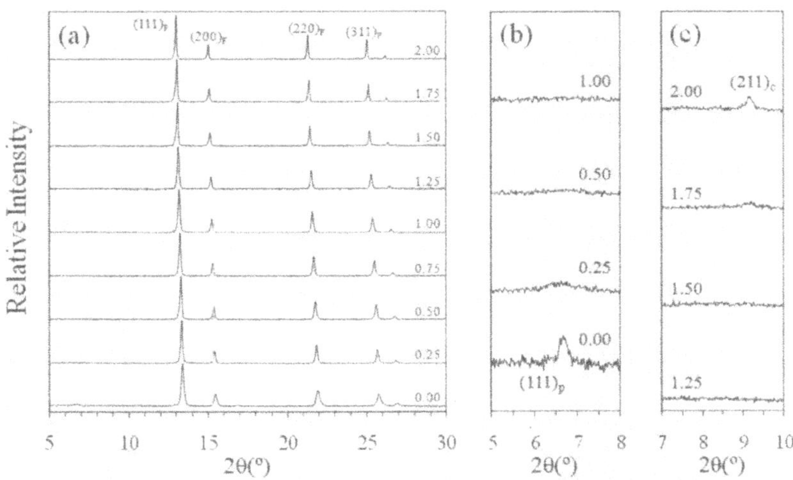

Figure 1: a) X-ray diffraction patterns of $Gd_2(Zr_{2-x}Ce_x)O_7$, b) (111) pyrochlore superlattice reflection and c) (211) C-type superlattice reflection.

At the other end of the system, a weak reflection was observed at $2\theta = 9.18°$ in the diffraction pattern of the $x = 2$ composition, Figure 1c. This reflection was indexed as the $(211)_C$ superlattice reflection of the C-type rare earth oxide structure. A similar very weak and diffuse reflection was observed in the diffraction pattern of the $x = 1.75$ composition, but was absent in the diffraction pattern of the $x = 1.50$ composition. From these XRD data, it could be concluded that the pyrochlore structure is formed for compositions in the range $0 \leq x \leq 0.25$ and the C-type structure in the range $1.75 \leq x \leq 2.00$. Intermediate compositions, between $0.25 < x < 1.75$, appear to adopt a defect fluorite structure as determined from the absence of additional superlattice reflections.

These findings contradict those of Patwe *et al.*, whose X-ray diffraction analysis of this system concluded that only 5 mol% Ce^{4+} could be contained in the pyrochlore lattice of $Gd_2Zr_2O_7$, determined by the absence of supercell reflections in the X-ray diffraction pattern of $Gd_2(Zr_{1.8}Ce_{0.2})O_7$[7]. Since anion vacancy ordering is understood to be sensitive to thermal history, the difference between these studies may be attributed to differences in the exact processing conditions applied.

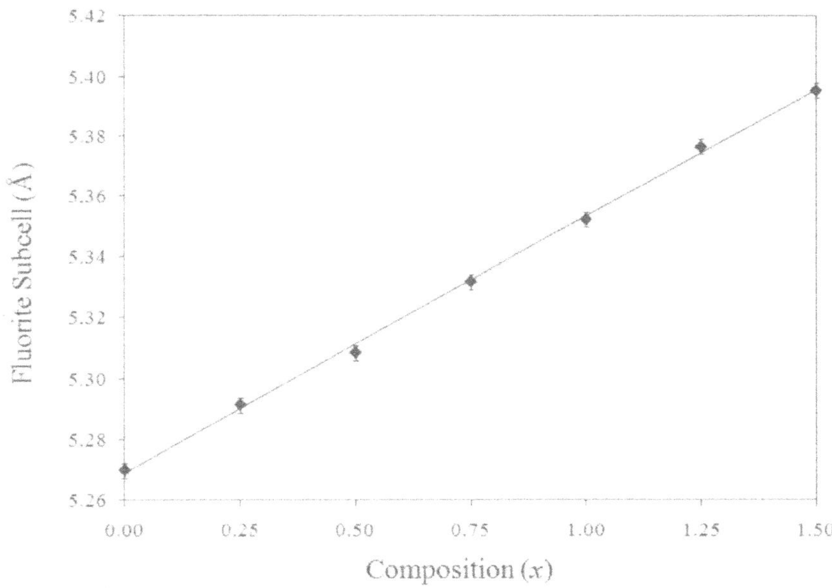

Figure 2: Dependence of cubic fluorite subcell parameter (a_F) on composition (x). Error bars represent 3σ.

SCANNING ELECTRON MICROSCOPY

Electron microscopy was used to confirm the single phase nature of the sintered specimens. Figure 3 shows electron micrographs obtained for the $Gd_2(Zr_{1.50}Ce_{0.50})O_7$ composition, which are representative of all compositions in the system. The secondary electron image in Figure 3(a) shows a uniform interlocking grain structure, with grain sizes ranging from 5 to 12 μm. Figure 3(b) shows the corresponding backscattered image, with uniformity in contrast indicating the presence of a single phase, without significant variation in composition. The porosity of the specimen, apparent as black circular features in the SEM images, is consistent with the experimentally determined density of 93 ± 1 % of theoretical.

Table I shows, as a representative example, the quantitative energy dispersive X-ray data from three grains of the $Gd_2(Zr_{1.50}Ce_{0.50})O_7$ composition. The mean composition of the three grains was $Gd_{1.96}Zr_{1.49}Ce_{0.56}O_7$, which was consistent with the target stoichiometry.

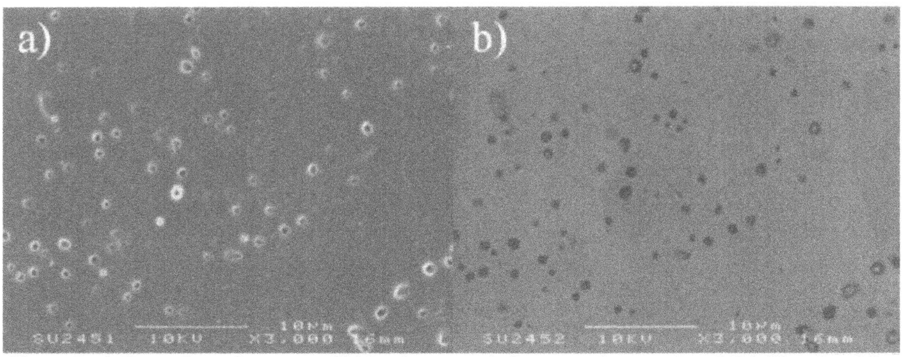

Figure 3: a) Secondary electron and b) backscattered electron micrographs of $Gd_2(Zr_{1.5}Ce_{0.5})O_7$.

Table I: Results from energy dispersive X-ray analysis of three grains of nominal $Gd_2(Zr_{1.5}Ce_{0.5})O_7$, normalised to full occupancy of cation sites.

Element	Grain 1		Grain 2		Grain 3		Mean	Std. Dev.
	at %	Norm. to 4 cations	at %	Norm. to 4 cations	at %	Norm. to 4 cations		
Gd	2.010	1.939	2.074	1.997	2.002	1.929	1.955	0.037
Zr	1.560	1.505	1.534	1.477	1.544	1.487	1.490	0.014
Ce	0.577	0.557	0.547	0.527	0.606	0.584	0.556	0.029

X-RAY ABSORPTION SPECTROSCOPY

Figure 4 shows Ce L_{III} edge XANES data for the $Gd_2(Zr_{1.75}Ce_{0.25})O_7$ composition, together with the data acquired for several cerium bearing standard compounds. Ce L_{III} edge XANES spectra of Ce^{3+} (e.g. $CePO_4$) and Ce^{4+} (e.g. CeO_2) species are characterised by a white line comprising one and two intense features, respectively. Several other more subtle features are also observed. The single white line feature in $CePO_4$ is attributed to the transition from an initial $2p^6 4f^1 5d^0$ state to a $2p^5 4f^1 5d^1$ final state, modified by local density of unoccupied states.

Following the notation convention employed by Bianconi, the consensus is that the two features, A and B, comprising the white line of CeO_2 are due to transition to: a) $2p^54f^05d^1$ and $2p^54f^25d^1\underline{L}^2$ final states (where $\underline{L}^n$ denotes a ligand hole); and b) the $2p^54f^15d^1\underline{L}$ final state[8]. Feature C arises from crystal field splitting of Ce 5d orbitals, whereas the weak feature D is assigned to final states with delocalised d-character[8, 9].

The XANES spectra shown in Figure 4 clearly allow easy fingerprinting of Ce oxidation state. From comparison of the spectra it is apparent that the XANES data of $Gd_2(Zr_{1.75}Ce_{0.25})O_7$ closely resemble that of CeO_2, demonstrating that Ce^{4+} is the dominant species. Similar spectra were obtained for all members of the $Gd_2(Zr_{2-x}Ce_x)O_7$ solid solution. Detailed comparison of data from $Gd_2(Zr_{1.75}Ce_{0.25})O_7$ and standards, shows that feature C is prominent and well resolved in XANES of $SrCeO_3$ and $CeTi_2O_6$ which contain Ce^{4+} in six-fold co-ordination. In contrast, this feature is less well resolved in the XANES of CeO_2 and $Gd_2(Zr_{1.75}Ce_{0.25})O_7$. This may indicate that Ce^{4+} in fact occupies the 8-fold co-ordinate Gd site, displacing Gd to occupy the 6-fold co-ordinate Zr-site.

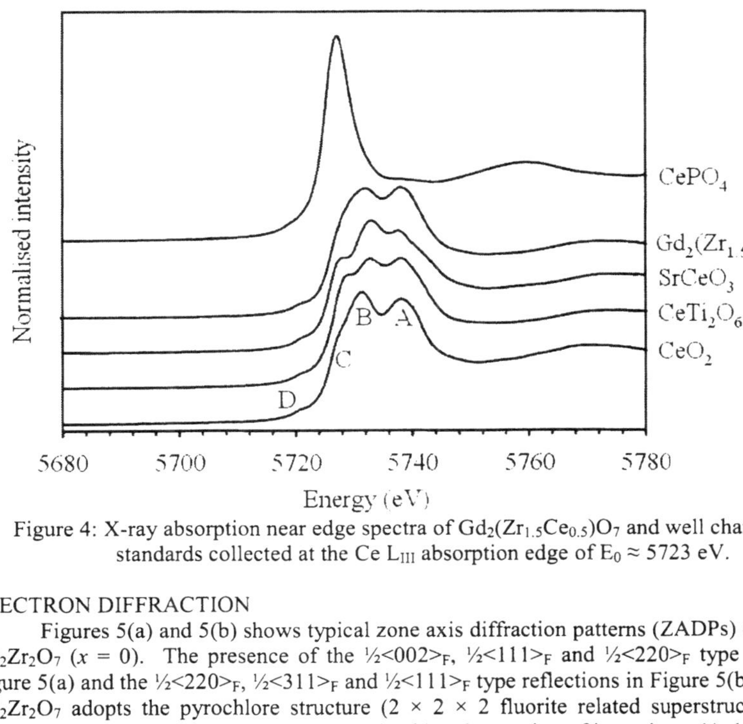

Figure 4: X-ray absorption near edge spectra of $Gd_2(Zr_{1.5}Ce_{0.5})O_7$ and well characterised standards collected at the Ce L_{III} absorption edge of $E_0 \approx 5723$ eV.

ELECTRON DIFFRACTION

Figures 5(a) and 5(b) shows typical zone axis diffraction patterns (ZADPs) obtained from $Gd_2Zr_2O_7$ ($x = 0$). The presence of the $\frac{1}{2}\langle 002\rangle_F$, $\frac{1}{2}\langle 111\rangle_F$ and $\frac{1}{2}\langle 220\rangle_F$ type reflections in Figure 5(a) and the $\frac{1}{2}\langle 220\rangle_F$, $\frac{1}{2}\langle 311\rangle_F$ and $\frac{1}{2}\langle 111\rangle_F$ type reflections in Figure 5(b) indicate that $Gd_2Zr_2O_7$ adopts the pyrochlore structure ($2 \times 2 \times 2$ fluorite related superstructure). Anion vacancy ordering in this system was determined by observation of intensity at $\frac{1}{2}\langle 111\rangle_F$ type and $\frac{1}{2}\langle 002\rangle_F$ type positions, whereas, cation ordering was determined by observation of intensity at $\frac{1}{2}\langle 111\rangle_F$ positions. Similar super-cell reflections were observed for $Gd_2(Zr_{1.75}Ce_{0.25})O_7$ ($x = 0.25$), however, intensity at $\frac{1}{2}\langle 002\rangle$ type positions was observed to be weak and diffuse.

Increased diffusivity of ½<002>$_F$ reflections, whilst maintaining strong reflections at ½<111>$_F$ type positions, is indicative of a reduction in the degree of ordering of anion vacancies but retention of cation ordering associated with the pyrochlore super-lattice. The limiting radius ratio (r_A/r_B), below which the pyrochlore structure ($A_2B_2O_7$) is destabilised, in favour of the defect fluorite structure ($A_{0.5}B_{0.5}O_{1.75}$), where the A and B cations are randomly distributed, is reported to be approximately 1.46 under ambient conditions[3]. The observed change in the super-lattice reflections with increasing cerium content is therefore consistent with an increase in the average B-site ionic radius arising from substitution of the larger Ce^{4+} cation for Zr^{4+}.

In the <110>$_F$ ZADP of $Gd_2(Zr_{1.50}Ce_{0.50})O_7$, Figure 5(c), the ½<002>$_F$ and ½<220>$_F$ type reflections were absent but satellite reflections present at the ½<111>$_F$ type positions were observed. This indicates a loss of anion vacancy ordering but retention of cation ordering associated with the pyrochlore supercell. In the <112>$_F$ ZADP, Figure 5(d), supercell reflections at ½<220> type positions are absent, but satellite reflections are present at the ½<111>$_F$ type positions. These satellite reflections are viewed as part of a ring of diffuse intensity with normal axis parallel to the [111]$_F$ directions, similar diffuse rings are present centred around ½<311>$_F$ type positions. The presence of diffuse rings such as these is indicative of ordering along the corresponding real space direction, with a degree of disorder in the orthogonal plane. The diffuse rings at ½<311>$_F$ positions are intersected by horizontal 'dark lines', highlighted by the arrows in Figure 5(d), and some asymmetry in the intensity across these dark lines is apparent. Effects such as these are characteristic of atomic relaxation displacements through 'size-effect-like' strain distortions[10]. These patterns are characteristic of a defect fluorite structure and have been observed in a number of other rare-earth sesquioxide stabilised cubic zirconias[11, 12, 13]. The radius ratio r_A/r_B for $Gd_2(Zr_{1.50}Ce_{0.50})O_7$ is 1.390 which is close to that of $Er_2Zr_2O_7$ (r_A/r_B = 1.394) which has been reported to crystallise with the defect fluorite structure[14]. In summary, $Gd_2(Zr_{1.50}Ce_{0.50})O_7$ adopts a defect fluorite structure characterised by disordered oxygen vacancies but retention of cation ordering with a reduced correlation length compared to the pyrochlore parent phase.

The $Gd_2(Zr_{1.00}Ce_{1.00})O_7$ sample exhibited complex ZADPs, Figures 5(e) and 5(f). In the <110>$_F$ ZADP, Figure 5(e), intensity is present as a diffuse square-wave distribution, running through the ½<111>$_F$ and ¼<220>$_F$ positions. The presence of the ¼<220>$_F$ type reflections is characteristic of the C-type structure of Gd_2O_3 although the ½<002>$_F$ type reflections, also characteristic of the C-type structure, are absent. In the <112>$_F$ ZADP, Figure 5(f), diffuse rings around ½<311>$_F$ type positions are consistent with those observed for the defect fluorite structure, however, satellite reflections at ½<111>$_F$ type positions are very diffuse. The absence of the ½<002>$_F$ type reflections and the existence of some residual diffuse scattering around the ½<111>$_F$ positions for $Gd_2(Zr_{1.00}Ce_{1.00})O_7$ suggests the formation of a hybrid structure with a cation and oxygen vacancy ordering pattern between the defect fluorite and the C-type structures. Similar patterns have been observed in a number of other closely related systems[15, 16, 17]. The structure of this hybrid phase is currently under investigation. Note: the presence of *some* but not *all* supercell reflections associated with the defect fluorite and C-type structures rules out the possibility that the electron diffraction patterns consist of a two phase mixture.

The ZADPs for $Gd_2Ce_2O_7$, Figures 5(g) and 5(h), display superlattice reflections consistent with the C-type structure. Similar ZADPs were observed for $Gd_2(Zr_{0.50}Ce_{1.50})O_7$ although the C-type super-cell reflections were weaker and more diffuse in character.

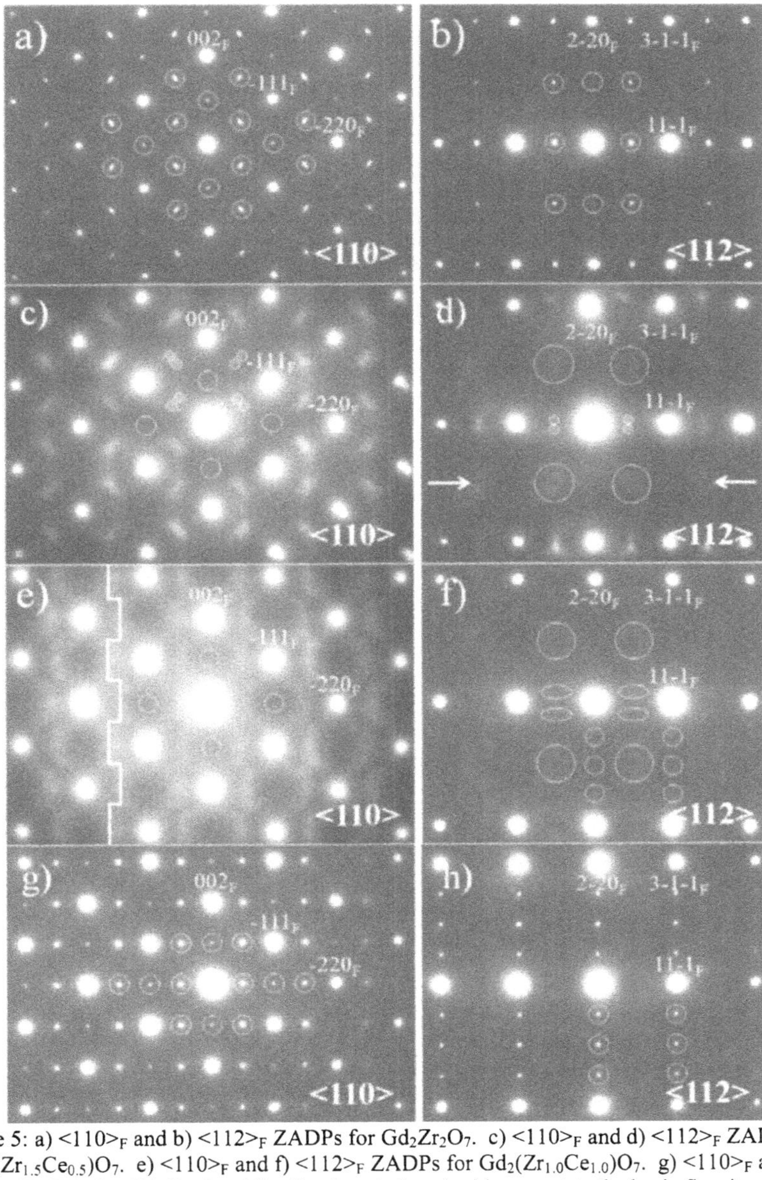

Figure 5: a) $\langle 110\rangle_F$ and b) $\langle 112\rangle_F$ ZADPs for $Gd_2Zr_2O_7$. c) $\langle 110\rangle_F$ and d) $\langle 112\rangle_F$ ZADPs for $Gd_2(Zr_{1.5}Ce_{0.5})O_7$. e) $\langle 110\rangle_F$ and f) $\langle 112\rangle_F$ ZADPs for $Gd_2(Zr_{1.0}Ce_{1.0})O_7$. g) $\langle 110\rangle_F$ and h) $\langle 112\rangle_F$ ZADPs for $Gd_2Ce_2O_7$. All reflections indexed with respect to the basic fluorite subcell.

SATELLITE DARK FIELD IMAGING

A satellite dark field image of the $x = 0.5$ sample was obtained using a ½<111>$_F$ type satellite reflection and is shown in Figure 6. A nanodomain texture is apparent on the scale of ~ 200 Å. One explanation for the existence of a nanodomain structure of this type is the presence of nanoscale phase separation. Withers *et al.*[15] observed a 'mottled' microstructure in the $(1-x)CeO_2.xYO_{1.5}$ system, similar to that observed in this study. They attributed this texture to compositional variation on the nanometer length scale. In the disposal environment, phase separation of this type could give rise to incongruent dissolution, compromising the performance of the wasteform, and potential risk of criticality arising from differential release and precipitation of fissile and neutron poison species. An alternative explanation for this phenomenon is the presence of short range ordering, such as that described by Tabira *et al.*[12] in the $(1-x)ZrO_2.xPrO_{1.5}$ system. Variations in contrast were attributed to fundamental changes in the local symmetry, as confirmed by the presence or absence of superlattice reflections in electron diffraction patterns acquired from different domains. High resolution electron microscopy studies are currently underway to investigate the origin of this nanodomain structure.

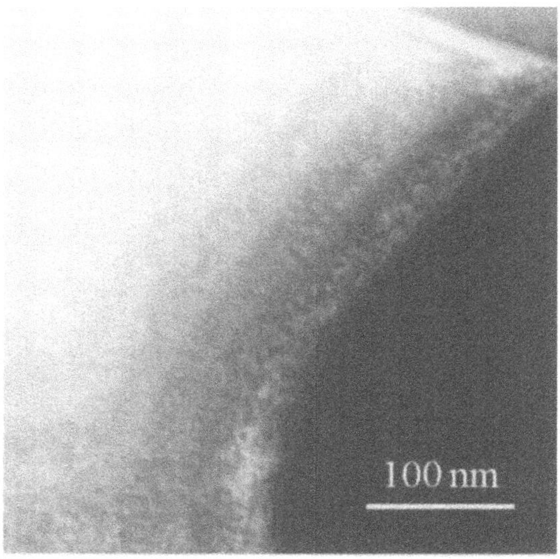

Figure 6: Satellite dark field image of $Gd_2(Zr_{1.5}Ce_{0.5})O_7$.

CONCLUSIONS

Phase boundaries were determined for the $Gd_2(Zr_{2-x}Ce_x)O_7$ system using a combination of X-ray and electron diffraction and are summarised in Table II. Comparison of Ce L_{III} edge XANES data from $Gd_2(Zr_{2-x}Ce_x)O_7$ and standard materials, confirmed the substitution of Ce^{4+} in this solid solution. Substitution of the larger Ce^{4+} cation for Zr^{4+} in the $Gd_2Zr_2O_7$ end member leads first to a loss of anion vacancy ordering in the defect fluorite phase field, but with retention of short range cation ordering characteristic of the pyrochlore supercell. Higher cerium content compositions display zone axis diffraction patterns with intensities at ¼, ½ and ¾ integer positions characteristic of a 4 x 4 x 4 repeat of the fluorite unit cell, consistent with the C-type structure. The electron diffraction patterns of $Gd_2(Zr_{1.0}Ce_{1.0})O_7$ revealed very complex, diffuse intensity distributions, displaying reflections consistent with both defect fluorite and C-type structures. However, not all allowed reflections associated with these structures were present, indicating the adoption of a complex hybrid structure with ordering between the defect fluorite and C-type structures.

Table II: Phase boundaries of the $Gd_2(Zr_{2-x}Ce_x)O_7$ system determined based on analysis of electron diffraction patterns.

Structure	Boundaries
Pyrochlore	$0.00 \leq x \leq 0.25$
Defect fluorite	$0.25 < x < 1.00$
Complex hybrid intermediate	$1.00 \leq x < 1.50$
C-type	$1.50 \leq x \leq 2.00$

This study has shown that up to 12.5 mol% Ce^{4+}, as a Pu^{4+} surrogate, can be accommodated by the pyrochlore structure of $Gd_2Zr_2O_7$. A key result of this study is the identification of ~ 200 Å nanodomains in the microstructure of $Gd_2(Zr_{1.5}Ce_{0.5})O_7$.. Domains of this type could be attributed to compositional heterogeneity at the nanoscale, or the presence of a short range ordered structure. Such compositional heterogeneity could adversely affect the performance of a Pu plutonium bearing $Gd_2Zr_2O_7$ wasteform, since incongruent dissolution of nanodomains would increase the potential risk of criticality arising from differential release and precipitation of fissile and neutron poison species. Further high resolution electron microscopy (HREM) are essential to determine the origin of these nanodomains and potential impact on wasteform performance.

ACKNOWLEDGEMENTS

This work was carried out as part of the TSEC programme KNOO and as such we are grateful to the EPSRC for funding under grant EP/C549465/1. Use of the National Synchrotron Light Source, Brookhaven National Laboratory, was supported by the U.S. Department of Energy, Office of Science, Office of Basic Energy Sciences, under Contract No. DE-AC02-98CH10886. We thank the National Nuclear Laboratory for part funding of a studentship for DPR. We are grateful to the Radioactive Waste Management Directorate for support to present the research published here (in part) at Materials Science & Technology 2009 Conference and Exhibition. We thank B. Ravel and J. Woicik for assistance in the acquisition of Ce L_{III} edge XANES data.

REFERENCES

[1]W. Donald, B. Metcalfe, R. Taylor, The Immobilisation of High Level Radioactive Wastes Using Ceramics and Glasses, *J. Mat. Sci.*, **32**, 5851-87, (1997).

[2]W. Lee, M. Ojovan, M. Stennett, N. Hyatt, Immobilisation of Radioactive Waste in Glasses, Glass Composite Materials and Ceramics, *Adv. App. Ceram.*, **105**, 3-12, (2006).

[3]R. Ewing, W. Weber, J. Lian, Nuclear Waste Disposal – Pyrochlore ($A_2B_2O_7$): Nuclear Waste form for the Immobilisation of Plutonium and "Minor" Actinides, *J. App. Phys.*, **95** (11), 5949-71, (2004).

[4]P. Bingham, R. Hand, M. Stennett, N. Hyatt, M. Harrison, The Use of Surrogates in Waste Immobilisation Studies: A Case Study of Plutonium, *Mater. Res. Soc. Symp. Proc.*, **1107**, 421-28, (2008).
[5]B. Ravel, M. Newville, ATHENA, ARTEMIS, HEPHASTUS: Data Analysis for X-ray Absorption Spectroscopy using IFEFFIT, *J. Synchrotron Rad.*, **12**, 537-41, (2005).
[6]R. Shannon, Revised Effective Ionic Radii and Systematic Studies of Interatomic Distances in Halides and Chalcogenides, *Acta Cryst.*, **A32**, 751-67, (1976).
[7]S. Patwe, B. Ambekar, A. Tyagi, Synthesis, Characterisation and Lattice Thermal Expansion of some Compounds in the System $Gd_2Ce_xZr_{2-x}O_7$, *J. Alloy. Comp.*, **389**, 243-46, (2005)
[8]A. Bianconi, A. Marcelli, H. Dexpert, R. Karnatak, A. Kotani, T. Jo, J. Petiau. 1987. Specific Intermediate-Valence State of Insulating 4*f* Compounds Detected by L-III X-ray Absorption, *Phys. Rev. B*, **35**, 806-12, (1987).
[9]A. Soldatov, T. Ivanchenko, S. Dellalonga, A. Kotani, Y. Iwamoto, A. Bianconi, Crystal-Structure Effects in the Ce L-III Edge X-ray Absorption Spectrum of CeO_2 Multiple Scattering Resonances and Many-Body Final States, *Phys. Rev. B*, **50**, 5074-80, (1994).
[10]B. Butler, R. Withers, T. Welberry, Diffuse Absences due to the Atomic Size Effect, Acta Cryst., **A48**, 737-46, (1992).
[11]Y. Tabira, R. Withers, Structure and Crystal Chemistry as a Function of Composition across the Wide Range Non-Stoichiometric $(1-\varepsilon)ZrO_2 \cdot SmO_{1.5}$, $0.38 < \varepsilon < 0.55$, Oxide Pyrochlore System, *J. Sol. Stat. Chem.*, **148**, 205-14, (1999).
[12]Y. Tabira, R. Withers, J. Barry, L. Elcoro, The Strain-Driven Pyrochlore to "Defect-Fluorite" Phase Transition in Rare Earth Sesquioxide Stabilized Cubic Zirconias, *J. Sol. Stat. Chem.*, **159**, 121-29, (2001).
[13]Y. Liu, R. Withers, L. Norén, The Pyrochlore to 'Defect Fluorite' Transition in the $Y_2(Zr_yTi_{1-y})_2O_7$ System and its Underlying Crystal Chemistry, *J. Sol. Stat. Chem.*, **177**, 4404-12, (2004).
[14]K. Sickafus, L. Minervini, R. Grimes, J. Valdez, M. Ishimaru, F. Li, K. McClellan, T. Hartmann, Radiation Tolerance of Complex Oxides, *Science*, **289**, 748-51, (2000).
[15]R. Withers, J. Thompson, N. Gabbitas, L. Wallenberg, T. Welberry, Microdomains, Solid Solutions and the "Defect Fluorite" to C-Type Sesquioxide Transition in CeO_2-$RO_{1.5}$ and $ZrO_{1.5}$ Systems, *J. Sol. Stat. Chem.*, **120**, 290-98, (1995).
[16]S. Garcia-Martin, M. Alario-Franco, D. Fagg, A. Feighery, J. Irvine, Modulated Fluorite-Type Structure of Materials from the $(1 - x)Y_{0.5}Zr_{0.5}O_{1.75} - xY_{0.75}Nb_{0.25}O_{1.75}$ $(0 \leq x \leq 1)$ System, **12**, 1729-37, (2000).
[17]N. Gabbitas, J. Thompson, R. Withers, A. Rae, A Single Crystal X-Ray Study of the Fluorite-Related "Solid Solutions" of $CeO_2 - YO_{1.5}$, *J. Sol. Stat Chem.*, **115**, 23-36, (1995).

WASTE FORM DEVELOPMENT FOR SRS MOX PLANT EFFLUENT

Erich Hansen, Timothy Jones, Tommy Edwards, and Alex Cozzi
Savannah River National Laboratory
Aiken, SC 29803

ABSTRACT

The Waste Solidification Building will process high activity waste (HAW) from the Mixed Oxide Fuel Fabrication plant. This work investigated the variability in the HAW waste stream and its impact on the ability to process the waste into a well mixed grout product using bench scale mixing.

A statistically designed text matrix was developed on the maximum acidic composition of the HAW waste inputs of nitric acid, gallium, silver and other minor metals. Substitution of the element aluminum for gallium and copper for silver was considered, yielding 40 tests. The resulting acidic solutions were neutralized and pH adjusted to a pH of 12 using sodium hydroxide. These streams were blended with cement and zircon flour at a water to cement ratio of 0.3 to make grout. The grouts were mixed and visually determined to be homogenous.

The fresh grout yield stress, 24 hour set time and 24 hour bleed liquid properties were measured. In all cases, there was no bleed liquid and the grouts set within 24 hours.

The yield stress results ranged between 99 to 1495 Pascal, providing an indication of the required torque to yield a mixable grout. The yield stress decreases with increasing nitric acid concentration and increased with increasing gallium and other minor metals. Silver, undissolved solids fraction, pH, and density had no impact on the yield stress. As the water content in the neutralized/pH adjusted stream increased, the yield stress increased.

INTRODUCTION

The Savannah River Site has begun construction of the Waste Solidification Building (WSB) for the treatment and solidification of the radioactive liquid waste streams generated by the Mixed Oxide (MOX) Fuel Fabrication Facility (MFFF). Cementation has been selected as the solidification method for waste streams generated in the MFFF.

The HAW is a moderately acidic aqueous stream generated in the following processes: the MOX acid recovery and recycle processes; the alkaline treatment process; and the aqueous purification process. This waste also includes rinse water from equipment and drain flushes and is expected to have a pH <1. The various waste streams will be evaporated to reduce the volume of the waste. Prior to the cementitious process, the waste streams are neutralized using concentrated NaOH solution. Additional NaOH solution is added to make the streams caustic, targeting a pH of 12.

In the WSB process, neutralized waste solutions will be added to stainless steel 55 gallon drums containing cementitious dry materials. The waste solutions are blended with the dry solids using mechanical agitation and upon completion of mixing the drums are secured and vented. The drums will then be stored until disposition at the appropriate waste facility.

The objective of this study was to perform a variability study of simulated HAW to determine the impact of feed variability on the quality of the final grout and on the mixability of the salt solution into the dry powders. The acidic feeds were processed through the neutralization/pH adjustment process. The adjusted feeds were then blended with the dry materials to make the final waste forms. A secondary objective was to determine if elemental substitution for cost prohibitive or toxic elements in the simulant affects the mixing response, thus providing a more economical simulant for use in full scale tests. Though not an objective, the HAW simulant used in the full scale tests was also tested and compared to the results from this task.

EXPERIMENTAL

Simulant Development

Based on the anticipated feed stream from the MFFF, a simulant composition was developed to simulate the WSB waste. The concentrations of these metals in this HAW stream were further increased by a factor of five in the WSB flowsheet to account for the uncertainty of the composition and the implementation of an evaporator in the WSB. Table I is the simulant of the acidic product. The following maximums were also specified for the HAW simulant, where the acid stream contained 7.39M HNO_3, an erbium concentration of 0.0021 M (substitute for Americium[1], equivalent to 0.507 g/L of Am) and a sodium concentration of 3.794 g/L. The acid and erbium concentrations will be maintained in this variability study.

Table I. Expected Compositional Range of the Acidic HAW Simulant

Metal	Concentration (g/L)		Metal	Concentration (g/L)	
	From MFFF	5X		From MFFF	5X
Ag	11.7	58.5	Au	0.10	0.50
Ba	0.8	4.00	Al	0.86	4.30
Cd	0.13	0.65	Cu	0.15	0.75
Pb	0.18	0.90	Ta	0.70	3.5
Hg	0.003	0.015	V	0.30	1.5
Cr	0.54	2.7	As	0.003	0.015
Ga	1.62	8.1			

The variables that were used in the factor space for this variability study are provided in Table II and are based on the starting composition of the acid solutions. The "other metals" are incorporated as a variable since their contribution is considerable. The composition of the "other metals" is provided in Table III. A lower limit of zero is placed on all of the variables, thereby incorporating water only as a condition. Note that Al and Cu are present in the "other metals" and will remain as such. Additional tests will be performed to determine the impact of the substitution of Al for Ga and Cu for Ag.

Given the constraints on four variables—gallium, silver, nitrates and "others"—and the substitution of aluminum and copper for gallium and silver, respectively, a statistical design was developed to investigate for main effects and possible interactions. The baseline tests (16 total) investigate pair wise interactions between Ga, Ag, HNO_3 and "others" in the HAW. The "others" are metals that are added as a group. Replacing Ga with Al and Ag with Cu resulted in 34 tests and was based on a linear model of the major components with both single and double substitutions. Incorporating all of the tests into one matrix after removing overlapping test points and adding replicates to determine variability in the measurement resulted in a total of 40 test points in the final test matrix.

Table II. Variable for the Initial Factor Space

Variable	Min Values	Max Values
Solution Molarity	0	7.39 Molar HNO_3
Ga	0	0.11618 Molar Ga
Ag	0	0.5423 Molar Ag
"Other metals"	0	1.1032 M (NO_3) See Table 3 for breakdown

Table III. Composition of "Other Metals."

Metal	g/L	Metal	g/L
Ba	4.00	K	5.95
Cd	0.65	Mg	2.21
Al	4.30	Ca	3.78
Cu	0.75	Am*	0.507

*Er will be used in place of Am, at equal molarity

To further limit variability in this study, the acidic solutions were neutralized/pH adjusted to a final pH value of 12, providing the same concentration (molar basis) of free hydroxides in the salt solution.

The response variable that was measured is the yield stress as determined using the vane method, performed in triplicate. This response variable, specifically the first measurement, was statistically analyzed against the input/measured/calculated variables as listed in Table IV. The calculated undissolved solids were determined from the measured dissolved and total solids. The molarity of the chemical or "others" was calculated by determining the number of moles of each prior to being neutralized/pH adjusted. Based on the dilution due to neutralization/pH adjustment and grout formulation, the molarity was determined by dividing the starting moles with the final grout volumes.

The acid solutions were made using reagent grade chemicals. Approximately 200 mL of de-ionized (DI) water was added to a 1-L volumetric flask. Chemicals having the largest to least mass contributions when then added sequentially. DI water is then added to the 1-L mark on the flask. Two subsamples of the acid solution are pulled and analyzed for total solids and density. Twelve acid simulants contained undissolved solids after processing was complete.

The neutralization/pH adjustment was performed using a two liter glass reactor with a cooling jacket. A water bath was used to control the temperature of the cooling water jacket to 10 C and to remove heat generated during the exothermic reaction between the sodium hydroxide, nitric acid, and nitrates. The targeted quantity of NaOH solution was based on the quantity required to neutralize the nitric acid and convert the metal nitrates to hydroxides in the acid solution. In all cases where there were undissolved solids in the acid solution during the neutralization/pH adjustment process, all of the solids dissolved prior to re-precipitating. The resulting solutions were analyzed for total solids in the slurry, soluble solids in the supernate, density and rheology of the slurry. Undissolved solids were calculated from the measured dissolved and total solids values.

For the two solutions which contained silver as the only metal nitrate in the acidic solution, the neutralization/pH adjustment process resulted in a black colored precipitated solids that quickly settled. This type of settling behavior was not observed in any of the other simulants that had any other metal nitrate blended with the silver nitrate.

Table IV. Variables Used to Determine Effect on Response Variable

Parameters from Input data used to make Acidic Solutions	
HNO_3 Batch Sheet Molarity	NO_3 Batch Sheet Molarity
Ag, Ga, Al, Cu, "Others" Batch Sheet Molarity	
Measured Properties of Acidic Solution	
Wt% total solids	Density
Measured Properties of Neutralized/pH Adjusted Solution	
Density	pH
Wt% soluble solids in supernate	Wt% total solids in solution
Calculated Variables in Neutralized/pH adjusted solution	
Wt% undissolved solids in solution	
Calculated Variables in grout	
Zircon flour fraction	Cement fraction
Water fraction	Neutralized pH adjusted solution fraction
Ag, Ga, Al, Er, Cu, Ca, Ba, K, Cd, Na, Mg molarity	Density of grout
NO_3 molarity	
Calculated Variables in grout – "Others" is grouped	
Ag, Ga, Al, Cu, "Others" Molarity	

Waste Form Preparation

The grout batches all consisted of a water to cement ratio of 0.30, and a density modifier (zircon flour) to cement ratio of 1/12. The quantity of simulant made was based on using the densities and mass fractions of the various components (simulant, cement, and zircon flour) and volume additively to determine the final grout density.

Grout Mixer Operation

The mixing tests were performed using a $1/6^{th}$ horsepower orbital mixer having three defined mixing speeds, low (#1), intermediate (#2) and high (#3). To begin each mixing test, approximately 75% of the liquid was placed into the mixer bowl. Next, the entire batch of cement and zircon flour blend was added to the mixing bowl and spread out evenly. Then the remaining liquid is poured over the top of the cement/zircon flour blend. The mixing bowl was raised into position and the mixer was started at its low speed and mixed for two minutes. Material adhered to the upper portion of the mixing bowl or paddle blade was scraped back into the grout mixture during the brief stoppages between mixing speed adjustments. The mixer was then set to speed #2 and allowed to mix for five minutes.

Upon completion of mixing, the mixing bowl and blade were removed from the mixer and the grout scraped from the blade back into the bowl.

Grout Properties

Freshly mixed grout was ladled into a 1000-ml plastic cup while the cup was tapped to release trapped air. The cup was lifted into the measurement position for the first vane measurement. At the conclusion of the first vane measurement, the sample cup was lowered and repositioned for a second measurement. A third vane measurement was also performed.

At the conclusion of the vane measurements, the grout sample was used to provide feed for subsequent measurements. Two 2-inch cylinder sample plastic containers and one 2-inch cube plastic container were loaded with grout, during which time the containers were tapped to remove as much entrained air as possible. The cylinder samples were used for bleed and density measurements. The square sample was used for set measurement.

Fresh grout samples were prepared in the same manner as that for grout density measurements, but were placed into 2-inch cubes rather than the 2-inch cylinders. The cubes were covered to eliminate or minimize evaporation. After 24 hours of curing, the free liquid was recovered and weighed. The volume of the fluid is then estimated using the density of the salt solution by dividing the measured mass by the density. A single bleed water sample was performed.

The fresh grout sample was prepared using the same method as described for bleed water. After three days of setting, a set measurement was performed using a modified Vicat Needle.[2] The needle (2 mm diameter) is placed and secured at the surface of the grout and the position indicator set to zero. The needle assembly, which has a total mass of 300 grams is released and the travel distance is measured. In these tests, a 1 mm travel distance was used as the basis for set.

The densities of the cured grouts were measured using cylindrical sample bottles having two inch diameters, an overall height of four inches and a flat bottom. The fresh grout samples were loaded into the sample bottles in the same manner as that for the vane measurements. The objective was to remove as much entrained air by tapping the sample bottle and leveling the top surface of the grout prior to curing. After curing for 3 to 4 days, the sample was removed from the sample bottle and weighed. Four diameter and three height measurements were made on the specimen. The measurements were averaged to calculate the volume. The density was then calculated by dividing the mass by the volume. Duplicate samples were run.

RESULTS

The cured grout samples did not contained any bleed water, and they satisfied the set requirements in every case. These data are not discussed further.

Several plots were prepared as part of the review of the measurements from these tests. Figure 1 provides a plot of the yield stress measurements by test number (#). The yield stress measurements are identified by a time indicator, which is set at values of 1, 4, and 6 reflecting the initial measurement, 2nd measurement (3 minutes later), and the 3rd measurement (3 minutes after the 2nd). The plot indicates that the yield stress measurements for a test run were increasing over the course of the measurements for most of the test runs, indicating the cementitious reactions may be occurring during the measurement. A strong linear correlation of yield stress for each measurement time was identified. Therefore, the data also indicate that any of the three yield stress data sets could have been used as the response variable. This data also reflects that if only a single yield stress measurement is obtained from multiple grout samples, the timing of the measurement is critical.

Table V summarizes the effect on the response variable as the input variable increases. As the concentration of input variables on the left side of Table V increase, the yield stress tends to decrease. The opposite is true for the variables on the right side of Table V. Note the variables maybe interrelated, hence yielding the same type of response. For instance, the Na and NO_3 molarity in the final grout show essentially the same type of decreasing trend in yield stress as these variables increase. This is because the vast majority (close to 100%) of the Na and NO_3 in the neutralized/pH adjusted solution is $NaNO_3$. The variables not listed in Table V did not show any linear trend, such as the concentration of undissolved solids in the neutralized/pH adjusted fluid.

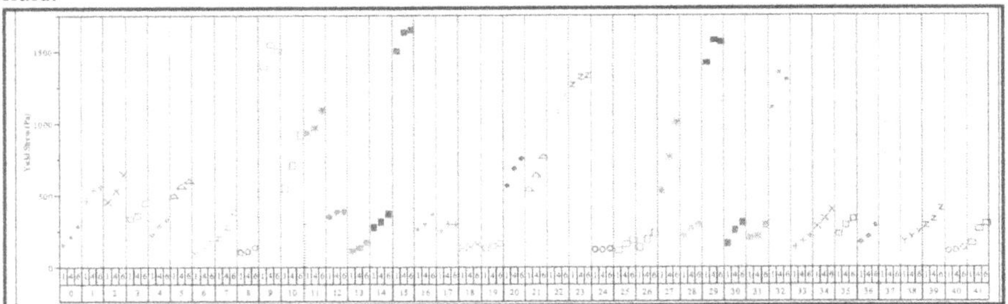

Figure 1. Yield stress measurements for each run as a function of time.

Table V. Response in Yield Stress to Increasing Input Variable.

Decrease in Yield Stress	Increase in Yield Stress
Batch Sheet Input Data – HNO_3	Final Grout – Cement Fraction
Acid Solution - density	Final Grout – Zircon Fraction
Neutralized-pH adjusted fluid – density	
Neutralized-pH adjusted fluid – wt% TS	
Neutralized-pH adjusted fluid – wt% SSS*	
Final Grout – Simulant Fraction	
Final Grout – Na Molarity	
Final Grout – NO_3 Molarity	

*soluble solids in supernate

The initial statistical analysis explored the relationship between the natural logarithm of the first yield stress measurement and the acid batching variables defined in the variability study. There are 40 experimental runs making up these results. The intent of the analysis is to get an initial impression of those variables that are seen to have a statistically significant impact on the response variable. For these tests, the estimated coefficients from this analysis suggest the relationship between the input variable and the variables HNO_3, Ga, Cu, and "Others" have a statistically significant influence on the response variable. Increases in the concentrations of HNO_3 and Cu lead to decreases in yield stress while increases in Ga and "Others" lead to increases in yield stress. The effects of the other two batching variables, Al and Ag, are not statistically significant.

As a consequence, if one were to substitute Al for Ga in future testing, it is likely that the yield stress would be less for the testing with Al as compared to the yield stress of corresponding tests containing Ga. The same general trend in results would be expected if Cu were used in place of Ag. This trend was statistically shown to result in a lower yield stress, provided all other parameters are maintained. This can also be observed in Figure 2 for the Cu for Ag substitution and Figure 3 for the Al for Ga substitution. The arrows shown in these figures indicate the effect of substituting, given all other parameters are held constant.

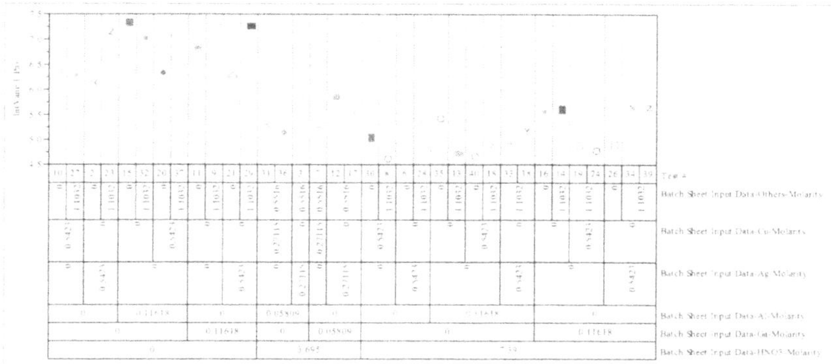

Figure 2. Plot of natural logarithm of first yield stress measurement (Pa) for trials investigating substitutability grouped as indicated by the information provided on the x-axis. Arrows show direction of yield stress change during substitution.

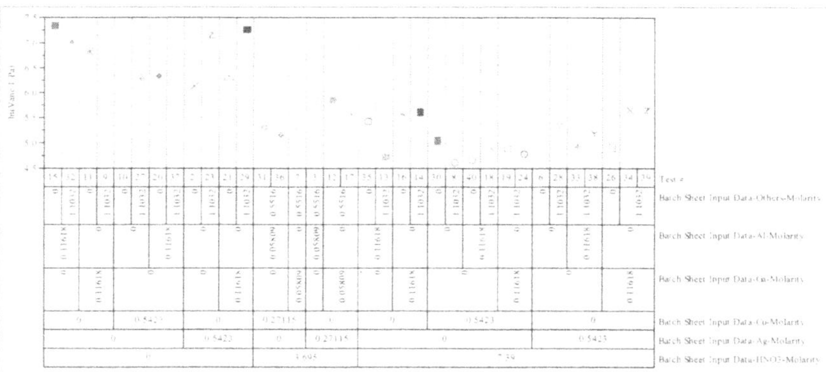

Figure 3. Plot of natural logarithm of first yield stress measurement (Pa) for trials investigating substitutability grouped as indicated by the information provided on the x-axis. Arrows show direction of yield stress change during substitution.

Excluding tests that involved Al and Cu from the data used to explore models for the yield stress response resulted in 20 remaining data points. Figure 4 provides the JMP results from modeling the natural logarithms of the first yield stress measurements versus a main-effects model for HNO_3, Ga, Ag, and "Others" in the acid batches. All of these candidate, explanatory variables are statistically significant except for Ag. Once again, increases in the concentrations of HNO_3 lead to decreases in yield stress while increases in Ga and "Others" lead to increases in yield stress. There is no indication of lack of fit for this modeling effort. Note that in the

Parameter Estimates section of Figure 4, for every molar increase in HNO_3 results in a log yield stress decrease of -0.209479 and covers a large range of molarity. For Ga and "Others", this estimated parameter is larger that than for HNO_3, but their molarity range is much smaller.

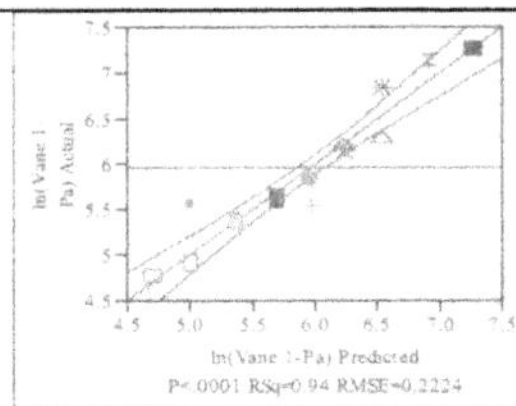

Response ln(1st Vane-Pa) Whole Model Actual by Predicted Plot

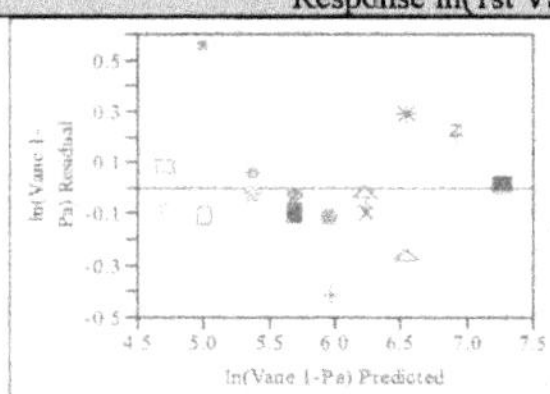

Residual by Predicted Plot

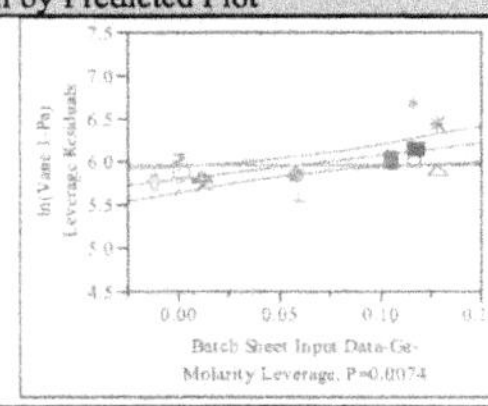

Leverage Plot – Ga Molarity

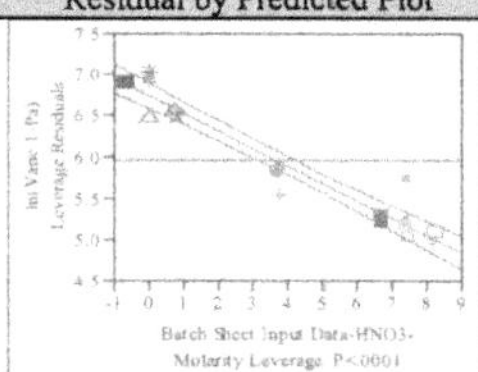

Leverage Plot – HNO_3 Molarity

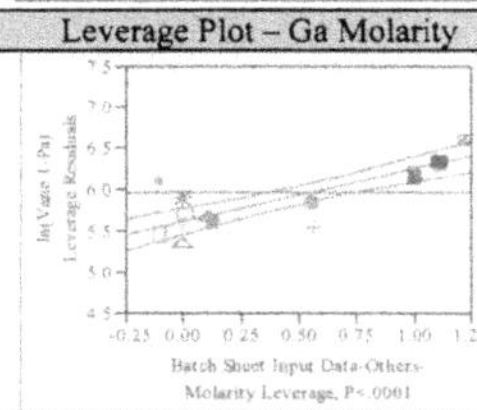

Leverage Plot – Others Molarity

RSquare	0.938505
RSquare Adj	0.926975
Root Mean Square Error	0.222376
Mean of Response	5.956817
Observations (or Sum Wgts)	20

Summary of Fit

Source	DF	Sum of Squares	Mean Square	F Ratio
Model	3	12.075108	4.02504	81.3944
Error	16	0.791216	0.04945	**Prob > F**
C. Total	19	12.866324		<.0001

Analysis of Variance

Source	DF	Sum of Squares	Mean Square	F Ratio
Lack Of Fit	5	0.34908912	0.069818	1.7370
Pure Error	11	0.44212701	0.040193	**Prob > F**
Total Error	16	0.79121612		0.2068
				Max RSq
				0.9656

Lack of Fit

Term	Estimate	Std Error	tRatio	Prob>|t|
Intercept	6.221287	0.095999	64.81	<.0001
Batch Sheet Input Data-HNO_3-Molarity	-0.209479	0.014346	-14.60	<.0001
Batch Sheet Input Data-Ga-Molarity	2.7983318	0.912494	3.07	0.0074
Batch Sheet Input Data-Others-Molarity	0.6290817	0.096096	6.55	<.0001

Parameter Estimates

Figure 4. Modeling of Natural Logarithm of First Yield Stress Measurements

CONCLUSIONS

Given the conditions of preparing the grout in this task, all of the grouts were visually well mixed prior to preparing the grouts for measurements. All of the cured grouts were measured for bleed and set. All of the cured grouts satisfied the bleed and set requirements, where no bleed water was observed on any of the grout samples after one day and all had set within three days of curing. This data indicates, for a well mixed product, bleed and set requirement are satisfied for the range of acidic feeds tested in this task.

The conclusions from this testing that impact the mixability/yield stress are:

- Statistically the yield stress decreases with increasing HNO_3 concentration and increases with increasing gallium and "Other" metals concentration.
- Statistically, silver had no impact on the yield stress.
- Increasing the sodium nitrate concentration causes a decrease in the yield stress.
- As the water content in the neutralized/pH adjusted solution increases, the yield stress increases.
- The undissolved solids fraction, pH, and density of the neutralized/pH adjusted solution do not cause the yield stress to trend.
- Table V lists the variables that show a linear trend to either increase or decrease the yield stress as the variable in the grout increases in concentration. Many of the variables which show the same type of trending are interrelated, such as Na and NO_3. Other notables are;
    - As the water content increases in the neutralized salt solution, the yield stress increases. This is due to the fraction cement (and/or zircon flour) increasing to maintain a constant water to cement ratio.
    - As the total solids concentration increased in the neutralized/pH adjusted solution, the yield stresses decreased. This is consistent with the above statement, since the quantity of water in the grout decreases as the total solids content increases in the neutralized/pH adjusted solution for a given grout volume.
    - The undissolved solids fraction, pH and density of the neutralized/pH adjusted solutions do not cause the yield stress to trend.

These tests showed that the surrogate elements, aluminum for gallium and copper for silver are not acceptable substitutes. Statistically, as the concentration of aluminum and copper increased the yield stress decreased. This effect is contrary to the gallium, where the yield stress increased with increasing gallium concentration. Silver was shown not to impact yield stress. The following observations were made:

- The substitution of Cu for Ag and Al for Ga was shown, statistically not to be adequate elemental substitutes. Substitution of these elements resulted in a lower response variable, resulting in a non-conservative condition of testing.
- Excluding tests involving Al and Cu from the "acid solution" data set resulted in HNO_3, Ga and "others" becoming statistically significant. The response function decreased with increasing HNO_3 concentration and increased with increasing Ga and "Others" concentration in the acid solution.

REFERENCES

[1] Villarreal, R. and Spall, D., "Selection of Actinide Chemical Analogues for WIPP Tests - Potential Nonradiactive Sorbing and Nonsorbing Tracers for Study of Ion Transport in the Environment" LA-13500-MS, Los Alamos National Laboratory, Los Alamos, NM August 1998.

[2] ASTM C807-08, "Standard Test Method for Time of Setting of Hydraulic Cement Mortar by Modified Vicat Needle"

FLAMMABLE GASSES IN THE SALTSTONE PROCESS FLOWSHEET

A.D. Cozzi
Savannah River National Laboratory
Aiken, SC, USA

ABSTRACT

The fate of flammables in the Saltstone process can be evaluated from the initial concentration in the process and the propensity to form or dissipate flammables in each operation of the Saltstone Production Facility (SPF) and Saltstone Disposal Facility (SDF).

Flammable compounds in the Saltstone process can originate either as flammable compounds in the incoming salt solution, flammable compounds in the premix materials, or flammable compounds formed by the decomposition of trace materials added either as grinding aids for the dry materials or admixtures used to destabilize foaming of the Saltstone slurry.

This work discusses the potential sources of volatiles that may be introduced into the Saltstone process and the mechanism for evolution of the volatiles in each of the process steps.

INTRODUCTION

The Saltstone feed tank receives waste from several influent streams. A simplified flowsheet of the process is shown in Figure 1. The waste salt solution is pumped to the salt feed tank (SFT) in the SPF. The premix materials; cement, slag and fly ash, are blended together prior to transfer to the grout mixer. The premix is gravity fed to the grout mixer in the SPF and the salt solution is incorporated into the premix in the grout mixer, yielding Saltstone slurry. The Saltstone slurry free falls into a hopper and then is pumped to the vault with a positive displacement pump.

Until recently, many volatile components present in Saltstone have been assumed to contribute a negligible amount to overall flammability inside the vault vapor space. Because of an increase in the total quantity of flammable fuel projected in future Saltstone waste streams; this assumption may no longer be valid. A review of the Saltstone process and the nature of the chemical reactions that take place during hydration of the dry material were conducted to determine if vapor/liquid equilibrium data established for use in the SRS high level waste (HLW) Tank Farms can be applied inside a Saltstone Vault. The results of this review indicate that Henry's Law coefficients developed for HLW salt solutions can be conservatively applied to grout being stored at Saltstone.

A comprehensive review of potential sources of flammable gasses in the Saltstone process was performed to support the revision of the safety analysis for the Saltstone Facility to accommodate changes in the Composite Lower Flammability Limit.

This paper discusses the potential sources of volatiles that may be introduced into the Saltstone process and the mechanism for evolution of the volatiles in each of the process steps.

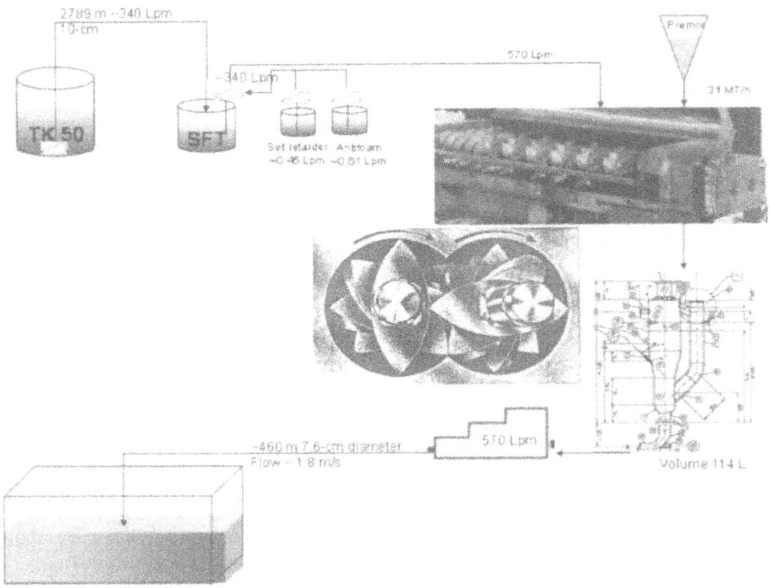

Figure 1. Simplified flowsheet of the Saltstone process.

SOURCES OF VOLATILE COMPONENTS

Tank 50

Tank 50 is the feed tank for the Saltstone process. Treatment of vessel vent system flushes from H-Canyon processes is the primary contributor of ammonia to the salt solution influents to tank 50. The primary source of Isopar L* is from the Modular Caustic Side Solvent Extraction (CSSX) Unit (MCU). Tank 50 may also have residual tetraphenylborate (TPB) present from previous Tank Farm activities that can decompose to benzene and other phenolic compounds.

The H-Canyon flowsheet for processing unirradiated fuel includes gel strikes. The gelatin is a collection of amino acids extracted from collagen inside animal connective tissue. Gelatin is not readily decomposed by caustic. The flammable contribution from degradation of non-flammable components is expected to be negligible. The primary flammable components in the material being fed to Tank 50 are ammonia and Isopar.

Premix

Premix components, cement, slag, and fly ash (nominally 10%, 45%, 45% by weight in the premix) are weighed and blended in the premix air blender hoppers in 2700 kg batches (~ 1 batch every 6 min) prior to pneumatic transfer to the premix feed bin.

---

* Isopar L – Exxon Corp. Isoparaffinic solvent

Portland Cement

The Ordinary Portland Cement (OPC) is added to the premix as a source of lime to ensure activation of the slag. The cement industry has adopted the use of hazardous and other wastes as a source of inexpensive fuel for production of clinker. In addition, the cement industry typically uses an assortment of organic constituents to aid in the grinding of the clinkers to make cement. For example, Holcim, the vendor for the SPF, uses 0.15 weight percent (wt %) diethylene glycol as a grinding aid. The flammable contribution from degradation of non-flammable components in the cement is expected to be negligible. An SPF cement sample collected in 2005 was tested and a loss on ignition (LOI) of 1 – 1.5 wt % was measured.[1] Analysis of a representative sample of cement shows ammonia concentrations to be < 2 mg/kg and concentrations < 0.4 mg/kg for benzene, toluene, and the xylenes. The contribution to the CLFL from carryover from the fuel and grinding aids is expected to be negligible.

Slag

To enhance the reductive capacity of the premix, slag, which contains reduced transition metals and sulfur compounds is one of the two primary components of the premix. The reductive capacity of the slag may produce ammonia via the reduction of nitrites and nitrates. Slag is the byproduct of the steel making process. The use of coke in the process introduces aromatic and aliphatic hydrocarbons, which can be present in the slag at measurable levels. Grinding aids used for slag size reduction are similar to those identified for grinding cements and may be a source for ammonia forming compounds. A slag sample collected in 2005 was tested and a LOI of approximately 0.6 wt % was measured between 150 to 800 °C.[1] Results of an analysis of a representative sample of slag show ammonia concentrations to be ~10 mg/kg. This is most likely due to the decomposition of the amine acetate in the grinding aid used by the supplier. Concentrations for benzene, toluene, and the xylenes were determined to be <0.4 mg/kg. Testing performed to determine the release of ammonia from Saltstone grout prepared from salt solution with and without ammonia was not quantified because the contribution from reduction of nitrates and nitrites in salt solution by the slag could not be deconvoluted from the contribution from the ammonia in the salt solution.[2]

Fly Ash

To reduce the temperature rise in the vault due to heat of hydration, ASTM Class F fly ash is added to the premix. Fly ash is a by product of the coal burning power generation industry and the composition and properties are greatly influenced by where the coal ore is mined and the operation of the coal burning power plant. Plant operating efficiency affects the amount of residual carbon remaining in the fly ash. Coal contains measurable quantities of aromatic and aliphatic hydrocarbons (benzene, toluene, alkylbenzes, heptane and octane). All of which are carried over, to some extent, into the fly ash. Ammonia in fly ash may be present as either ammonium salts or as chemisorbed ammonia. Carbon Burn Out (CBO) fly ash has been developed to limit the amount of ammonia in fly ash for construction of occupied facilities. In the CBO process, fly ash that does not meet the ASTM Class F specification is reheated to approximately 700 °C to reduce the LOI to values typically less than 2 wt %. Results of analysis of representative samples of both CBO and non-CBO fly ash show for the non-CBO fly ash show ammonia concentrations to be ~180 mg/kg and concentrations < 0.4 mg/kg for benzene, toluene, and the xylenes. The CBO fly ash had an ammonia concentration of < 2 mg/kg and concentrations < 0.4 mg/kg for benzene, toluene, and the xylenes.

Admixtures

Operating conditions in the SPF sometimes require the use of admixtures to adjust the fresh properties of the Saltstone slurry to stay within the operating parameters of the SPF. In the recent campaigns, the SPF used both a set retarder and defoamer to meet the slurry properties for optimum production.

Set Retarder

Set retarder is added to the salt solution to temporarily inhibit the cementitious reactions that occur when the salt solution is blended with the premix materials. The current set retarder used by the SPF is Daratard 17[†]. Daratard 17 is a mixture of approximately 25 wt % corn syrup and 25 wt % calcium lignosulfonate in water. The Daratard 17 is added to the Saltstone slurry at 0.45 L/m. The potential contribution to the flammability limit from the degradation of non-flammable components in Daratard 17 is expected to be negligible.

Defoamer

Defoamer is added to the Saltstone slurry to destabilize the foam that may occur when the salt solution is blended with the premix materials. One defoamer used by the SPF is Clear Air 100[‡]. Clear Air 100 is 100% tributylphosphate (TBP). TBP can decompose to produce butanol. The Clear Air 100 is added to the Saltstone slurry at 0.6 L/m. The potential contribution to the flammability limit from degradation of non-flammable components in Clear Air 100 can be calculated.

Recently, a siloxane based defoamer[§] was introduced as a replacement for Clear Air 100. This antifoam contains xylene. The potential contribution to the flammability limit from the xylene in Q2-1383-A can be calculated.

Process Flowsheet

The fate of flammables in the Saltstone process can be evaluated from the initial concentration in Tank 50, the flammables introduced by the premix and admixture materials, the production of ammonia via the reduction of nitrates ands nitrites by slag, and the propensity to form or dissipate flammables in each operation of the SPF and SDF.

Saltstone Process Facility

Transfer to Salt Feed Tank

Slurry pumps in Tank 50 may or may not be operating during salt solution transfers. The telescoping transfer pump in Tank 50 is currently located proximal to the bottom of the tank. The salt solution is pumped at approximately 340 liters per minute (gpm) through 2790 meters of 10-cm pipe to the salt feed tank. The residence time in the transfer line is approximately sixty minutes during normal operations. It is unlikely that volatilization of flammables from the salt solution will occur in the transfer line.

† W.R. Grace, Daratard 17 ASTM C494 Type B and D admixture
‡ Clearwater International LLC, Clear Air 100 ASTM C494 Type B and D admixture
§ Dow Corning Q2 1383-A

Salt Feed Tank

The SFT is an underground 25000 liter tank that is procedurally limited to 14000 liters. The behavior of ammonia in the SFT would be governed by a Henry's law constant that is a function of parameters related to the vessel/impeller characteristics, the air/liquid properties as well as the process conditions. A sparge introduces air into a salt solution with dissolved ammonia. A portion of the ammonia—limited by the appropriate Henry's law constant—is released to the bubble. This carries the now entrained ammonia to the ventilated head space effectively reducing the ammonia concentration in the salt solution. With a working volume of 11000 liters and an in/out flow rate of 340 L/m, the residence time for salt solution in the SFT is approximately 30 minutes. It is not anticipated that significant quantities of flammable vapors will evolve from the salt solution in the SFT. Flammable vapors released in the SFT would be removed to the vessel vent system.

Salt solution is drawn from the SFT and pumped to the grout mixer between 340 to 375 L/m, depending on the formulation. Admixtures are introduced into the pump suction. The salt solution transfer rate and SFT working volume require that the SPF operate under "bleed and feed" conditions where the SFT is taking salt solution transfers from Tank 50 at a similar rate as the salt solution is being pumped to the grout mixer.

Grout Mixer

Premix enters the mixer at approximately 31 metric tons per hour. The salt solution, containing set retarder and defoamer, is added to the premix. The mixer is 0.18 m long with two 25-cm barrels forming a figure "8"—see inset in Figure 1. The working volume in the mixer is approximately 115 liters and the residence time is between 5 to 15 seconds. In the mixer, the salt solution and premix are combined using a low-speed, high-shear mixing process (150 rpm; 3-mm radial clearance between blades and barrel). The mixer breaks up clumps of dry materials rapidly mixes it with the salt solution to wet the particles. Here, ammonium salts that are adhered to the fly ash can begin to be released into the Saltstone slurry and convert to ammonia. The added defoamer reduces the surface tension to improve the release of air entrained in the premix and to mitigate foaming attributed by the salt solution when rapidly mixed. The added set retarder inhibits the initial hydration reactions of the premix to ensure that there is sufficient time after a process set back (The SPF Distributed Control System (DCS) determines that a process parameter is not within allowable limits and shuts the process down.) to remedy the cause of the set back and restart the plant before the grout can begin to gel enough to require excessive pump pressure to initiate flow in the transfer line. The mixer discharges the Saltstone slurry is into the grout hopper.

It is not anticipated that significant quantities of flammable vapors will evolve from the salt solution interaction with premix in the grout mixer. It is also unlikely that any reaction would occur in the mixer to produce flammable components from degradation of non-flammable components in the premix or admixtures. Flammable vapors released in the grout mixer would be removed by the vessel vent system.

Grout Hopper

The Saltstone slurry free falls from the grout mixer into the larger leg of the grout hopper. The working capacity of the hopper is approximately 115 liters. This is similar in volume to the mixer, so it can be estimated that the residence time is also between 5 to 15 seconds. The grout pump is operated to maintain a constant level in the large (measuring) leg of the hopper. As a matter of procedure, the hopper is emptied and flushed with water every 30 minutes.

It is not anticipated that significant quantities of flammable vapors will evolve from the Saltstone slurry in the grout hopper. It is also unlikely that any reactions would occur in the grout hopper to produce flammable components from degradation of non-flammable components in the premix or admixtures. Any flammable vapors released would be removed by the vessel vent system. The Saltstone grout may contain entrained air due to the processing in the grout mixer and from the fall.

Grout Pump and Transfer Line

The grout pump draws the Saltstone slurry from the hopper. The grout pump is a duplex hose pump and is considered a positive displacement pump. The slurry is split into two lines at the inlet to the pump and is recombined at the discharge of the pump prior to exiting the SPF for the SDF. The grout pump transfers the Saltstone slurry to the SDF in 7.5-cm pipe of approximately 460 m of piping.

It is not anticipated that significant quantities of flammable vapors will evolve from the Saltstone slurry in the grout pump/grout line. It is also unlikely that any reactions would occur in the grout pump/grout line to produce flammable components from degradation of non-flammable components in the premix or admixtures. Any flammable vapors released would be released upon entry into the vault. The residence time of the Saltstone slurry in the 460 meters of piping for a transfer rate of 570 L/m is approximately 4 minutes.

Saltstone Disposal Facility

The Saltstone Disposal Facility (vault) is comprised of twelve individual cells each 30 m x 30 m x 7.5 m high. A schematic of an individual cell is shown in Figure 2.

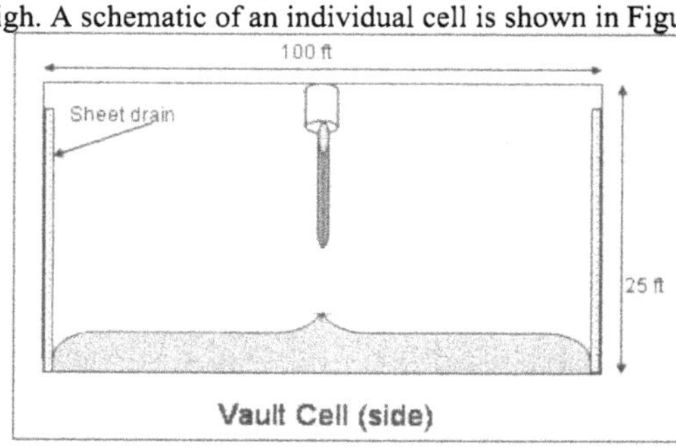

Figure 2. Schematic of Cross-Section of an Individual Cell

Filling

The Saltstone slurry enters the vault in the center of the cell. The slurry free falls between 1.5 to 7.5 meters prior to striking the free surface of the cell or previously poured grout. The impact of the Saltstone slurry striking the free surfaces generates significant aerosol and mist in the cell. The mechanism for the release of ammonia—and other flammables that have considerable solubility in salt solution—to the vapor space of the cell during pouring is governed by Henry's law. Although the action of active pouring into the cell may be kinetic, the equilibrium condition described by Henry's law applies. That is, the concentration of ammonia—and other soluble species such as butanol—in the vapor space is represented by the Henry's Law constant for that species in the specific salt solution at the temperature of interest.

As the slurry enters the cell, small pools of Saltstone slurries form and a "finger" "breaks out" into a stream to release the pool. Either due to the topography of the base (or cured grout) or the constantly changing rheology of the slurry, or a combination of both, flow is arrested or another "finger" is formed. The thickness of the "fingers" has not been measured or estimated. Figure 3 is a schematic of the flow pattern in the cell that is influenced by both yield strength and plastic viscosity.

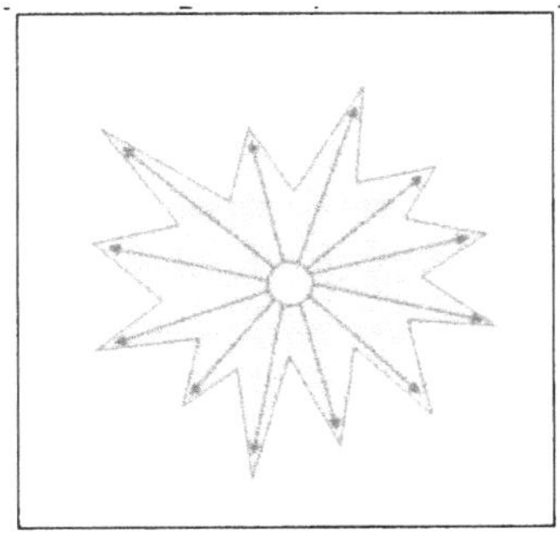

Inside Cell (top)

Figure 3. Flow Pattern of Saltstone slurry in Cell

Saltstone slurries prepared in the laboratory behave like a Bingham plastic fluid. It has been demonstrated that the rheological properties are dependent upon the mixing method so discussion is limited to laboratory work, as no rheological analysis has been performed on samples from the actual saltstone unit operations or similar equipment. The yield stress measured on laboratory samples prepared with simulants is in the range of 2 to10 Pa. The yield stress component of the fluid governs several properties that pertain to plant operation and grout performance. During setback, the development of yield stress dictates the pressure required to initial flow of the saltstone slurry in the transfer line. In the vault, the yield stress is instrumental in maintaining suspension of the premix particles in the salt solution during flow. The development of yield stress is an indication that the cement in the saltstone grout is beginning to hydrate and that the slag is being activated by the salt solution. The increase of yield stress also contributes to the halting of flow. During this time, the remainder of the ammonium salts on the fly ash and slag will be released to the salt solution. The additional ammonia will contribute to the concentration of ammonia in the salt solution and subsequently, increase the equilibrium concentration expected in the vapor. Further increases in the ammonia concentration in the salt solution will occur due to the hydration of the cement. At this time the temperature of the saltstone slurry is governed by the temperature of salt solution transferred from Tank 50 (procedurally limited to 40 °C) and the temperature of the surface of the vault floor—typically 20 °C below the maximum temperature (maximum temperature is procedurally limited to 95 °C). It has been proposed that Henry's law can be used to bound the concentration of ammonia vapor in the head space. The concentration of ammonia in the vapor space may be calculated from the concentration of ammonia in the salt solution—the concentration of ammonia in the Tank 50 salt feed, corrected for the ammonia contribution from the premix—and the Henry's law constant for the salt solution at the temperature of the saltstone slurry, which can be estimated.[3]

Curing

As lifts of saltstone are poured and the premix materials hydrate, heat is generated. For Saltstone slurries currently being produced, the maximum heat generation occurs 12 hours after mixing.

The activation of the slag, and to some extent the fly ash, is comparable to glass dissolution. That is, slag and fly ash are primarily alkaline earth silicates and, when exposed to a caustic environment, begin to dissolve and form hydroxylated silicon which reacts with the alkaline earths and form cementitious compounds. During these reactions, bound or chemisorbed compounds are released to the salt solution. At this time the temperature of the saltstone slurry is governed by the temperature of the previous lift and the energy generated by the hydration reactions occurring. As with the ammonia release during pouring, the concentration of ammonia in the vapor space may be calculated from the concentration of ammonia in the salt solution—adjusted for the concentration of ammonia in the salt solution due to hydration reactions and evaporation, and the Henry's law constant for the salt solution at the temperature of the curing saltstone—procedurally limited to 95 °C. Recent testing has shown that Henry's law bounds the release of ammonia from curing saltstone. Although the ammonia production from the reduction of nitrates and nitrites by the blast furnace slag has not been quantified, it was determined that the quantity of ammonia released from saltstone prepared with CBO fly ash and salt solution spiked with 200 mg/L ammonia and cured at 95 °C will be bounded by Henry's law using the constant determined in the study.

CONCLUSIONS

Flammable compounds in the saltstone process can originate either as flammable compounds in the incoming salt solution, flammable compounds in the premix materials, flammable compounds formed by the decomposition of trace materials added either as grinding aids for the dry materials or admixtures used to destabilize foaming of the saltstone slurry, or flammable compounds due to chemical reactions among the slurry components. The flammable compounds that have been identified from the above sources are:

- Salt solution
    - Ammonia
    - Isopar
- Premix materials
    - Ammonia from blast furnace slag—most likely from grinding aid
- Admixtures
    - Xylene from the Q2-1383a antifoam
- Reactions
    - Ammonia from reduction of nitrates an nitrites by blast furnace slag

- The release of flammable compounds from the Tank 50 transfer pump through the Saltstone Process Facility to the Saltstone Disposal Facility is expected to be minimal.

- The release of ammonia and other flammables to the vapor space of the Saltstone Disposal Facility during pouring is expected to be governed by Henry's law.

- The concentration of ammonia compounds in the Saltstone Disposal Facility is expected to be governed by Henry's law. This has been demonstrated in for saltstone prepared with salt solutions spiked with 200 mg/L ammonia, although the actual concentration of ammonia in the salt solution is not known due to the production of ammonia from reduction of nitrates an nitrites by blast furnace slag, the ammonia measured was comparable to the amount predicted using the Henry's law constant determined for the study.

REFERENCES

1. Harbour, J.R., Hansen, E.K., Edwards, T.B., Williams, V.J., Eibling, R.E., Best, D.R., Missimer, D.M., "Characterization of Slag, Fly Ash and Portland Cement for Saltstone," WSRC-TR-2006-00067.

2. Zamecnik, J.R., Cozzi, A.D., "Ammonia Concentration in Saltstone Headspace – Summary Report," SRNS-STI-2008-00119 Revision 0, September 2008.

3. Swingle, R.F., Porier, M.R., Berry, C.J., Edwards, T.B., and Young, J.E., "Ammonia Henry's Law Constants in SRS High Level Waste Pump Tanks," WSRC-TR-2000-00226.

DEVELOPMENT OF CRYSTAL-TOLERANT WASTE GLASSES

J. Matyáš, J.D. Vienna, A. Kimura, M. Schaible, and R.M. Tate
Pacific Northwest National Laboratory
Richland, WA, USA

ABSTRACT

Crystal-tolerant high-level waste glasses are being developed to allow high waste loading without decreasing melter lifetime by keeping the spinel crystals small and suspended in the molten glass. Adding ~ 1 wt% NiO to the baseline glass resulted in formation of large spinel crystals, up to 210 μm in size, and a crystal accumulation rate of ~ 227 mm/year, the highest of all tested glasses. When noble metals were added to this glass, large spinel crystals did not form and the accumulation rate decreased ~ 8.5 times. Adding ~ 5 wt% Fe_2O_3 to the baseline glass resulted in a high number density of ~ 10-μm spinel crystals that remained suspended in the glass melt even after 17 days at 850°C. The accumulation rate of spinel crystals in high-chromia crucibles was only slightly higher than that in double crucibles, except for baseline glass that was ~ 2.6 times faster. The enrichment of this glass with chromium that leached from the wall resulted in an increased number density of ~ 50-μm crystals.

INTRODUCTION

The main risk for the continuous operation of radioactive-waste-glass melters is that noble metals and spinel crystals will accumulate on the melter bottom and in the glass discharge riser of the melter during extended idling periods. The excessive accumulation of the particle sludge can disrupt the passage of alternating electrical current among submerged electrodes and obstruct the discharge of molten glass, hindering the melter's performance and shortening its lifetime. To avoid or minimize the accumulation of crystals, high-level waste (HLW) glasses have been formulated with a low liquidus temperature (T_L), or temperature T_l at which the equilibrium fraction of spinel crystals in the melt is below 1 vol %, nominally below 1050°C. However, these constraints cannot prevent the formation, growth, and accumulation of spinel crystals in considerably cooler regions (~ 850°C) of the glass discharge riser during melter idling. The slurry-fed ceramic melters that were installed at the Pamela vitrification plant in Belgium in 1985 experienced pluggages of the bottom and overflow nozzles of the glass discharge riser after a few months of operation.[1] In addition, the constraints significantly limit the waste loading, which is reflected in a high volume of waste glass, and thus result in high capital, production, and disposal costs.

The behavior of spinel crystals, such as the dynamics of spinel nucleation, growth, and dissolution in the velocity, temperature, and redox fields that exist in the glass-melting process has been studied in great detail over the past 15 years. Izak et al.[2] investigated the feed-to-glass conversion and showed that submicron high-iron spinel crystals precipitate from the primary (nitrate) melt in the cold cap. These crystals eventually dissolve in the borosilicate melt, but some can survive and act as nucleation sites for crystals that form in the molten glass. Additionally, the nucleation of spinel crystals in the bulk of the melt is affected by temperature and the presence of nucleation agents. The number density of crystals increased by several orders of magnitude as the temperature approached the glass-transition region.[3] The noble metals (Rh, Ru, Pd, and Pt) increased the number density of crystals by up to four orders of magnitude compared with baseline glass.[4] Alton et al.[5] studied the kinetics of the growth and dissolution of spinel crystals over a range of temperatures. The crystal growth rate ranged from 5 to 22 μm/h at temperatures of 750 to 1050°C. Wilson et al.[6] measured the equilibrium concentration of spinel as a function of temperature and composition and showed that additions of 3 wt% NiO to baseline glass increased the theoretical maximum fraction of spinel in glass from 0.08 to 0.14. The settling of spinel crystals in simulated HLW glass was studied in the laboratory using crucible scale experiments.[7,8] The settling distance was measured as the distance between the glass level and the uppermost crystals in the centerline of the crucible.[8] Klouzek et al.[9] observed the settling of monodisperse alumina particles in the liquid of the same viscosity as the viscosity of HLW glass just above the T_L. Mathematical modeling of spinel settling revealed that the crystal size is the main factor determining the rate of settling of individual crystals.[10] The rate of settling increased with the crystal size to the power of 2.3.

Pacific Northwest National Laboratory (PNNL) is studying variations in composition to develop crystal-tolerant HLW glasses. By keeping the spinel crystals small and therefore limiting spinel deposition in the melter, these glasses will allow high waste loading without decreasing melter lifetime. A series of double crucible and high-chromia crucible tests were performed on designed glasses of different compositions to investigate the effect of Cr, Ni, Fe, and noble metals on the accumulation rate of spinel crystals in the glass discharge riser of the HLW melter. In addition, the processing properties of glasses, such as the viscosity and T_L, were measured as a function of temperature and composition.

EXPERIMENTAL

Glass Matrix Design and Fabrication

Baseline glass composition (EMSP-BL) was formulated with a composition data of AZ-101 simulant (see Table I). Fractions of additives (composed of B_2O_3, Li_2O, Na_2O, and SiO_2) and waste loading were adjusted by maximizing the waste loading in the glass while satisfying the glass property and component constraints.[11] A test matrix of eight compositions was designed by changing component fractions of Cr_2O_3, NiO, Fe_2O_3, and noble metals (Rh_2O_3 and RuO_2) in the baseline glass while decreasing the mass fraction of all other components proportionally. Table II shows the composition of designed glasses, including the baseline glass.

Glass batches were prepared from AZ-101 simulant, boric acid, carbonates of Li and Na, SiO_2, Cr_2O_3, NiO, and Fe_2O_3. All the components were hand-mixed in plastic bags, mixed in an agate mill for 5 min, and melted at 1200°C for 1 h in a Pt-10%Rh crucible covered with a Pt lid. Then the glass melt was quenched on a clean steel plate, and the glass was crushed in a tungsten carbide mill for 2 min, remelted for 1 h at the same temperature, and again quenched and crushed.

Table I. Composition of HLW AZ-101 Simulant in Mass Fraction of Oxides and Halogens.

Component		Component		Component		Component	
Al_2O_3	0.2263	Fe_2O_3	0.3999	Rh_2O_3	NA	Cl	0.0006
B_2O_3	0.0159	K_2O	0.0095	RuO_2	NA	Ce_2O_3	0.0056
BaO	0.0026	MgO	0.0036	SiO_2	0.0467	CoO	0.0003
CaO	0.0158	MnO	0.0097	SO_3	0.0023	CuO	0.0011
CdO	0.0178	Na_2O	0.0786	TiO_2	0.0008	La_2O_3	0.0061
Cr_2O_3	0.0047	NiO	0.0175	ZnO	0.0006	Nd_2O_3	0.005
F	0.0002	P_2O_5	0.0088	ZrO_2	0.1148	SnO_2	0.0027

NA: Noble metals were not added.

Table II. Composition of Designed Glasses in Mass Fraction of Oxides and Halogens.

Component	EMSP-BL-(AZ-101)	EMSP-Cr0.6-(AZ-101)	EMSP-Cr1.2-(AZ-101)	EMSP-Ni1.07-(AZ-101)	EMSP-Ni1.5-(AZ-101)	EMSP-Ni1.5 w/noble metals-(AZ-101)[a]	EMSP-Fe20-(AZ-101)	EMSP-Fe20/Ni1.5-(AZ-101)
Al_2O_3	0.0821	0.0818	0.0813	0.0818	0.0814	0.0814	0.0768	0.0759
B_2O_3	0.0799	0.0796	0.0791	0.0796	0.0792	0.0792	0.0748	0.0738
BaO	0.0009	0.0009	0.0009	0.0009	0.0009	0.0009	0.0009	0.0009
CaO	0.0057	0.0057	0.0057	0.0057	0.0057	0.0057	0.0054	0.0053
CdO	0.0064	0.0064	0.0064	0.0064	0.0064	0.0064	0.0060	0.0060
Cr_2O_3	0.0017	**0.0060**	**0.0120**	0.0017	0.0017	0.0017	0.0016	0.0016
F	0.0001	0.0001	0.0001	0.0001	0.0001	0.0001	0.0001	0.0001
Fe_2O_3	0.1451	0.1444	0.1435	0.1444	0.1438	0.1437	**0.2000**	**0.2000**
K_2O	0.0034	0.0034	0.0034	0.0034	0.0034	0.0034	0.0032	0.0032
Li_2O	0.0199	0.0198	0.0197	0.0198	0.0197	0.0197	0.0186	0.0184
MgO	0.0013	0.0013	0.0013	0.0013	0.0013	0.0013	0.0012	0.0012
MnO	0.0035	0.0035	0.0035	0.0035	0.0035	0.0035	0.0033	0.0033
Na_2O	0.1866	0.1858	0.1846	0.1858	0.1850	0.1849	0.1745	0.1725
NiO	0.0064	0.0063	0.0063	**0.0107**	**0.0150**	**0.0150**	0.0060	**0.0150**
P_2O_5	0.0032	0.0032	0.0031	0.0032	0.0032	0.0032	0.0030	0.0029
SiO_2	0.4031	0.4013	0.3988	0.4013	0.3995	0.3994	0.3771	0.3727
SO_3	0.0008	0.0008	0.0008	0.0008	0.0008	0.0008	0.0008	0.0008
TiO_2	0.0003	0.0003	0.0003	0.0003	0.0003	0.0003	0.0003	0.0003
ZnO	0.0002	0.0002	0.0002	0.0002	0.0002	0.0002	0.0002	0.0002
ZrO_2	0.0416	0.0414	0.0412	0.0414	0.0412	0.0412	0.0389	0.0385
Cl	0.0002	0.0002	0.0002	0.0002	0.0002	0.0002	0.0002	0.0002
Ce_2O_3	0.0020	0.0020	0.0020	0.0020	0.0020	0.0020	0.0019	0.0019
CoO	0.0001	0.0001	0.0001	0.0001	0.0001	0.0001	0.0001	0.0001
CuO	0.0004	0.0004	0.0004	0.0004	0.0004	0.0004	0.0004	0.0004
La_2O_3	0.0022	0.0022	0.0022	0.0022	0.0022	0.0022	0.0021	0.0020
Nd_2O_3	0.0018	0.0018	0.0018	0.0018	0.0018	0.0018	0.0017	0.0017
SnO_2	0.0010	0.0010	0.0010	0.0010	0.0010	0.0010	0.0009	0.0009
Waste loading, wt %	36.24	36.09	35.86	36.09	35.93	35.90	33.91	33.52

[a]Added noble metals {0.0003 of Rh_2O_3 and 2.9E-5 of RuO_2 in the form of $Ru(NO)(NO3)_x(OH)_y$}.

Double Crucible and High-Chromia Crucible Tests

Figure 1 shows the cross-section of the double crucible assembly and a high-chromia crucible. In the double crucible test, the alumina crucible was nested in a larger silica crucible and covered with molten glass to eliminate Marangoni convection in the meniscus and bubble generation at the bottom of silica crucibles (undisturbed settling).[7] The core-drilled second silica crucible held the alumina crucible in place and filled the space, saving more than 0.5 kg of glass per test. To minimize the effect of the surface crystallization and to eliminate the impact of temperature history on crystal nucleation and growth, powdered glass was melted in Pt-crucible at 1200°C for 1 h. Then the crucible was removed from the melting furnace, and molten glass was poured into three double crucibles or six high-chromia crucibles that were rested inside the furnace at 850°C, mimicking the temperature in the glass discharge riser. Double and high-chromia crucibles were removed at various times and cross-sectioned. The marked rectangular pieces shown in Figure 1 were cut out from the crucibles, thin-sectioned, and analyzed with scanning electron microscopy-electron dispersive spectroscopy (SEM-EDS) and Clemex image analysis to determine the thickness of the spinel sludge layer and the size and surface fraction of spinel crystals in the accumulated layer.

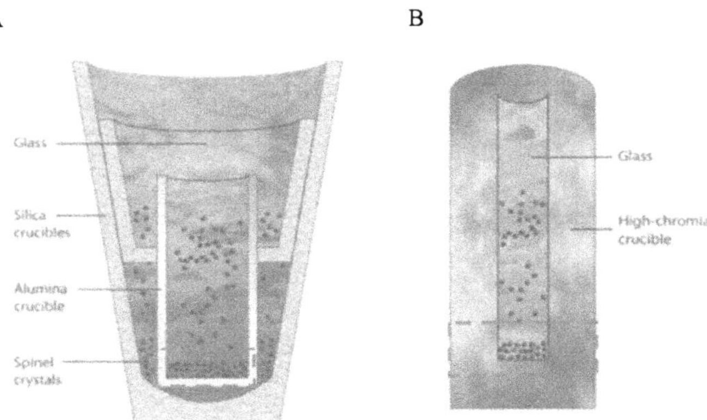

Figure 1. Cross-section of double crucible assembly (A) and high-chromia crucible (B) (red-marked area shows the location of thin-sectioned samples).

Liquidus Temperature

The T_L of EMSP glasses was obtained with heat-treatments of glasses at selected temperatures for 24 ± 2 h in Pt-alloy boxes with tight fitting lids (to reduce volatility). Glass samples were quenched, cross-sectioned, polished, and analyzed with a high-magnification optical microscope for the presence of crystals. The T_L was determined within 5°C of the temperature range between the highest and the lowest temperature at which a glass sample contains or does not contain spinel crystals in the glass, respectively.

Viscosity

The viscosity of selected EMSP glasses was measured as a function of temperature with a fully automated rotating-spindle Brookfield Digital Viscometer (DV-III). The Pt spindle was calibrated with a National Bureau of Standards borosilicate glass (Standard Reference Material 717). Approximately 50 mL of glass, as-measured by EtOH displacement in a volumetric cylinder, was transferred into a Pt/10%Rh crucible and put through a ramp/soak temperature schedule in a high-temperature Deltech® furnace. The viscosity measurement started at ~1150°C and was continued at 50°C intervals during stepwise cooling until the temperature reached 950°C; the viscosity measurement was completed by stepwise heating at 50°C intervals from 950°C to 1150°C. About a 30-min soak time was allowed at each temperature. This hysteresis approach allowed the potential impacts of crystallization at lower temperatures and volatization at higher temperatures via duplicate measurements to be assessed.

RESULTS AND DISCUSSION

Figure 2 shows the growth of a spinel layer on the bottom of the crucible for four glasses that were heat-treated at 850°C for various times. Adding ~1 wt% of NiO or ~ 0.4 wt% of Cr_2O_3 to the baseline glass increased the accumulation rate of spinel crystals by a factor of ~ 9 and ~ 2, respectively. The high accumulation rate (~ 227 mm/year) of spinel crystals in the glass containing 1.5 wt% of NiO is not surprising if we consider the settling of up to 210-μm-sized crystals. Figure 3 shows the time-sequence SEM images of high-Ni glass after heat-treatments at 850°C, visualizing the growth of the spinel layer as well as the change in size and concentration of crystals in this layer over time. Figure 4 details the average crystal size distribution in the top ~ 2.6-mm spinel layer (Figure 3D, A) and in the bottom ~ 0.36-mm spinel layer (Figure 3D, B) for the same glass after heat-treatment at 850°C for 194 h. The average size of crystals in the bottom layer is about 30 μm smaller compared with the average size of crystals in the top part of the settling layer. Figure 5 shows the growth of spinel crystals, as the change of average crystal size, and the layer compaction over the time in high-Ni glass heat-treated at 850°C. The nucleated spinel crystals grew fast, reaching ~ 41 μm in just about 67 h. Addition of noble metals to high-Ni glass {EMSP-Ni1.5 w/noble metals (AZ-101)} decreased the spinel layer thickness from ~ 2.70 mm to ~ 0.32 mm after 7-day heat-treatment at 850°C. The nucleation agents (noble metals) increased the number density of crystals, generating small crystals that settle slowly.

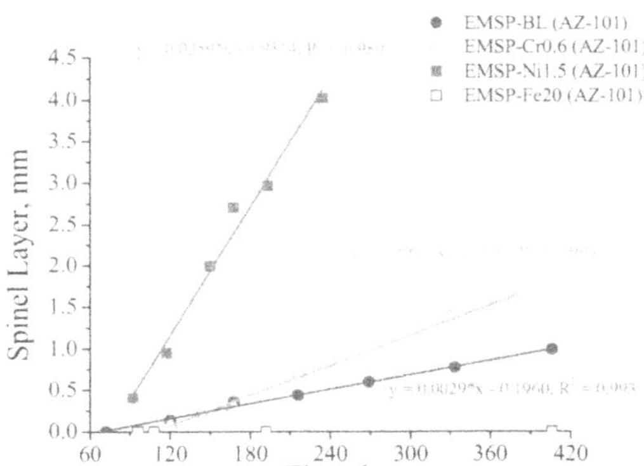

Figure 2. Thickness of spinel layer as a function of time for selected EMSP glasses that were heat-treated in double crucibles at 850°C.

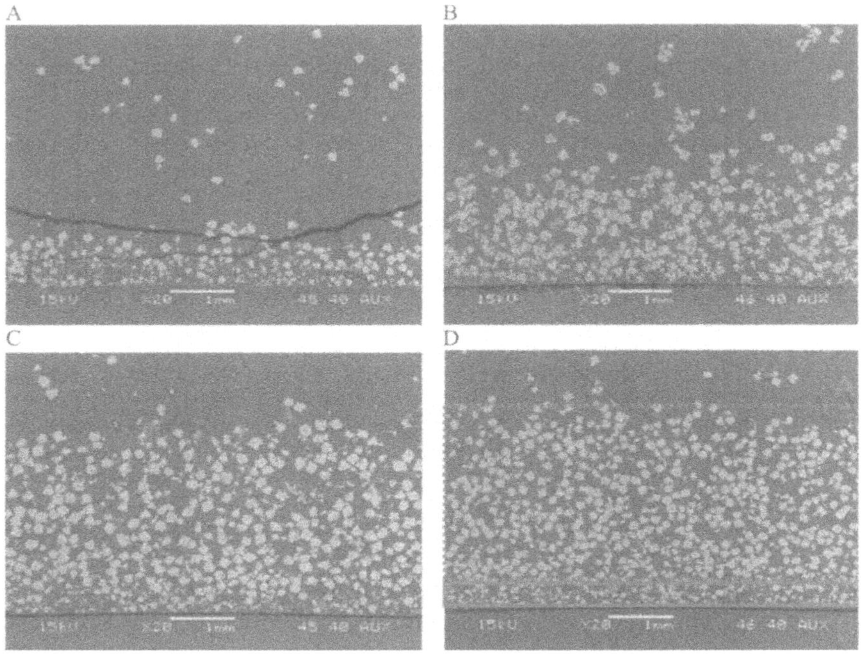

Figure 3. SEM images of spinel layer in EMSP-Ni 1.5 (AZ-101) glass after heat-treatment at 850°C for 92 h (A), 150 h (B), 168 h (C), and 194 h (D).

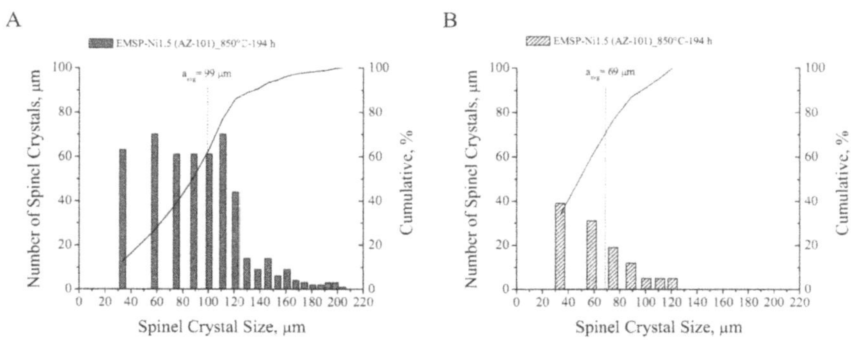

Figure 4. Distribution of average crystal size in the layer of settled spinel in EMSP-Ni 1.5 (AZ-101) glass after heat-treatment at 850°C for 194 h (Figure 3D): top ~ 2.6 mm (A) and bottom ~ 0.36 mm (B).

A question arises whether it would be possible to control the accumulation rate of spinel crystals through optimization of noble metals concentration in radioactive wastes. A small concentration of noble metals may be sufficient to nucleate enough spinel crystals to limit their growth to a size of approximately 10 μm. High concentrations of noble metals are not favorable because of their tendency to form large agglomerates.[12] These agglomerates settle rapidly regardless of phase equilibria. In contrast, the glass containing 20 wt% of Fe_2O_3 did not form the spinel layer even after 17 days at 850°C; see Figure 2. Adding ~ 5 wt% of Fe_2O_3 to the baseline glass significantly increased the number density of spinel crystals (700 to 850 crystals/mm^2) and decreased their size to ~ 10 μm in average; see Figure 6.

Figure 7 shows the thickness of the spinel layer as a function of time for three glasses that were heat-treated in high-chromia crucibles at 850°C. The accumulation rate of spinel crystals in high-chromia crucibles was only slightly higher than that in double crucibles, except for baseline glass that was ~ 2.6 times faster. The enrichment of this glass with chromium that was leached out from the wall resulted in increased number of ~ 50-μm chromium-enriched spinel crystals. Table III summarizes the concentration of Cr, Fe, and Ni in two glasses at the center of the cross-sectioned crucible ~ 2 mm from the left (L) and the right (R) walls and in the middle (M). Glasses were heat-treated in high-chromia crucibles at 850°C for 4 days. Significantly higher concentrations of Cr_2O_3 were detected in areas close to the refractory wall because of the dissolution of high-chromia refractory by molten glasses.

Figure 8 shows the measured viscosity curves and predicted viscosities at 850°C if the linear fit is made to viscosity data for several glasses. Predicted viscosities in the glass discharge riser during idling are all well above 150 Pa.s. Table IV summarizes determined T_L's of the glasses. The T_L of high-Ni glass containing noble metals is the highest and is about 170°C higher compared with baseline glass. The T_L's of glasses increased with Cr>Ni>Fe>noble metals.

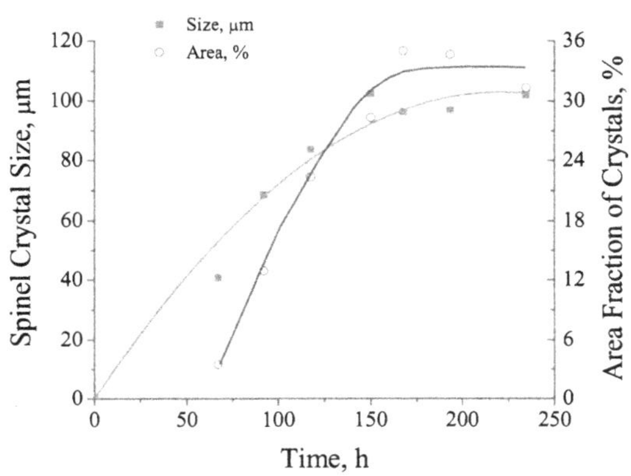

Figure 5. Change of average spinel crystal size and layer compaction over time in EMSP-Ni 1.5 (AZ-101) glass heat-treated at 850°C.

Figure 6. SEM image of EMSP-Fe20 (AZ-101) glass heat-treated at 850°C for 27 days.

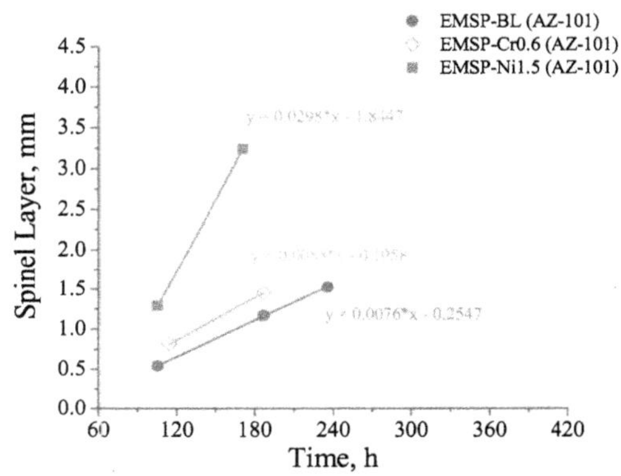

Figure 7. Thickness of spinel layer as a function of time for selected EMSP glasses that were heat-treated in high-chromia crucibles at 850°C.

Table III. Concentration of Cr, Fe, and Ni in glass in mass % of oxides at the center of the cross-sectioned crucible ~ 2 mm from the left (L) and the right (R) walls and in the middle (M). Glasses were heat-treated in high-chromia crucibles at 850°C for 4 days.

Glass ID	Location	Cr_2O_3	Fe_2O_3	NiO
EMSP-BL (AZ-101)	Glass L	0.56	3.32	0.05
	Glass M	0.18	10.7	0.44
	Glass R	1.29	8.48	0.17
EMSP-Ni1.5 (AZ-101	Glass L	1.17	5.83	0.23
	Glass M	0.15	9.5	0.53
	Glass R	0.9	7.05	0.13

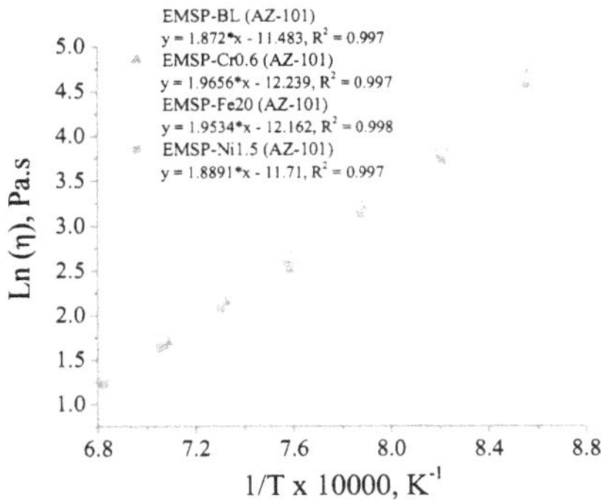

Glass ID	$\eta_{850°C}$, Pa.s
EMSP-BL (AZ-101)	178.5
EMSP-Cr0.6 (AZ-101)	192.8
EMSP-Fe20 (AZ-101)	186.8
EMSP-Ni1.5 (AZ-101)	165.6

Figure 8. Viscosity curves and predicted viscosities of selected EMSP glasses at 850°C.

Table IV. Liquidus temperatures of EMSP glasses.

Glass ID	T_L, °C
EMSP-BL (AZ-101)	964 - 969
EMSP-Cr0.6 (AZ-101)	1017 - 1022
EMSP-Cr1.2 (AZ-101)	1095 - 1100
EMSP-Fe20 (AZ-101)	1105 - 1109
EMSP-Fe20/Ni1.5 (AZ-101)	1210 - 1215
EMSP-Ni1.07 (AZ-101)	1015 - 1018
EMSP-Ni1.5 (AZ-101)	1077 - 1082
EMSP-Ni1.5/noble metals (AZ-101)	1136-1140

CONCLUSION

Spinel, a solid solution of magnetite (Fe_3O_4), trevorite ($NiFe_2O_4$), and nichromite ($NiCr_2O_4$), was the primary crystalline phase of designed EMSP glasses. The main factor that determined the accumulation rate of spinel crystals was the crystal size. Large crystals formed with an increased concentration of NiO and settled rapidly regardless of phase equilibria. A small concentration of noble metals in the molten glasses prevented the formation of large spinel crystals. Increased concentration of Fe_2O_3 resulted in a high number density of ~ 10-μm spinel crystals that remained suspended in the glass. The enrichment of glass with chromium resulted in increased number of ~ 50-μm chromium-enriched spinel crystals.

Our future goal is to evaluate the effect of spinel forming components on size and concentration of crystals, and on accumulation rate. In addition, we will identify minor components that most affect the size of spinel crystals and build an experimental model that can predict crystal accumulation as a function of feed composition. This model would ultimately allow a substantial increase in the waste loading.

ACKNOWLEDGEMENT

This work was funded by the U.S. Department of Energy's Environmental Management Program EM20. Pacific Northwest National Laboratory is operated by Battelle for the U.S. Department of Energy under Contract DE-AC05-76RL01830.

REFERENCES

[1]L.F. Ermold and D.A. Knecht, Foreign travel report: Visits to UK, Belgium, Germany, and France to Benchmark European Spent Fuel and Waste Management Technology, Westinghouse Idaho Nuclear Company, Inc., WINCO-1140, August 1993.

[2]P. Izak, P. Hrma, B.W. Arey, and T.J. Plaisted, Effect of Feed Melting, Temperature History, and Minor Component Addition on Spinel Crystallization in High-Level Waste Glass, *Journal of Non-Crystalline Solids*, 289, 17-29 (2001).

[3]T. Plaisted, J. Alton, B. Wilson, and P. Hrma, Effect of Minor Component Addition on Spinel Crystallization in Simulated High-Level Waste Glass, *Ceramic Transactions*, 119, 291-299 (2001).

[4]J. Alton, T.J. Plaisted, and P. Hrma, Kinetics of Growth of Spinel Crystals in a Borosilicate Glass, *Chemical Engineering Science*, 57, 2503-2509 (2002).

[5]J. Alton, T.J. Plaisted, and P. Hrma, Dissolution and Growth of Spinel Crystals in a Borosilicate Glass, *Journal of Non-Crystalline Solids*, 311, 24-35 (2002).

[6]B.K.Wilson, P. Hrma, J. Alton, T.J. Plaisted, and J.D. Vienna, The Effect of Composition on Spinel Equilibrium and Crystal Size in High-Level Waste Glass, *Journal of Materials Science*, 37, 5327-5331 (2002).

[7]J. LaMont and P. Hrma, A Crucible Study of Spinel Settling in a High-Level Waste Glass, *Ceramic Transactions*, 23, 343-348 (1998).

[8]J. Kloužek, J. Alton, T.J. Plaisted, and P. Hrma, Crucible Study of Spinel Settling in High-Level Waste Glass, *Ceramic. Transactions*. 119, 301-308 (2001).

[9]J. Kloužek and A. Franek, Modeling of Particle settling in high viscosity liquid, *Ceramics-Silikáty*, 45 (2), 70-75 (2001).

[10]J. Matyáš, J. Kloužek, L. Němec, and M. Trochta, Spinel settling in HLW melters, *ICEM'01 The 8th International Conference On Radioactive Waste Management and Environmental Remediation*, Bruges, Belgium (2001).

[11]J.M. Perez, JR, D.F. Bickford, D.E. Day, D.S. Kim, S.L. Lambert, S.L. Marra, D.K. Peeler, D.M. Strachan, M.B. Triplett, J.D. Vienna, and R.S. Wittman, High-Level Waste Melter Study Report, Pacific Northwest National Laboratory, Richland, WA, PNNL-13582 (2001).

[12]K.M. Fox, D.K. Peeler, T.B. Edwards, D.R. Best, I. A. Reamer, R.J. Workman, J.C. Marra, B.J. Riley, J.D. Vienna, J.V. Crum, J. Matyas, A.B. Edmondson, J.B. Lang, N.M. Ibarra, A. Fluegel, A. Aloy, A.V. Trofimenko, R. Soshnikov, International Study of Aluminum Impacts on Crystallization in U.S. High Level Waste Glass, SRNS-STI-2008-00057, Savannah River National Laboratory, Aiken, SC,2008.

PHASE STABILITY OF DEFENSE WASTE PROCESSING FACILITY (DWPF) TYPE HIGH LEVEL NUCLEAR WASTE GLASSES

Amanda L. Billings, Carol M. Jantzen, Connie C. Herman and Sharon L. Marra
Savannah River National Laboratory
Aiken, SC, USA

ABSTRACT

The Defense Waste Processing Facility (DWPF) at the Savannah River Site (SRS) has been vitrifying high-level radioactive waste (HLW) through a Joule heated melter since 1996. As part of waste quality acceptance, phase stability of the borosilicate glass waste forms must be determined. From the commencement of waste processing into the present-day, waste form compositions have varied and phase stability has continued to be determined by creating Time-Temperature-Transformation (TTT) diagrams. Crystallization behavior and TTT diagrams have been examined for varying glass waste compositions using isothermal heat treatments and X-ray diffraction (XRD) phase analysis. Typically, transition metal ferrite based spinel, nepheline, acmite, and lithium silicate crystalline phases are precipitated. Glass composition-crystalline phase stability relationships will be discussed for a recently processed composition and historical compositions.

INTRODUCTION

Glasses, which are amorphous by definition, can transform into more ordered crystalline phases at temperatures below the liquidus and above the glass transition temperature. By annealing a glass within this glass transformation region, the potential for crystal nucleation and growth is high. In nuclear waste glasses, this transformation region is relatively large, encompassing the region of approximately 400 °C to 1050 °C. The crystallization kinetics of nuclear waste glasses can be determined by creating time-temperature-transformation (TTT) curves. This provides information regarding the processing conditions necessary to avoid significant devitrification, which can alter the long-term durability of the waste glass. Generating these curves is also important for nuclear waste glasses as they provide the effects of different accident scenarios (low temperature melter idling, transportation fires) and the effects of long-term (10^6-10^8 years) repository storage.

The Defense Waste Processing Facility (DWPF) is currently immobilizing high-level liquid waste by combining sludge (waste) with glass frit, melting the slurry mix of sludge and frit and pouring the molten glass into stainless steel canisters to create the final waste form at the Savannah River Site (SRS). Eventually, these canistered waste forms will be sent to the Civilian Radioactive Waste Management System (repository) for final disposal. Prior to plant start-up and radioactive operations, TTT curves were completed for projected waste compositions in two separate studies. Beginning in 1982, Jantzen et al. published experiments relating three different waste blend scenarios for two different frit compositions and examined the effects of melt insolubles and oxygen fugacity on devitrification kinetics.[1-4] In 1992, TTT curves were published by Cicero et al. based on updated composition projections just prior to DWPF start-up.[5] Those glasses are commonly referred to as the waste-form compliance plan glasses (WCP).[5] Both studies were anticipated to compositionally bound the DWPF glass waste forms. Since that time, higher waste loadings, less sludge washing, different tank blending scenarios and salt processing strategies have been implemented. In addition, frit development strategies have significantly changed, resulting in higher alkali glass systems being processed to improve melt rate, waste loading, and waste throughput.

Beginning with Sludge Batch 3 (SB3), higher waste loadings, new washing and frit development strategies, as well as the incorporation of excess material streams, pushed projected waste form compositions to a region that appeared to be compositionally different than the previously established regions.[1-5] Compositional differences can be observed in Table I. More specifically, during processing of SB3, DWPF targeted a nominal waste loading (WL) of 38 wt %, much higher than the planned 28 wt % sludge WL. This was done in an effort to maximize waste throughput, which is a function of melt rate and waste loading. In addition, the Liquid Waste Organization (LWO) Tank Farm operations utilized a target end point for sludge washing for SB3 to approximately 1.2M Na^+, as compared to 0.5M Na^+ for previous sludge batches in order to minimize water sent to the evaporator system and to accommodate the excess material streams from the high-level waste canyon. The result was a higher sodium-concentrated sludge, which was compensated by the strategic development of Frit 418. The combination of all of the changes ultimately transitioned the DWPF glass forming system outside the compositional region evaluated prior to radioactive operations. Specifically, note the lower B_2O_3, CuO, K_2O, and TiO_2 (due to abandonment of the Precipitate Hydrolysis process for salt treatment) and the higher Li_2O and Na_2O (due to frit and washing strategy changes) concentrations for the SB3-Frit 418 system in Table I. Therefore, a new TTT diagram for this system was warranted in order to understand the impact of these changes.

Table I. Approximate Waste Form Compositions of Various Glasses Used to Generate TTT Diagrams[1-6]

Oxide (Wt %)	SB3-Frit 418	Range of WCP Glasses	Range of 165 Glasses	Range of 131 Glasses
Al_2O_3	5.96	2.99-7.15	2.59-10.15	3.04-10.51
B_2O_3	5.16	7.03-10.33	6.44-8.23	9.44-11.09
BaO	0.06	0.11-0.25	0.1-0.33	0.04-0.07
CaO	1.13	0.84-1.09	1.14-1.42	1.49-1.63
Ce_2O_3	0.09	0	0	0
Cr_2O_3	0.09	0.09-0.15	0.14-0.19	0.17-0.20
Cs_2O	0	0.02-0.09	0-0.03	0.03
CuO	0.04	0.25-0.45	0-0.06	0.06-0.08
Fe_2O_3	12.75	7.78-13.25	7.53-15.52	11.36-13.52
K_2O	0.08	2.21-3.86	0	0
La_2O_3	0.04	0	0.05-0.11	0.14-0.21
Li_2O	5.16	3.22-4.62	3.96-4.51	3.09-3.53
MgO	1.39	1.41-1.49	0.87-0.95	1.40-1.59
MnO	2.59	1.41-2.54	1.45-2.71	1.96-2.59
MoO_3	0	0.08-0.22	0	0
Na_2O	13.88	8.56-12.62	8.26-9.88	8.90-12.29
Nd_2O_3	0	0.06-0.55	0	0
NiO	0.69	0.41-1.06	0.97-1.94	0.94-1.26
P_2O_5	0	0	0.09-0.16	0.12-0.14
PbO	0.06	0	0.26-0.51	0.42-0.46
RuO_2	0.02	0.01-0.05	0.05-0.10	0.05-0.06

SiO_2	50.20	46.50-55.80	50.26-52.52	43.25-46.00
SO_4	0.43	0	0	0
SrO	0	0	0.05-0.08	0.05-0.06
ThO_2	0	0	0.03-1.07	0.03-1.07
TiO_2	0.01	0.56-1.03	0.07-0.16	0.72-0.75
U_3O_8	0	0	0.91-1.91	1.93-2.16
ZnO	0.06	0	0.04-0.08	0.05-0.09
ZrO_2	0.11	0.05-0.33	0.77-1.09	0.47-0.50

EXPERIMENTAL

Glass Formulation

A single stock of simulated (non-radioactive) SB3-Frit 418 glass was used to support the chemical and physical characterization associated with developing a TTT diagram representative of the "current" compositional region. The current study was based on a non-radioactive version of the waste glasses in order to be consistent with the WCP glass TTT study.[5] More specifically, the sludge composition was normalized without U_3O_8 and ThO_2 and was combined at a waste loading of 38 wt % with Frit 418. Frit 418 has a composition in weight percent of 8 Na_2O, 8 Li_2O, 8 B_2O_3, and 76 SiO_2. The SB3 glass composition used for experimental testing can be found in Table I. RuO_2 was added to the glass at 0.02 wt % because it is present as a minor component in the sludge. It was included in this testing because noble metals such as RuO_2 are not fully soluble in the glass and will often serve as nucleation sites for crystallization, as was observed in an earlier study by Bickford et al.[2]

Glass Fabrication

It should be noted that kilogram quantities of glass were required to support the SB3-Frit 418 TTT development task. This required multiple individual batches targeting the same composition to be melted and then combined into one large cache of glass. The glass was prepared from the proper proportions of reagent-grade oxides, carbonates, boric acid, and salts in fourteen 300 g batches (numbered 1-14). The raw materials were thoroughly mixed and placed into 95% platinum/5% gold, 600 mL crucibles. The crucibles were placed into an electrically heated, high-temperature furnace at the target melt temperature of 1150 °C. After an isothermal hold at 1150 °C for 1.0 hour, the crucibles were removed from the furnace and poured onto a clean, stainless steel plate and allowed to air cool (quench). The glasses were then crushed and pairs of the fourteen 300 g batches were combined into seven 600 g batches (named A-G) and re-melted at 1150 °C for 1.0 hour of homogenization following the same procedure as above.

Chemical Analysis

To confirm that the as-fabricated glasses corresponded to the target composition, representative samples of each of the 600 g glass batches were analyzed by means of two dissolution techniques, sodium peroxide fusion (PF) and lithium-metaborate fusion (LM). The samples prepared by LM were used to measure Al, Ba, Ca, Ce, Cr, Cu, Fe, K, La, Mg, Mn, Na, Ni, Pb, Ru, S, Si, Ti Zn, and Zr concentrations. Samples prepared by PF were used to measure B and Li concentrations. Each glass was prepared in duplicate for each cation dissolution technique. All of the prepared samples were analyzed by Inductively Coupled Plasma – Atomic Emission Spectroscopy (ICP-AES) with the instrumentation being re-calibrated between the duplicate analyses. Glass standards were also intermittently measured to assess the performance of the ICP-AES instrument over the course of the analyses. Specifically, several samples of WCP Batch 1 (BCH)[7] were included in the sample analysis.

Results of the chemical analyses were reviewed before initiation of any measurements or heat treatments.[8] The objective of the review was to facilitate a decision on the use of all or only a subset of the available batches of glass (A-G). The individual batches determined to be compositionally consistent could then be combined in order to create a cache of compositionally consistent glass that could be used to support the glass transition measurement and TTT study. The result of the statistical review of the measured compositions revealed some differences between the average measured concentrations of MnO for the SB3-TTT-G batch and the rest of the batches. Due to this irreconcilable difference, it was recommended that this batch be excluded from the cache of glass that was to be created.[8] Eliminating batch G from the cache ensured that a compositionally consistent glass would be used to support programmatic objectives. Batches A-F were then combined and mixed thoroughly in one container where samples could subsequently be drawn for heat treatments.

Glass Transition Temperature Determination

A Netzsch 409 PC Luxx differential scanning calorimeter (DSC) was used to measure the glass transition temperature (T_g) of the glass. The ASTM procedure E 1356 was also used as a guide for determining the T_g.[9] A portion of the quenched glass was ground and sieved to a particle size between 100 and 200 mesh. A heating rate of 20 K/min was used to heat three separate samples from room temperature to 1000 °C in a platinum pan. The value of T_g reported in this document was determined by the slope intercept method at the onset of the endothermic inflection and is the average of three separate tests.

Heat Treatments

In order to develop a TTT diagram for the SB3 Frit-418 glass composition, glasses were heat treated as follows. Approximately 30 g of glass from the chemically verified cache were placed in open platinum alloy crucibles and re-melted at 1150 °C for 45 minutes. Platinum alloy crucibles were used instead of alumina crucibles in order to minimize the possibility of alumina dissolution during long-term heat treatments. Durations shorter than one hour were used in order to minimize any volatilization of the chemically verified glass, yet still maintain a homogenized melt. After 45 minutes, the crucibles were removed from the furnace and placed into separate furnaces preheated to the desired isothermal hold temperature. Once the desired duration of heat treatment was reached, crucibles were removed from the furnace and immersed in a shallow water bath to quickly cool or "quench" the glass. This process only exposed the exterior of the crucible to the water (i.e., water did not contact the glass). Each glass sample was heat treated at a unique time and temperature in order to cover the anticipated crystallization region. An independent thermocouple and temperature recorder were used during heat treatments to ensure that the temperature remained relatively constant during the treatments. A sample of the glass was also heat treated according to the centerline canister cooling (CCC) temperature profile in order to simulate DWPF pouring conditions.[10] The CCC glass, along with a sample of the original "quenched" glass (non-heat treated glass from the glass cache), was submitted along with all other heat treated samples for further analysis.

Phase Analysis (Qualitative and Quantitative)

Glass samples were evaluated for crystallization using X-ray Diffraction (XRD). Depending on the type and extent of crystallization, the existence of crystalline phases in the glass can impact properties such as durability. The crystallization information may also provide insight into the potential liquidus temperature (T_L) of the glass since crystals should not be identified above the T_L. Samples were run under conditions providing a detection limit of approximately 0.5 vol %. That is, if crystals (or undissolved solids) were present at 0.5 vol % or greater, the diffractometer would not only be capable of detecting the crystals but would also allow a qualitative determination of the type of crystal(s) present. Jade 9.0© software was used for phase analysis.

For those samples which exhibited a crystalline phase, quantitative XRD measurements were also conducted. A whole pattern Rietveld least squared refinement was conducted on samples using 10 wt % high purity alumina as an internal intensity standard. The Archimedes method was used to measure the density of the original glass[11] and standard densities of the crystalline phases were referenced. Once both densities were determined, total volume percent crystallinity was calculated.

RESULTS AND DISCUSSION

Chemical Composition

The targeted and measured compositions of the final SB3-Frit 418 glass cache, which was used for heat treatments, are found in Table II. No significant discrepancies were found between the targeted and analyzed compositions. This indicates that the batching, melting and batch down-select process were successful and any sample taken from the glass cache would be representative of the measured composition. A more detailed description of the statistical analysis of the glass batches is discussed in a report by T.B. Edwards.[8]

Table II. Target and Overall Measured Concentrations by Weight Percent Oxide for the Cache of SB3/Frit-418 Glass[8]

Oxide	Target	Measured
Al_2O_3	5.96	6.11
B_2O_3	5.16	5.12
BaO	0.06	0.05
CaO	1.14	1.09
Ce_2O_3	0.09	0.11
Cr_2O_3	0.09	0.09
CuO	0.04	0.04
Fe_2O_3	12.75	12.39
K_2O	0.08	0.11
La_2O_3	0.05	0.04
Li_2O	5.16	5.24
MgO	1.39	1.29
MnO	2.59	2.77
Na_2O	13.88	13.95
NiO	0.69	0.63

PbO	0.06	0.05
RuO_2	0.02	0.01
SiO_2	50.20	50.46
SO_4	0.43	0.40
TiO_2	0.01	0.01
ZnO	0.06	0.06
ZrO_2	0.11	0.10
Sum	100.00	100.12

Glass Transition Temperature (T_g)

The glass transition onset temperature for the SB3-Frit 418 glass was measured using DSC. The onset of T_g was determined to occur at 443 ±3 °C. The T_g is an indicator of the approximate temperature where the supercooled liquid converts to a solid on cooling or conversely, where the solid begins to behave as a viscoelastic solid on heating. This is extremely important because crystallization cannot occur below the onset of T_g. These results are consistent with the XRD data for this study, as no crystallization was detected in any of the samples treated at 400 °C. The T_g of the SB3 glass is also comparable to the measured glass transition ranges for the seven WCP glasses. The glass transition range for those glasses was between 418 °C and 460 °C.[12]

Phase Analysis (X-ray Diffraction)

XRD was completed on every heat treated sample as well as quenched and CCC heat treated samples of the SB3 glass. Table III lists the results of the XRD analysis on all of the samples which contained a crystalline phase. Only eleven samples were found to contain a crystalline phase or multiple crystalline phases. Crystallization only occurred at extended durations of heat treatment time (48 hours and longer) at temperatures between 600 °C and 900 °C. No crystallization was detected for this glass composition at temperatures below 600 °C, above 900 °C, at heat treatment times less than 48 hours, or in the quenched or CCC treated glass samples. Phases that could potentially affect the durability of the residual glass, such as nepheline ($NaAlSiO_4$) and lithium metasilicate (Li_2SiO_3), were present in samples treated at 600 °C for 192 and 768 hours and samples treated at 700 °C for 48, 192 and 785 hours. Sodium transition metal silicate-based phases, which included aegirine ($NaFe^{3+}(SiO_3)_2$) (synonymous to acmite) and krinovite ($NaMg_2CrSi_3O_{10}$), sodium aluminosilicate-based phases such as albite ($NaAlSi_3O_8$) and nepheline, as well as lithium metasilicate were detected in samples treated at 600 °C for 192 and 768 hours and samples treated at 700 °C and 800 °C for 48, 192 and 785 hours. Samples treated at 800 °C and 900 °C, where the spinel-type phase trevorite ($NiFe_2O_4$) crystallized, are not expected to exhibit a significant decrease in durability for the remaining glass.

As compared to the phases precipitated in the WCP glasses and earlier waste glass studies by Jantzen and Bickford (the 165 and 131 glasses listed in Table I),[1-5] all crystalline phases detected were the same, with the exception of krinovite and albite. These two phases were only detected in one sample treated at 700 °C for 48 hours. These phases do however remain in the same crystalline families and are similar in composition to aegirine and nepheline. This confirms that no new phase end-members were formed in the SB3-Frit 418 glass even though new waste processing strategies were implemented.

For the eleven samples which contained crystalline phases, quantitative measurements of each of the phases were completed. The results of the quantitative XRD analysis can also be found in Table III. In this table, volume percent crystallinity for each phase detected by XRD, as well as total crystallinity, is listed. The quantitative results show that the maximum amount of crystallization (for those times and temperatures tested) occurred at 600 °C for 768 hours of heat treatment. After that duration, approximately 53 vol % of the sample was crystalline. The second most crystallized region occurred at 700 °C with approximately 29 vol % crystals after 785 hours of treatment. Heat treatments at 800 and 900 °C resulted in fewer crystals being formed than at 600 and 700 °C.

Table III. Quantitative X-ray Diffraction Results

Heat Treatment Conditions	Phases	Vol % Crystalline	Total Vol % Crystalline	Vol % Amorphous
600 °C-192 Hrs	Lithium Silicate (Li_2SiO_3)	8.29	17.33	82.67
	Aegirine ($NaFe^{3+}(SiO_3)_2$)	3.06		
	Nepheline ($NaAlSiO_4$)	5.97		
600 °C -768 Hrs	Aegirine ($NaFe^{3+}(SiO_3)_2$)	32.07	52.95	47.05
	Lithium Silicate (Li_2SiO_3)	10.86		
	Nepheline ($NaAlSiO_4$)	10.01		
700 °C -48 Hrs	Aegirine ($NaFe^{3+}(SiO_3)_2$) and Krinovite ($NaMg_2CrSi_3O_{10}$)*	13.49	14.96	85.04
	Lithium Silicate (Li_2SiO_3)	1.47		
700 °C -192 Hrs	Lithium Silicate (Li_2SiO_3)	7.41	26.07	73.93
	Aegirine ($NaFe^{3+}(SiO_3)_2$)	18.66		
700 °C -785 Hrs	Lithium Silicate (Li_2SiO_3)	9.67	28.84	71.16
	Aegirine ($NaFe^{3+}(SiO_3)_2$)	19.17		
800 °C -48 Hrs	Aegirine ($NaFe^{3+}(SiO_3)_2$)	0.67	0.67	99.33
800 °C -192 Hrs	Trevorite ($NiFe_2O_4$)	0.99	0.99	99.01
800 °C -785 Hrs	Trevorite ($NiFe_2O_4$)	2.55	3.60	96.40
	Aegirine ($NaFe^{3+}(SiO_3)_2$)	1.05		
900 °C -48 Hrs	Trevorite ($NiFe_2O_4$)	0.60	0.60	99.40
900 °C -192 Hrs	Trevorite ($NiFe_2O_4$)	1.39	1.39	98.61
900 °C -768 Hrs	Trevorite ($NiFe_2O_4$)	0.60	0.60	99.40

* The crystalline phases formed in this sample were complex and formed a solid-solution which made quantitative analyses difficult and volume percentages had to be estimated.

After evaluating the XRD data, a TTT diagram could be developed (Figure 1). Qualitative phase analysis results were used to determine the preferred regions of crystallization which are plotted on a log time (in hours) versus temperature (°C) scale. All samples which remained amorphous after the specified heat treatment temperature and duration are represented by a solid square. The location and arrangement of the crystalline phase data points are used to define the estimated location of a phase transformation line, which is represented by the dashed line and indicates the time and temperature where the phase transitions begin to occur. The profile of the CCC curve is also plotted on the TTT diagram for reference.

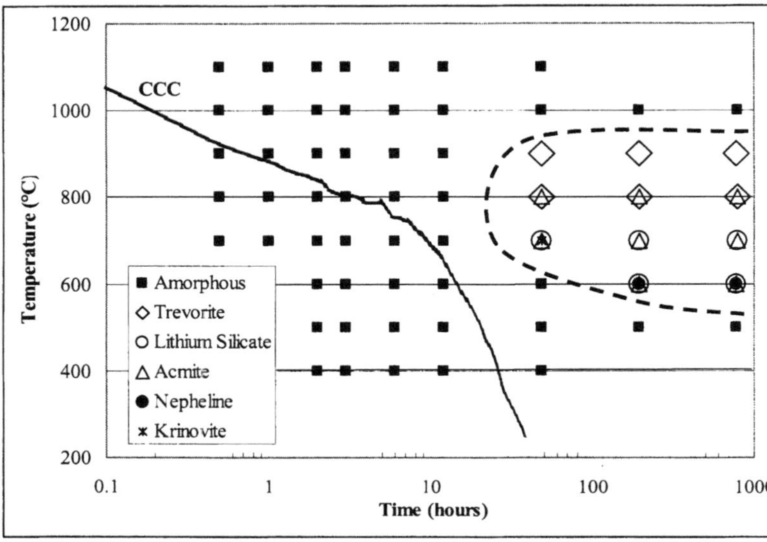

Figure 1. Time-Temperature-Transformation Diagram for SB3-Frit 418 Glass

The TTT diagram for SB3-Frit 418 at a 38 wt % WL has four main crystalline phases: nepheline, lithium silicate, aegirine, and trevorite, and two minor phases: krinovite and a very small amount of albite. The entire crystallization region was encompassed by the spinel-type phase, trevorite, or iron silicate-based aegirine. At lower temperatures (700 °C and 600 °C), lithium silicate and nepheline formed as well as aegirine. One sample after 48 hours of treatment at 700 °C, contained a significant concentration of complex crystalline solutions (> 14 vol % total) consisting of krinovite, aegirine and albite as well as lithium silicate.

It is important to note that all crystallization occurred at times significantly beyond the heat-treatment profile represented by the CCC curve. This suggests that this glass is extremely stable and resistant to crystallization upon simulation of prototypic melter and pouring operations in DWPF represented by a CCC profile-treated glass. Also, the CCC treated sample complements the results of the above statement and TTT diagram, and exhibits no detectable crystallization. Also note that the two potentially durability affecting phases, nepheline and lithium silicate, did not occur autonomously. Those phase regions occurred in the major phase region of aegirine and were also present with albite and krinovite. Although not observed in past studies, this suggests that major silicate phases such as lithium silicate and nepheline could possibly be nucleated by aegirine at these lower temperatures.

The phases formed under the treatment conditions discussed are similar to those established by the previous WCP glass study (Figure 2).[5] Trevorite was found to occur most often at the higher treatment temperatures, while silicate and sodium aluminosilicate phase regions (lithium metasilicate and nepheline specifically) occurred around 600-700 °C most frequently. Even with the different blending, washing and frit development strategies, it appears that crystallization regions are very similar to those previously studied.[5]

The two major differences in the TTT diagrams of the WCP glasses, 165 and 131 glasses and the SB3-Frit 418 glass are the shape and location of the nose of the TTT curve. In Figures 2a and 2c there is a noticeable double nose signifying a separation between two or more different crystallization regions. This appeared when crystallization occurred more frequently at ≥1000 °C at shorter times. The location of the nose of the SB3-Frit 418 TTT curve (Figure 1) also shifted from shorter heat treatment times to longer heat treatment times as compared to the WCP TTT diagrams (Figure 2). This suggests that the more recent glass waste form is more stable (or less prone to crystallization) at those shorter times than the previously examined glass compositions.

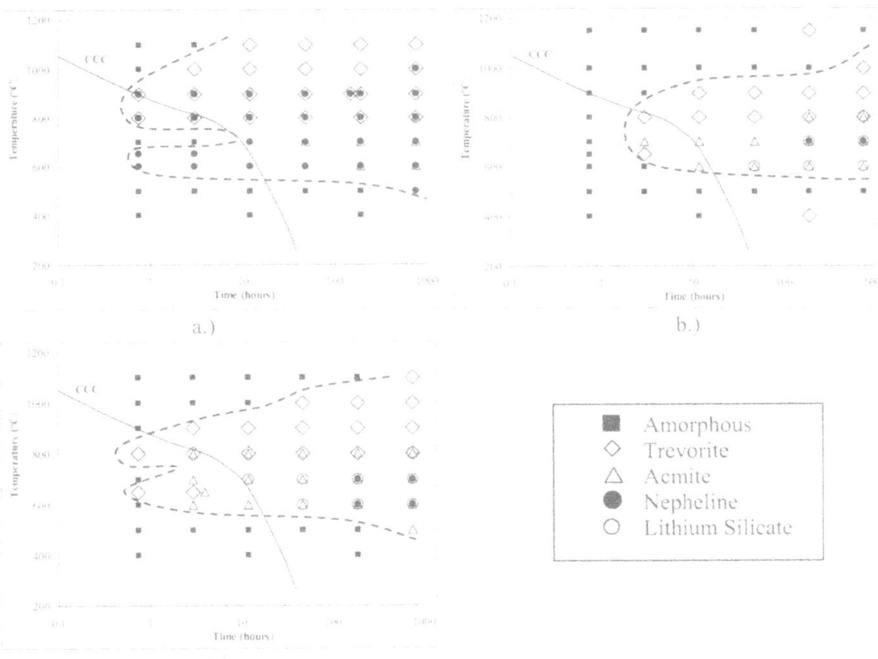

Figure 2. TTT diagrams from the WCP glass study. These diagrams were generated from a.) Purex type waste glass, b.) HM type waste glass, and c.) glass from a blend of the two types of wastes.

CONCLUSIONS

A TTT diagram was completed for a simulated waste glass based on SB3-Frit 418 at a 38 wt % WL. The diagram was generated using one glass composition heat treated at various temperatures and durations of time. Crystallization only occurred in the temperature region of 600 - 900 °C at times greater than 48 hours. For those samples in which crystallization was detected, trevorite and/or aegirine (acmite) were primarily identified by XRD. Lithium metasilicate and nepheline were major phases formed in addition to aegirine at the 600 and 700 °C temperatures. Crystallization was most prevalent at 600 °C as approximately 53 vol % of a sample heat treated for 768 hours at that temperature was crystallized.

The onset T_g of this particular glass was measured at 443 °C ±3, therefore this glass will never exhibit a change in morphology below that temperature, no matter the duration. Of the matrix of times and temperatures tested, crystallization did not occur until far beyond the temperatures and times represented by the CCC curve. The T_g of this glass fell within the region of measured temperatures of the WCP glasses.[5] Phases remained the same or remained within the same crystalline families as those previously observed in the studies of the WCP, 165 and 131 glasses.[1-5] Regions of maximum crystallization also remained within the same temperature regions (500 °C - 700 °C) found in the previous studies.[1-5] This supports that even under extremely different processing strategies, and varying glass compositions, the waste form still behaved similarly with respect to phase transformation when heat treated under abnormal temperature conditions. The conditions imposed on the glass during this study are highly unlikely to occur to any waste glass currently being stored at DWPF.

REFERENCES

[1]D.F. Bickford and C.M. Jantzen, "Vitrification of SRL Defense Waste Glass," pp. 557-65 in Vol. VII, *Scientific Bases for Nuclear Waste Management.* Edited by G. L. McVay. Elsevier, NY, 1985.

[2]D.F. Bickford and C.M. Jantzen, "Devitrification of Defense Nuclear Waste Glasses: Role of Melt Insolubles," *J. Non-Cryst. Solids,* **84** [1-3] 299-307 (1986).

[3]C.M. Jantzen and D.F. Bickford, "Leaching of Devitrified Glass Containing Simulated SRP Nuclear Waste," pp. 135-46 in Vol. VII, *Scientific Bases for Nuclear Waste Management.* Edited by G. L. McVay. Elsevier, NY, 1985.

[4]C.M. Jantzen, D.F. Bickford, and D.G. Karraker, "Time-Temperature-Transformation [TTT] Kinetics in SRL Waste Glass," *Advances in Ceramics,* **8** 30-8 (1984).

[5]C.A. Cicero, S.L. Marra, and M.K. Andrews, "Phase Stability Determinations of DWPF Waste Glasses (U)," Westinghouse Savannah River Company, Aiken, SC, WSRC-TR-93-00227, Revision 0, May, 1993.

[6]A.L. Billings and T.B. Edwards, "Time-Temperature-Transformation (TTT) Diagrams for the Sludge Batch 3 - Frit 418 Glass System," Savannah River National Laboratory, Aiken, SC, SRNL-STI-2009-00025, 2009.

[7]C.M. Jantzen, M.A. Pickett, K.G. Brown, T.B. Edwards, and D.C. Beam, "Process/Product Models for the Defense Waste Processing Facility (DWPF): Part I. Predicting Glass Durability from Composition Using a Thermodynamic Hydration Energy Reaction Model (THERMO)," Savannah River Technology Center, Aiken,SC, WSRC-TR-93-0672, 1993.

[8]T.B. Edwards, "A Statistical Review of the Chemical Compositions of the Batched Glass for the SB3 TTT Study," Savannah River National Laboratory, Aiken, SC, Inter-Office Memorandum SRNL-SCS-2007-00027, 2007.
[9]"Standard Test Method for Assignment of the Glass Transition Temperatures by Differential Scanning Calorimetry," American Society for Testing and Materials (ASTM), West Conshohocken, PA, ASTM Standard E 1356, 2003.
[10]S.L. Marra and C.M. Jantzen, "Characterization of Projected DWPF Glass Heat Treated to Simulate Canister Centerline Cooling," Westinghouse Savannah River Company, Aiken, SC, WSRC-TR-92-142, Rev. 1, 1993.
[11]"Glass Density Using the Mettler AT400 (or Equivalent Balance)," Savannah River National Laboratory, Aiken, SC, SRTC Procedure Manual, L29, ITS-0057, 2006.
[12]B.H. Culbertson, S.L. Marra, M.K. Andrews, C.A. Cicero, B.J. Hardy, and M.J. Plodinec, "Phase. Stability and Control of the Temperature of the DWPF Product," Savannah River National Laboratory, Aiken, SC, WSRC-IM-91-116-7, Rev. 1, 2006.

ALKALI/ AKALINE-EARTH CONTENT EFFECTS ON PROPERTIES OF HIGH-ALUMINA NUCLEAR WASTE GLASSES

J. S. McCloy, C. Rodriguez, C. Windisch, C. Leslie, M.J. Schweiger, B. R. Riley, J. D. Vienna
Pacific Northwest National Laboratory
Richland, Washington 99352, United States

ABSTRACT

A series of high alumina (>20 mass %) borosilicate glasses were made and characterized based on the assumption that the primary modifier cation field strength plays a significant role in mediating glass structure of nuclear waste glasses. Any crystallization upon quenching or after heat treatment at 950°C for 24 hours was identified and quantified by X-ray diffraction. Particular note was taken of any aluminosilicates formed, such as those in the nepheline group ($MAlSiO_4$ where M=K, Na, Li), as these remove multiple glass-formers from the network upon crystallization. The relative roles of potassium, sodium, lithium, calcium, and magnesium on glass structure and crystallization in high alumina glasses were explored using Raman and infrared vibrational spectroscopy. Strong evidence was found for the importance of four-membered rings in glasses with 10 mol % alkaline earths (Ca, Mg).

INTRODUCTION

Due to the life-cycle costs associated with fabrication, storage, and disposal of high-level nuclear waste, it is desirable to maximize the waste loading to decrease the overall volume of vitrified waste for storage. One of the significant limitations to waste loading is the aluminum concentration for which elaborate separation steps have been devised to lower its concentration in waste feed. It is hoped that an understanding of the glass structure would help clarify the susceptibility of high-aluminum glasses to nepheline ($NaAlSiO_4$) crystallization, which results in reduced chemical durability as measured by the Product Consistency Test (PCT).[1] It would thus be very beneficial to understand the limitations of high-aluminum glasses and be able to predict crystallization and processing properties.

The goal of this project is to develop new models and refine those currently existing for glasses with high-aluminum (high waste loading) for nuclear waste vitrification. A focus was made on high-aluminum waste compositions, rather than high iron or bismuth, because > 50% of Hanford wastes and >30% of Savannah River wastes are high in aluminum.

Previously, thermochemical models based on freezing-point depression of quasi-crystalline precursors have been proposed to explain the nepheline-spinel liquidus,[2] but no structural model has been put forward that describes the relationship between glass structure and nepheline formation. High-waste-loaded glasses are particularly vulnerable to nepheline formation because of their high sodium and aluminum contents, which increase the liquidus temperature.[3] Previously, it had been shown that nepheline was unlikely to form outside of its primary phase field in the ternary Na_2O-Al_2O_3-SiO_2 submixture phase diagram, and the "nepheline discriminator" was introduced, which showed that nepheline formation was highly unlikely when $SiO_2/(SiO_2+Al_2O_3+Na_2O) > 0.62$, where oxide components are in mass fraction.[1;4] While very useful, this criterion is very restrictive for glasses with high waste-loading, which tend to contain more alumina and less silica. Thus, there was much interest in pursuing ways to push the acceptable glass formulation space beyond this criterion and into composition regions where higher alumina concentrations are acceptable.

The focus of this current study was to explore the effect of boron and the alkali/ alkaline-earth modifier on the crystal structure and propensity for aluminosilicate (i.e., nepheline) crystallization in representative high-alumina nuclear waste alumino-borosilicate glasses.

EXPERIMENTAL

Initially, experimentation began with a baseline waste composition that was representative of a high-alumina waste stream from Hanford, but renormalized without the minor components (BaO, CdO, SO_3, TiO_2, and ZnO) or the radioactive components (see Table I). The low concentrations of radioactive components of the original baseline waste were thought to be inconsequential for the properties of interest and so and were eliminated from the study. In general, we fabricated the glasses using oxides and carbonates, melted them at 1150 to 1400°C for ~1 hour (as indicated in Table II), crushed and remelted them at the same or slightly lower temperature for ~1 hour, and then quenched them. Target compositions for all the glasses are listed in Table III. Glasses were then heat treated at 950°C for 24 hours and quenched again.

X-ray diffraction was used to identify crystalline phases using JADE® and quantify them using RIQAS®. A Scintag Pad V X-ray diffractometer (XRD) with Cu Kα radiation (λ = 1.5406 Å, 45 kV, and 40 mA) and a Peltier-cooled Si(Li) solid-state detector was used for diffraction analysis. The measurements were done with θ to 2θ geometry in a step-scan approach from 5° to 110° 2θ using a step size of 0.04° 2θ and a dwell time of 4 seconds per step. Raman spectra were collected using a Raman spectrometer (SPEX 1877), a liquid nitrogen-cooled silicon charged coupled device (CCD) detector (Princeton instruments), and an argon ion laser using the 488-nm excitation line. Data were collected in backscatter geometry using a 400-µm slit, with approximately 100 mW of laser power at the sample, and a 100-second acquisition time. Infrared absorption was collected with a Fourier Transform Infrared Spectrometer (Thermo Nicolet 6700) using a deuterated tri- glycine sulfate (DTGS) detector. Data were averaged over 16 scans collected from 4000 to 400 cm^{-1} at 2 cm^{-1} resolution in transmission mode using 0.5 mass % of the powdered glass samples pressed into KBr pellets, and a blank KBr pellet was used as a background. The Beer-Lambert law was used for conversion to absorbance.

The first set of glasses was then formulated to test the effect of the aluminum-to-boron ratio in alumino-borosilicate waste glasses. To the baseline waste composition was added only Na_2O, B_2O_3, and SiO_2. Enough Na_2O was added to bring the concentration of these glasses to approximately 10 mol %, and then various amounts of B_2O_3 and SiO_2 were added to achieve different Al/B ratios (e.g., Na-Lo, Na-Lo2, and NaHi2). Having very low modifier content, these glasses were difficult to work with and when heat treated crystallized many iron compounds. It was decided, then, to decrease the alumina content and increase the Na_2O content to 20 mol %. Again, the B_2O_3 and Al_2O_3 ratios were modified, this time to three levels, keeping the other constituents constant (e.g., Na20m+B, Na20m, Na20m-B). The rationale for exploring high boron contents was that relatively few high boron glasses were found in the database,[5] and boron is believed to have an inhibiting effect on nepheline formation due to a lowering of the activity of the Na_2O in the glass melt.[6; 7; 8] Upon heat treatment, only the last glass (Na20m-B) with the least boron (highest Al/B) formed nepheline (28.6 mass %). It was then decided that this composition would be the baseline for further investigation.

A third series of glasses was then made, adding the boron and silicon to achieve the aforementioned Al/B ratio. Compositions were formulated so that there was 20 mol % of either K_2O or Li_2O (e.g., K20m-B, Li20m-B), and no extra Na_2O was added apart from what was defined in the waste. Since magnesium and calcium do not have direct chemical analogues of nepheline, despite their ability to substitute into the nepheline structure to some extent,[9] it was decided to make a final series of glasses with 10 mol % Na_2O with an additional 10 mol % of either K_2O, Li_2O, CaO, or MgO.

A summary of the waste loading, aluminum loading, nepheline discriminator, Al/B ratio, and M_2O/MO ratio (where $M_2O = Li_2O + Na_2O + K_2O$ and MO = MgO + CaO) is listed in Table IV. With the exception of the first set of glasses (Na-Lo, Na-Lo2, Na-Hi2), all the glasses studied here were fully compensated (i.e., peralkaline), that is, the aluminum tetrahedra were charge compensated by alkali and (in the case of Ca10 and Mg10) alkaline-earth modifiers.

RESULTS

Crystal fraction determination by XRD

Samples of quenched and heat-treated-glasses were crushed and doped with ~5 mass % crushed CaF_2 and scanned by XRD. Crystalline phases and their mass fractions are listed in Table III.

As previously mentioned, the first glass that successfully formed nepheline ($Na_{6.65}Al_{6.24}Si_{9.76}O_{32}$, 28.6 mass %) was Na20m-B, while other glasses formed mainly various transition-metal spinels. Upon quenching, the K20m-B would not form a glass at all, but instead formed a porous "rock" that was found by XRD to consist mostly of kalsilite ($KAlSiO_4$, 56.5 mass %), the potassium analogue of nepheline, with a small quantity of carnegieite ($NaAlSiO_4$, 3 mass %), a high-temperature form of nepheline. The lithium analogue, on the other hand, formed no aluminosilicates on quenching, but after heat treatment consisted mostly of $LiAlSiO_4$ (49.1 mass %). There was some difficulty in identifying this lithium aluminosilicate, as the XRD powder pattern did not list a space group or further references. Upon comparison of the Raman spectrum (see below), the assignment was determined to be β-eucryptite, which has also been called γ-spodumene in the literature.[10; 11] In the 10-mol % substituted glasses, the potassium glass (K10m-B) formed nepheline ($KNa_3Al_4Si_4O_{16}$, 14.9 mass %) on heat treatment. The lithium (Li10m-B) and magnesium (Mg10m-B) glasses formed only spinel. The calcium (Ca10m-B) glass formed an aluminosilicate phase like labradorite [$(Ca_{0.86}Na_{0.14})(Al_{1.86}Si_{0.14})Si_2O_8$, 6.7 mass

%], which would presumably also be deleterious to glass durability because of its removal of network-forming cations.

Raman spectra

For the 20 mol % series, the K glass formed kalsilite on quenching, the Na glass formed nepheline, and the Li glass formed eucryptite on heat treatment. The Raman spectrum for the Na glass did not change much with crystallization of nepheline while the spectrum for the Li glass changed dramatically (see Figure 1a). Additionally, a comparison of the Raman spectra for the heat-treated Na glass with mineral nepheline shows no signs of it, while the eucryptite and kalsilite spectra show strong features similar to that of mineral nepheline (Ward's #46E5580, Bancroft, Ontario, Canada, see Figure 1b). (Note: the Raman spectrum of this specimen of nepheline resembled that shown for α-Carnegieite $NaAlSiO_4$, slightly more than that shown for nepheline[12]). All tectosilicates, or three-dimensional framework silicates, have their strongest Raman active band below 600 cm^{-1}, as shown in Figure 1b, which distinguishes them from sheet, ring, chain, dimer, or isolated tetrahedra silicates.[13]

Inspecting the Raman spectra of the quenched series (all suffix "m–B") K10, Na20, Li10, Ca10, and Mg10 illustrates a progression, particularly in the region around 650 to 750 cm^{-1} (see Figure 1c). To the authors' knowledge, the importance of the low-frequency spectral region has not been commented on before in its relation to waste glasses, as the 900 to 1200 cm^{-1} region of T-O stretching inside tetrahedra (T=Al(IV), B(IV), F(IV), Si) is usually considered. The area that peaked around 486 cm^{-1} is a delocalized vibration of the symmetric stretch of bridging oxygens between tetrahedra indicative of six-membered rings of tetrahedra,[12; 14] while the region around 570 cm^{-1} is likely indicative of four-membered rings involving aluminum.[14; 15] It can be seen that the intensity of the latter band increases and broadens the overall first band as the field strength is increased, as would be expected with the increase of four-membered rings. An additional peak starts at 690 cm^{-1} for K, dampens, shifts to larger wavenumbers (up to 741 cm^{-1}), and strengthens again as the field strength is increased. A band near 750 to 770 cm^{-1} was noted in previous studies of nepheline precipitating glasses, and this was interpreted as being caused by six-membered borate rings with one or two B(IV) units.[16] No 850-cm^{-1} band[6; 16] was noted in any of the quenched glasses, and all the "-B" series glasses contained 8 to 10 mass % boron, although this may be about the top end of boron concentrations where the band would be visible.

A broad band at this location was evident in the heat-treated K10 glass (see Figure 1d). The 690 cm^{-1} peak is sharp, and both K10 and Na20 form nepheline with heat treatment. Intensities in the 480 to 500 and 700 to 720 cm^{-1} regions has also been assigned to $Al\text{-}O^-$ or Al(IV)-O-B(III) in lanthanum aluminoborates.[17] In the 900 to 1200 cm^{-1} region, the lower field strength cations show more intensity in the lower coordinated (Q^0, Q^1) tetrahedra at lower wavenumbers relative to higher ones. All spectra show a similar weak B(III) related peak at 1200 to 1500 cm^{-1}.

Infrared spectra

The infrared spectra of the quenched 10-mol % series show six general bands (see Figure 2a). The band at 446 cm^{-1} can be assigned to (Si,Al)-O-Si bending, the band at 691 cm^{-1} to T-O-T stretching (T=Si,Al), and the band at 940-1120 to T-O stretching in tetrahedra (T=Si,Al,B,Fe).[18] Three coordinated boron bands are at 1269 and 1380 cm^{-1}.[19] The weak band at 604 cm^{-1}, which is most notable in the Na20 and K10 glasses, is probably due to six-membered rings,[20] but other interpretations are possible, including Al(VI)[21] (unlikely for K) or B-O bending.[19]

Heat-treated K10 and Na20 glasses showed several additional absorption features in the region of 620 cm^{-1} and below, which may be due to nepheline (see Figure 2b), and the Li10 glass showed a pronounced dip around 604 cm^{-1} only after heat treatment (not shown). Changes in the Ca10 sample upon heat treatment consisted of additional broad absorption in the 500 to 650 cm^{-1} region, while the Mg sample showed no infrared absorption change after heat treatment (not shown). In the 20-mol % glasses, as the field strength increases, there is less resolved structure both in the quenched (see Figure 2 for Na20, Li20 and Figure 2b for K20) and unquenched glasses. This is also observed in the heat-treated Li20 glass, which showed considerably less structure than the heat-treated Na20 glass (see Figure 2b), despite having a higher crystalline content. The K20 glass, which crystallized kalsilite, has bands at 685 cm^{-1}, which has been assigned to Si-O-Al bending, and at 603 cm^{-1}, which has been assigned to vibrations of six-membered rings.[22; 23] These bands can also be identified in the samples containing nepheline (mineral, K10, and Na20) and, to a lesser extent and shifted to higher wave numbers, in the sample containing eucryptite (Li20). An additional band around 507 cm^{-1} can be identified in nepheline-containing glasses, which shifts higher or lower in wavenumber, presumably based on the relative Na/K contributions in the nepheline structure.[24] The overall structure of the same bands in the crystallized samples (Figure 2b) can be seen in the quenched ones (Figure 2a).

DISCUSSION

The crystal structure of nepheline (space group $P6_3$) is composed of six-membered rings of tetrahedra with two types of rings, one puckered.[9] The rings are composed of two types of tetrahedra, one with Al > Si and the other with Al < Si. The puckered rings are smaller and tend to hold the sodium ions while the unpuckered sites are larger and hold the potassium ions, when present. Natural nepheline has a tendency to incorporate various other cations into its structure, notably potassium, calcium, and iron. Potassium, calcium, and sodium lie in the middle of the rings, while aluminum, silicon, and iron are in the center of the oxygen tetrahedra. Thus it is not surprising that nepheline with its capability to easily incorporate substantial amounts of these other atoms would be of primary importance in many-component nuclear waste glasses. Feldspars, like labradorite found in the Ca10 glass, consist of four-membered rings of tetrahedra.[13]

Structurally, alumino-tectosilicates can be thought of as being composed of SiO_2 tetrahedra in six-membered rings, $Al_2Si_2O_8^{2-}$ in four-membered rings, and AlO_2^- in six-membered rings, and charge balancing cation plays a primarily role in determining the type of ring structure (four- or six-membered, see Figure 3).[14] It has been shown that in mixed alkali and alkaline-earth aluminosilicates, glasses with more alkali have predominantly $(Al,Si)_4O_8$ six-membered rings while with more alkaline earth, there will be some $Al_2Si_2O_8^{2-}$ four-membered rings. The distinct infrared and Raman vibrations associated with four- and six-membered rings is discussed by Mozgawa and colleagues.[23; 25]

It is known that as field strength increases (Mg>Ca>Li>Na>K), 1) aluminosilicate sheets and chains become more puckered, 2) intertetrahedral angles become more similar among various coexisting structural units, 3) Al preference for a particular site (usually Q^4) is diminished, and 4) higher coordinated Al becomes more favorable.[26] Other studies based on ^{27}Al magic angle spinning (MAS) nuclear magnetic resonance (NMR) have also shown that higher-field-strength cations (Mg, La) allow more Al(V) and Al(VI) than lower-field-strength cations (Ba, Ca) because the higher coordinated Al cations are neutralized by the high-field-strength cations.[17; 27] Additionally, molecular orbital calculations have shown that higher-field-strength cations compete more with Si or Al for bonding of oxygen (i.e., tetrahedral Mg).[28] Larger-field-strength cations perturb the tetrahedral framework and change the average T-O bond length. It has been shown in aluminoborates that higher-field-strength cations (Mg), can shift the normally encountered "chemical ordering" (i.e., Al, B tetrahedra avoidance) seen in Na and to a lesser extent in Ca aluminoborates to the "statistical random mixing" of tetrahedra seen in Mg aluminoborate.[29]

All these observations point to the expectation that alumino-borosilicates with Ca and Mg should be more disordered and thus have more diffuse Raman bands than their Na and K counterparts. This is in fact what is observed in the current dataset. With a more ordered glassy state, it is more likely that a crystalline phase will form, either during quenching or subsequent heat treatment. There is evidence from the infrared and Raman data that the quenched glasses, at least of the 10-mol % series, consist mainly of six-membered rings of tetrahedra. The Raman band at 486 cm^{-1} and the IR band at 603 cm^{-1} indicate six-membered rings while the Raman band at 570 cm^{-1} is assigned to four-membered rings. In the K10 and Na20 quenched glasses, the distinction between the Raman bands for six- and four-membered rings is fairly sharp but blurs and broadens together for the Li20, Li10, Ca10, and Mg10 glasses. Crystallization in K20, K10, Na20, and Li20 occurs into aluminosilicates with six-membered ring structures (kalsilite, nepheline, and eucryptite), while crystallization in Ca10 occurs into a four-membered ring structure (labradorite, a feldspar). This is further evidence that at the higher field strengths, at least in Ca10 and Mg10, there exist some local four tetrahedra structures in these glasses as is predicted by theory.[14]

CONCLUSIONS

Substitution of higher field strength cations for Na in high alumina waste glasses affects the crystallization of nepheline and other aluminosilicates (kalsilite, eucryptite, labradorite). Low–field-strength cations (K), when present in sufficient quantities, form large fractions of aluminosilicates, even upon quenching (i.e., K20). Upon heat treatment, smaller quantities of low–field-strength cations or larger quantities of higher-field-strength cations can also form aluminosilicates (K10, Na20, Li20). High-field-strength cations, when present at modest levels, do not form aluminosilicates (Li10, Mg10), or they form only small amounts of aluminosilicates (Ca10) upon heat treatment.

Raman spectra show more changes in the quenched glasses when going from low to high field strength than do infrared spectra. Infrared spectra are very indicative of nepheline formation, especially in the 500 to 700 cm^{-1} region where additional absorption structure appears. Infrared spectra of crystallized glasses are easier to interpret than corresponding Raman spectra because of low signal and large fluorescence in the Raman spectra of crystallized glasses. Both Raman and IR indicate the increasing presence of four-membered rings of tetrahedra in glasses with higher field strength.

Further work will involve studying the electrical conductivity and viscosity of these glasses to probe the relationship between these important practical properties and the glass structure. Additionally, an exploration of the role of Fe^{3+} in tetrahedra building in the presence of Al and B with different alkali and alkaline-earths will be investigated, leading to improved understanding of high waste-loading with high-iron wastes.

ACKNOWLEDGEMENTS

Pacific Northwest National Laboratory is operated for the U.S. Department of Energy by Battelle under Contract DE-AC05-76RL01830.

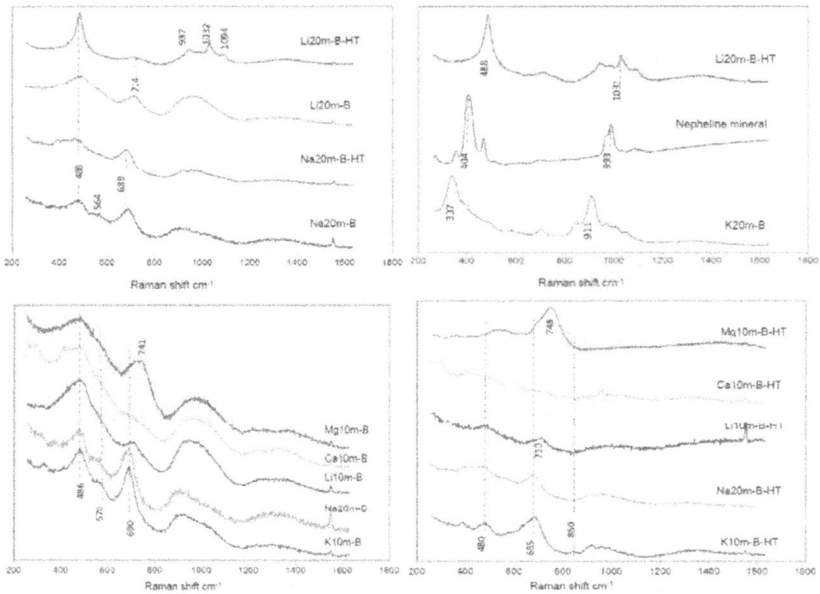

Figure 1: Raman spectra of studied glasses; (a,UL): quenched and heat-treated 20 mol% Na and Ca with the low Al/B ratio; (b, UR): comparison of nepheline-like $MAlSiO_4$ compounds where M=K, Na, Li; (c,LL): comparison of 10-mol % series quenched glasses, normalized; (d,LR): comparison of 10-mol % series heat-treated glasses, not normalized because of variable shaped backgrounds and weak Raman signals in some samples

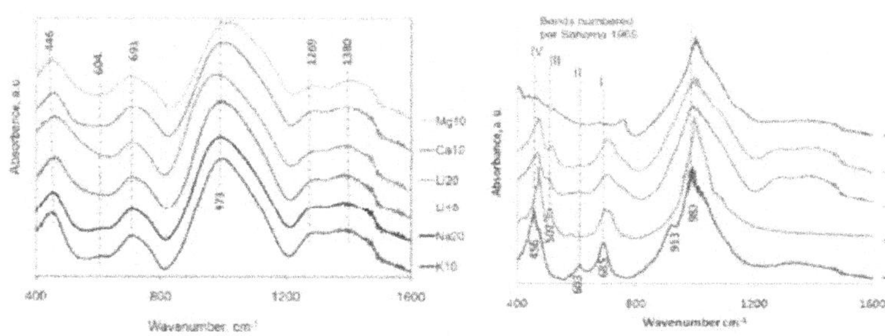

Figure 2: Infrared spectra of studied glasses; (a) comparison of quenched glasses that did not crystallize, normalized; (b) glasses forming nepheline or nepheline-like compounds on quenching or heat treatment, normalized

Figure 3: Schematic of four- and six-ring tetrahedral ring structures in alumino-silicates. Light colored tetrahedra are Al(IV) while dark tetrahedra are Si(IV). Tetrahedral corners contain oxygen atoms.

Table I: Baseline waste composition (High Aluminum Waste)

Waste Component	Mass%	Mol%
Al_2O_3	53.27	46.43
B_2O_3	0.42	0.54
Bi_2O_3	0.12	0.07
CaO	2.54	0.49
Cr_2O_3	2.39	3.79
F	0.05	0.04
Fe_2O_3	1.16	0.68
K_2O	1.48	6.94
Li_2O	13.11	7.30
MgO	0.31	0.30
Na_2O	0.38	1.13
NiO	0.26	0.57
P_2O_5	7.96	11.41
PbO	0.89	1.06
SiO_2	2.34	1.46
ZrO_2	0.91	0.36

Table II: Some metrics on the glasses tested

	Waste loading (mass%)	Al_2O_3 (mass%)	Nepheline discriminator (mass% ratio)	Al/B ratio (moles)	Si/(Al+B) (moles)	M_2O/MO (moles alk, alk.ea)	Melt T (t)/ Remelt T (t) °C (h)
Na-Lo	51.55	27.46	0.47	1.0	0.48	5.57	1300(1)/1325(0.3)/ 1350(1)
Na-Lo2	37.38	19.91	0.53	0.48	0.45	8.03	1300(1)/1300(1)
Na-Hi2	43.67	23.26	0.56	1.03	0.75	6.88	1350(1)/1400(0.5)/ 1300(1)
Na20m	44.05	23.56	0.43	1.01	0.56	13.04	1250(1)/1250(1)
Na20m+B	36.90	19.73	0.41	0.51	0.38	15.65	1200(1)/1150(1)
Na20m-B	48.78	26.08	0.44	1.99	0.75	11.74	1300(1)/1330(1)
K20m-B	42.74	22.95	0.53	1.99	0.75	14.95	1250(1)
Li20m-B	51.15	27.47	0.54	1.99	0.75	14.77	1150(1)/1200(1)
K10m-B	51.15	25.10	0.50	1.99	0.75	11.72	1350(1)/1350(1)
Li10m-B	51.15	27.47	0.50	1.99	0.75	11.44	1250(1)/1250(1)
Ca10m-B	51.15	26.86	0.50	1.99	0.75	1.03	1250(1)/1250(1)
Mg10m-B	51.15	27.06	0.50	1.99	0.75	0.93	1350(1)/1350(1)

Table III: Target compositions of glasses and baseline waste (mol%)

	Al_2O_3	B_2O_3	Bi_2O_3	CaO	Cr_2O_3	F	Fe_2O_3	K_2O	Li_2O	MgO	Na_2O	NiO	P_2O_5	PbO	SiO_2	ZrO_2
Baseline waste	46.82	0.54	0.49	3.82	0.68	7.00	7.36	0.30	1.41	0.58	11.50	1.07	1.48	0.36	16.23	0.64
Na-Lo	20.14	20.17	0.21	1.64	0.29	3.01	3.16	0.13	0.49	0.25	9.92	0.46	0.63	0.16	39.06	0.27
Na-Lo2	14.03	29.49	0.15	1.15	0.20	2.10	2.21	0.09	0.34	0.17	10.17	0.32	0.44	0.11	38.84	0.19
Na-Hi2	16.44	16.02	0.17	1.34	0.24	2.46	2.58	0.10	0.40	0.20	10.13	0.37	0.52	0.13	48.66	0.22
Na20m	16.77	16.66	0.18	1.37	0.24	2.51	2.64	0.11	0.41	0.21	20.04	0.38	0.53	0.13	37.61	0.23
Na20m+B	13.91	27.47	0.15	1.14	0.20	2.08	2.19	0.09	0.34	0.17	20.03	0.32	0.44	0.11	31.19	0.19
Na20m-B	18.69	9.39	0.20	1.53	0.27	2.79	2.94	0.12	0.45	0.23	20.05	0.43	0.59	0.15	41.92	0.25
K20m-B	17.70	8.89	0.18	1.44	0.26	2.64	2.78	20.08	0.43	0.22	4.35	0.40	0.56	0.14	39.68	0.24
Li20m-B	17.75	8.92	0.19	1.45	0.26	2.65	2.79	0.11	20.16	0.22	4.36	0.40	0.56	0.14	39.80	0.24
K10m-B	18.70	9.40	0.20	1.53	0.27	2.79	2.94	10.11	0.45	0.23	10.04	0.43	0.59	0.15	41.93	0.25
Li10m-B	18.79	9.44	0.20	1.53	0.27	2.81	2.95	0.12	10.01	0.23	10.09	0.43	0.59	0.15	42.14	0.26
Ca10m-B	19.09	9.59	0.20	9.96	0.28	2.85	3.00	0.12	0.46	0.24	9.95	0.43	0.60	0.15	42.81	0.26
Mg10m-B	18.76	9.42	0.20	1.53	0.27	2.80	2.95	0.12	0.46	9.95	10.07	0.43	0.59	0.15	42.06	0.26

Table IV: Crystalline phases and fractions identified by x-ray diffraction. Glasses denoted with * had only the phases identified. All phases that were identified are listed.

GLASS ID	Quenched	950°C -24 hour heat treated
EM09-Na-Lo *	Trevorite, ulvospinel, wilkinsonite, sodium aluminum silicate, sarcopside	Trevorite, mullite, sodalite, hematite
EM09-Na-Lo2 *	Magnesium aluminum iron oxide, trevorite	Nicromite, hematite
EM09-Na-Hi2 *	Magnetite, Na_2CaSiO_4	Magnetite, sillmanite, chromium iron dioxide
EM09-Na20m *	Trevorite, $MgFeAlO_4$ spinel	Magnetite (1.3 wt%), Hematite (1.2 wt%)
EM09-Na20m+B *	Nichromite	Chromium iron oxide $Cr_{1.3}Fe_{0.7}O_3$
EM09-Na20m-B *	Magnetite	Nepheline ($Na_{6.65}Al_{6.24}Si_{9.76}O_{32}$) (28.6 wt%), Hematite (2.5 wt%), Magnetite (0.7 wt%)
EM09-K20m-B	Kalsilite ($KAlSiO_4$) (56.5 wt%), Carnegieite ($NaAlSiO_4$) (3 wt%), Quartz (SiO_2) (5 wt%)	N/A
EM09-Li20m-B	Lithium iron zinc oxide $(Li_{0.75}Fe_{3.25})(Fe_{1.5}ZnO_5)O_8$ (2.1wt%)	Lithium Aluminum Silicate ($LiAlSiO_4$) (49.1 wt%), Chromite spinel ($LiFe_3Cr_2O_8$) (3.8 wt%)
EM09-K10m-B	Not tested	Nepheline ($KNa_3Al_4Si_4O_{16}$) (14.9 wt%), Magnetite or nichromite (2.2 wt%)
EM09-Li10m-B	Not tested	Magnetite or $LiFe_5CrO_{16}$ spinel (4.3 wt%)
EM09-Ca10m-B	Not tested	$(Ca_{0.86}Na_{0.14})(Al_{1.86}\ Si_{0.14})Si_2O_8$ (6.7 wt%), Magnetite or hercynite $(Fe,Al)(Al,Fe)O_4$ spinel (2.7 wt%), Hematite (2.5wt%)
EM09-Mg10m-B	Not tested	Magnetite or $(Mg,Fe)Al_2O_4$ spinel (8.2 wt%)

REFERENCES

[1]H. Li, J. D. Vienna, P. Hrma, D. E. Smith, and M. J. Schweiger, Nepheline precipitation in high-level waste glasses: compositional effects and impact on the waste form acceptability, 261-268 in *Proc. Materials Research Society*. **465**. (1997).

[2]C. M. Jantzen and K. G. Brown, Predicting the Spinel-Nepheline Liquidus for Application to Nuclear Waste Glass Processing. Part II: Quasicrystalline Freezing Point Depression Model, *J. Amer. Ceram. Soc.*, **90[6]**, 1880-1891 (2007).

[3]H. Li, B. Jones, P. Hrma, and J. D. Vienna, Compositional effects on liquidus temperature of Hanford simulated high-level waste glasses precipitating nepheline ($NaAlSiO_4$), 279-288. In *Ceramic Transactions*, **87**. American Ceramic Society, Westerville, OH, 1998.

[4]T. J. Menkhaus, P. Hrma, and H. Li, Kinetics of nepheline crystallization from high-level waste glass, 461-468. In *Ceramic Transactions*, **107**. American Ceramic Society, Westerville, OH, 2000.

[5]J. D. Vienna, A. Fluegel, D. S. Kim, and P. Hrma, Glass Property Data and Models for Estimating High-Level Waste Glass Volume. Pacific Northwest National Laboratory, PNNL-18501, Richland, Washington, 2009.

[6]H. Li, P. Hrma, J. D. Vienna, M. Qian, Y. Su, and D. E. Smith, Effects of Al_2O_3, B_2O_3, Na_2O, and SiO_2 on nepheline formation in borosilicate glasses: chemical and physical correlations, *J. Non-Cryst. Solids*, **331[1-3]**, 202-216 (2003).

[7]T. M. Besmann, K. E. Spear, and E. C. Beahm, Assessment of nepheline precipitation in nuclear waste glass via thermochemical modeling, in *Proc. Materials Research Society*. Fall symposium QQ, (1999).

[8]K. M. Fox, T. B. Edwards, and D. K. Peeler, Control of nepheline crystallization in nuclear waste glass, *Int. J. Appl. Ceram. Technol.*, **5[6]**, 666-673 (2008).

[9]K. T. Tait, E. Sokolova, and F. C. Hawthorne, The Crystal Chemistry of Nepheline, *Canad. Mineral.*, **41**, 61-70 (2003).

[10]S. K. Sharma and B. Simons, Raman study of crystalline polymorphs and glasses of spodumene composition quenched from various pressures, *Amer. Mineral.*, **66**, 118-126 (1980).

[11]M. Zhang, H. Xu, E. K. H. Salje, and P. J. Heaney, Vibrational spectroscopy of beta-eucryptite ($LiAlSiO_4$): optical phonons and phase transition(s), *Phys. Chem. Miner.*, **30[8]**, 457-462 (2003).

[12]D. W. Matson, S. K. Sharma, and J. A. Philpotts, Raman spectra of some tectosilicates and of glasses along the orthoclase-anorthite and nepheline-anorthite joins, *Amer. Mineral.*, **71[5-6]**, 694-704 (1986).

[13]J. J. Freeman, A. Wang, K. E. Kuebler, B. L. Jolliff, and L. A. Haskin, Characterization of natural feldspars by Raman spectroscopy for future planetary exploration, *Canad. Mineral.*, **46**, 1477-1500 (2008).

[14]F. Seifert, B. O. Mysen, and D. Virgo, Three-dimensional network structure of quenched melts (glass) in the systems SiO_2-$NaAlO_2$, SiO_2-$CaAl_2O_4$ and SiO_2-$MgAl_2O_4$, *Amer. Mineral.*, **67**, 696-717 (1982).

[15]D. Sykes and C. M. Scarfe, Melt Structure in the System Nepheline-Diopside, *J. Geophys. Res.*, **95**, (1990).

[16]H. Li, Y. Su, J. D. Vienna, and P. Hrma, Raman spectroscopic study - effects of B_2O_3, Na_2O, and SiO_2 on nepheline ($NaAlSiO_4$) crystallization in simulated high level waste glasses,

469-477. In *Environmental Issues and Waste Mangement Techniques V, Ceramic Transactions,* **107**. Edited by G. T.Chandler and X. Feng. American Ceramic Society, Westerville, OH, 2000.

[17]R. K. Brow, D. R. Tallant, and G. L. Turner, Polyhedral Arrangements in Lanthanum Aluminoborate Glasses, *J. Amer. Ceram. Soc.,* **80[5]**, 1239-1244 (1997).

[18]S. A. MacDonald, C. R. Schardt, D. J. Masiello, and J. H. Simmons, Dispersion analysis of FTIR reflection measurements in silicate glasses, *J. Non-Cryst. Solids,* **275[1-2]**, 72-82 (2000).

[19]E. I. Kamitsos, A. P. Patsis, M. A. Karakassides, and G. D. Chryssikos, Infrared reflectance spectra of lithium borate glasses, *J. Non-Cryst. Solids,* **126[1-2]**, 52-67 (1990).

[20]M. Sitarz, W. Mozgawa, and M. Handke, Vibrational spectra of complex ring silicate anions -- method of recognition, *J. Mol. Struct.,* **404[1-2]**, 193-197 (1997).

[21]W. R. Taylor, Application of infrared spectroscopy to studies of silicate glass structure: Examples from the melilite glasses and the systems Na_2O-SiO_2 and Na_2O-Al_2O_3-SiO_2, *Proc. Indian Acad. Scie. (Earth Planet. Sci.),* **99[1]**, 99-117 (1990).

[22]M. Sitarz, W. Mozgawa, and M. Handke, Rings in the structure of silicate glasses, *J. Mol. Struct.,* **511-512**, 281-285 (1999).

[23]W. Mozgawa and M. Sitarz, Vibrational spectra of aluminosilicate ring structures, *J. Mol. Struct.,* **614[1-3]**, 273-279 (2002).

[24]T. H. G. Sahama, Infrared absorption of nepheline, *Comptes Rendus de la Societe geologique de Finlande,* **XXXVII**, 107-117 (1965).

[25]W. Mozgawa, M. Handke, and W. Jastrzebski, Vibrational spectra of aluminosilicate structural clusters, *J. Mol. Struct.,* **704[1-3]**, 247-257 (2004).

[26]B. O. Mysen, Role of Al in depolymerized, peralkaline aluminosilicate melts in the systems Li_2O-Al_2O_3-SiO_2 , Na_2O-Al_2O_3-SiO_2, and K_2O-Al_2O_3-SiO_2, *Amer. Mineral.,* **75[1-2]**, 120-134 (1990).

[27]B. C. Bunker, R. J. Kirkpatrick, R. K. Brow, G. L. Turner, and C. Nelson, Local Structure of Alkaline-Earth Boroaluminate Crystals and Glasses: II, ^{11}B and ^{27}Al MAS NMR Spectroscopy of Alkaline-Earth Boroaluminate Glasses, *J. Amer. Cer. Soc.,* **74[6]**, 1430-1438 (1991).

[28]A. Navrotsky, K. L. Geisinger, P. McMillan, and G. V. Gibbs, The tetrahedral framework in glasses and melts — inferences from molecular orbital calculations and implications for structure, thermodynamics, and physical properties, *Phys. Chem. Miner.,* **11[6]**, 284-298 (1985).

[29]J. C. C. Chan, M. Bertmer, and H. Eckert, Site Connectivities in Amorphous Materials Studied by Double-Resonance NMR of Quadrupolar Nuclei: High-Resolution ^{11}B and ^{27}Al Spectroscopy of Aluminoborate Glasses, *JACS,* **121[22]**, 5238-5248 (1999).

NEPHELINE CRYSTALLIZATION IN NUCLEAR WASTE GLASSES

Kevin M. Fox, David K. Peeler and Thomas B. Edwards
Savannah River National Laboratory
Aiken, South Carolina, USA

ABSTRACT

This study was undertaken in an attempt to reduce known conservatism in a model that relates the composition of a high level waste glass to its propensity for nepheline crystallization upon slow cooling. Nepheline crystallization is to be avoided since it can reduce the chemical durability of a waste glass. A series of test glass compositions was selected, fabricated and characterized in support of this study. The results continued to identify compositional regions where the model remains conservative; however there was difficulty in identifying relationships among these compositions. An alternative nepheline discriminator was also evaluated but was shown to have difficulty in predicting nepheline crystallization in some of the slowly cooled glasses studied here. Further refinement of the current nepheline discriminator model may be possible using the data collected.

INTRODUCTION

Nepheline crystallization is of concern for high-level nuclear waste (HLW) glasses because its formation can reduce the chemical durability of the waste form. Waste compositions containing higher concentrations of Al_2O_3 and Na_2O are more prone to nepheline ($NaAlSiO_4$) crystallization,[1] particularly upon slow cooling. Currently, a nepheline discriminator is included as a process control constraint at the Defense Waste Processing Facility (DWPF) at the Savannah River Site (SRS).[2] This constraint uses the inequality that was proposed by Li[1] and later verified as applicable to DWPF-type glasses.[3-7] The nepheline discriminator is given as Equation 1, and relates the concentrations of SiO_2, Na_2O and Al_2O_3 (as weight percentages in glass) to a critical value of 0.62.

$$\frac{SiO_2}{SiO_2 + Na_2O + Al_2O_3} > 0.62 \qquad (1)$$

This equation defines a boundary line on the SiO_2-Na_2O-Al_2O_3 ternary diagram above which (or toward the SiO_2 corner of the ternary) nepheline is not predicted to crystallize in the glass upon quenching or slow cooling.

An earlier, Phase I study confirmed that some conservatism exists in the current nepheline discriminator.[8] Several glass compositions, particularly compositions that targeted higher Al_2O_3 concentrations, were shown to be very durable while their nepheline discriminator values were well below the current nepheline discriminator limit of 0.62. Increased concentrations of B_2O_3 and CaO were shown to improve durability responses and suppress the formation of nepheline. These results provided incentive to revise the nepheline discriminator to reduce some of this conservatism and incorporate the influence of B_2O_3. The Phase I study recommended that a second phase be undertaken to provide additional data in support of this revision.[8]

Twenty five glass compositions were subsequently selected for a Phase II study on reduction of conservatism in the nepheline discriminator.[9] The glass compositions were selected to identify any linear effects of composition on nepheline crystallization and were restricted to regions that fell within the validation ranges of the DWPF Product Composition Control System (PCCS) models.[10] The glasses were fabricated in the laboratory and characterized for crystallization and chemical durability after both quenching and slow cooling. Nepheline was identified in one of the quenched glasses and several of the slowly cooled glasses. A partitioning algorithm was used to identify trends in crystallization behavior based on glass composition. Generally, for the slowly cooled glasses MnO concentration influenced the crystallization of spinels and B_2O_3 and SiO_2 concentrations influenced the crystallization of nepheline. Durability responses varied from acceptable to unacceptable (with respect to that of the Environmental Assessment (EA) benchmark glass[11]) depending on the glass composition and type and extent of crystallization that occurred. It was not possible to identify any linear effects of composition on chemical durability performance for this set of study glasses. The results of the Phase II study alone were not sufficient to recommend modification of the current nepheline discriminator. It was recommended that the next series of experiments continue to focus not only on compositional regions where the PCCS models are considered applicable (i.e., the model validation ranges), but also be restricted to compositional regions where acceptable glasses are predicted to be produced but are prohibited by the current nepheline discriminator.[9]

The intent of this Phase III study is to investigate whether there are compositional regions available, particularly glasses with higher aluminum concentrations to support higher waste loadings, that are acceptable by all of the PCCS models with the exception of the nepheline discriminator. Two routes were considered in selecting glasses for this phase of the nepheline study: one based on the relationship of the current nepheline discriminator model to the other DWPF PCCS models[10] and the other based on the theory of crystallization in mineral and glass melts.[12]

EXPERIMENTAL PROCEDURE

A detailed description of the glass selection process has been provided elsewhere.[12] Briefly, a series of 29 test glass compositions were selected for this study using a combination of two approaches. The first approach was based on evaluating the glass composition region allowable by all of the DWPF PCCS models[10] with the exception of the current nepheline discriminator.[2] This approach was taken to determine whether there are glass compositions that, while predicted to crystallize nepheline upon slow cooling, would otherwise be acceptable for processing in the DWPF. The second approach was based on quasicrystalline theory of glass structure. A detailed description of this methodology is forthcoming, but in brief, ACT™ was used to relate simple ratios of the cation components of a series of glasses to the quasicrystalline phases calculated using NORMCALC™.*

Each of the study glasses was prepared from the proper proportions of reagent-grade metal oxides, carbonates, and boric acid in 150 g batches. The batch was melted in a high-temperature furnace at 1150 °C for one hour.[13] The glass was then poured onto a clean, stainless steel plate and allowed to air cool (quench). Approximately 25 g of each glass was heat-treated to simulate cooling along the centerline of a DWPF-type canister[14] to gauge the effects of thermal history on the product performance. This cooling schedule is referred to as the ccc heat treatment.

Chemical analysis was used to verify the compositions of the study glasses. Sodium peroxide fusion and lithium-metaborate fusion were used to prepare the glass samples for analysis. Each of the samples was analyzed, twice for each element of interest, by Inductively Coupled Plasma – Atomic Emission Spectroscopy (ICP-AES). Glass standards were also intermittently measured to assess the performance of the ICP-AES instrument over the course of the analyses. Representative samples of each quenched and ccc glass analyzed by X-ray diffraction (XRD, D8 Advance, Bruker AXS Inc., Madison, WI) to identify any crystalline phases present.

The Product Consistency Test (PCT) Method-A[15] was performed in triplicate on each quenched and ccc glass to assess chemical durability. Also included in the experimental test matrix was the EA benchmark glass,[11] the Approved Reference Material (ARM) glass, and blanks from the sample cleaning batch. Samples were ground, washed, and prepared according to the standard procedure.[15] Fifteen milliliters of Type-I ASTM water were added to 1.5 g of glass in stainless steel vessels. The vessels were closed, sealed, and placed in an oven at 90 ± 2 °C where the samples were maintained at temperature for seven days. Once cooled, the resulting solutions were sampled (filtered and acidified) and analyzed. Samples of a multi-element, standard solution were also included in the analysis as a check on the accuracy of the ICP-AES instrument used for these measurements. Normalized release rates were calculated based on the measured compositions using the average of the common logarithms of the leachate concentrations.

RESULTS AND DISCUSSION

The crystalline phases identified by XRD in the study glasses are provided in Table I. Phases described as "possible" were difficult to identify due to very low peak intensities. Glasses that were found to contain nepheline are highlighted in the table. Further discussion of these glasses will be provided later below.

Table I. Summary of XRD results for the study glasses

Glass ID	Heat Treatment	XRD Results
NE3-01	quenched	Trevorite
	ccc	Magnetite, possible Nepheline
NE3-02	quenched	Amorphous
	ccc	Amorphous
NE3-03	quenched	Amorphous
	ccc	Magnetite, Nepheline
NE3-04	quenched	Amorphous
	ccc	Nepheline, Lithium Silicate
NE3-05	quenched	Amorphous
	ccc	Amorphous
NE3-06	quenched	Amorphous
	ccc	Magnetite, Trevorite, Nepheline
NE3-07	quenched	Amorphous
	ccc	Magnetite, Nepheline
NE3-08	quenched	Amorphous
	ccc	Trevorite, Nepheline
NE3-09	quenched	Amorphous
	ccc	Nepheline, possible Trevorite
NE3-10	quenched	Amorphous
	ccc	Nepheline, Lithium Silicate, Magnetite, Hematite
NE3-11	quenched	Magnetite
	ccc	Nepheline, Lithium Silicate, Magnetite
NE3-12	quenched	Amorphous
	ccc	Nepheline
NE3-13	quenched	Amorphous
	ccc	Nepheline, Lithium Silicate, Magnetite
NE3-14	quenched	Amorphous
	ccc	Nepheline, Lithium Silicate, Trevorite
NE3-15	quenched	Amorphous
	ccc	Amorphous
NE3-16	quenched	Amorphous
	ccc	Amorphous
NE3-17	quenched	Amorphous
	ccc	Trevorite, Magnetite
NE3-18	quenched	Amorphous
	ccc	Possible Trevorite
NE3-19	quenched	Amorphous
	ccc	Amorphous
NE3-20	quenched	Amorphous
	ccc	Amorphous

Table I. Summary of XRD results for the study glasses (continued)

Glass ID	Heat Treatment	XRD Results
NE3-21	quenched	Amorphous
	ccc	**Nepheline, Lithium Silicate, Franklinite ($ZnFe_2O_4$)**
NE3-22	quenched	Amorphous
	ccc	Magnetite
NE3-23	quenched	Amorphous
	ccc	Amorphous
NE3-24	quenched	Amorphous
	ccc	**Nepheline, Lithium Silicate**
NE3-25	quenched	Amorphous
	ccc	**Nepheline, Lithium Silicate, Magnetite**
NE3-26	quenched	Trevorite
	ccc	**Nepheline, Lithium Silicate, Magnetite, possible Kyanite (Al_2SiO_5)**
NE3-27	quenched	Possible Trevorite
	ccc	**Nepheline, Lithium Silicate, Magnetite**
NE3-28	quenched	Amorphous
	ccc	**Bunsenite, Trevorite, Nepheline**
NE3-29	quenched	Amorphous
	ccc	**Nepheline, Franklinite ($ZnFe_2O_4$)**

All of the ICP-AES measurements for each oxide were averaged to determine a representative chemical composition for each glass, as given in Table II. A review of the measured[16] versus target[12] compositions showed that in general, there appear to have been only minor difficulties in meeting the targeted concentrations for the study glasses, none of which should impact the outcome of the study since the analyses will be performed using the measured, rather than target compositions.

Table II. Measured Compositions of the Study Glasses

	NE3-01	NE3-02	NE3-03	NE3-04	NE3-05	NE3-06	NE3-07	NE3-08	NE3-09	NE3-10	NE3-11	NE3-12	NE3-13	NE3-14	NE3-15
Al_2O_3	10.54	12.92	12.86	13.40	14.58	14.46	14.68	14.00	16.21	14.83	16.07	16.69	17.29	17.99	10.76
B_2O_3	4.74	7.08	8.54	4.60	12.34	11.21	11.35	8.94	6.50	7.45	7.41	7.74	8.05	5.24	6.30
BaO	0.01	0.07	0.07	0.01	0.08	0.08	0.01	0.01	0.08	0.08	0.01	0.01	0.06	0.07	0.04
CaO	4.59	0.01	4.70	0.24	0.01	0.01	1.52	2.96	4.70	1.01	0.01	0.70	0.04	4.26	0.01
CdO	0.01	0.24	0.25	0.01	0.27	0.25	0.01	0.01	0.28	0.25	0.01	0.01	0.21	0.24	0.14
Ce_2O_3	0.01	0.35	0.36	0.01	0.38	0.36	0.01	0.01	0.39	0.35	0.01	0.01	0.16	0.33	0.20
Cr_2O_3	0.17	0.01	0.09	0.01	0.19	0.03	0.01	0.01	0.01	0.02	0.13	0.14	0.12	0.02	0.18
CuO	0.01	0.13	0.13	0.01	0.14	0.13	0.01	0.01	0.14	0.13	0.01	0.01	0.11	0.12	0.08
Fe_2O_3	8.65	8.21	10.37	11.36	5.39	13.89	13.78	7.16	6.43	11.58	12.37	8.80	4.84	11.64	6.82
La_2O_3	0.01	0.09	0.09	0.01	0.10	0.09	0.01	0.01	0.10	0.09	0.01	0.01	0.07	0.08	0.05
Li_2O	3.95	3.94	3.91	4.88	5.37	4.20	4.17	6.11	4.60	4.30	7.00	3.94	4.79	6.84	4.89
MgO	1.38	0.01	0.01	0.01	1.54	0.05	0.31	0.87	0.01	1.43	0.01	0.01	0.18	0.04	0.54
MnO	3.50	5.54	3.47	0.20	0.22	0.58	2.63	0.60	0.24	2.01	4.18	0.56	4.42	0.18	1.39
Na_2O	16.18	15.27	13.62	17.12	14.93	11.34	11.08	13.67	13.01	13.12	10.74	13.64	16.28	10.96	16.55
NiO	1.67	0.01	0.39	0.01	0.01	0.01	0.01	2.25	2.19	0.01	0.40	0.35	1.93	0.01	2.18
PbO	0.01	0.19	0.19	0.01	0.21	0.20	0.01	0.01	0.22	0.20	0.01	0.01	0.16	0.19	0.11
SiO_2	40.65	41.50	38.94	45.94	41.77	37.44	36.58	42.04	43.59	41.88	40.81	45.03	38.19	37.60	47.33
SO_4^{2-}	0.07	0.45	0.47	0.07	0.47	0.43	0.07	0.09	0.48	0.45	0.07	0.07	0.38	0.41	0.27
TiO_2	1.91	1.89	0.29	0.01	0.78	1.94	1.91	0.19	0.01	0.01	0.01	0.52	0.84	1.76	0.15
ZnO	0.01	0.13	0.13	0.01	0.14	0.13	0.01	0.01	0.15	0.13	0.01	0.01	0.12	0.12	0.08
ZrO_2	0.01	0.20	0.21	0.01	0.22	0.20	0.01	0.01	0.22	0.19	0.01	0.01	0.17	0.20	0.12

Table II. Measured Compositions of the Study Glasses (continued)

	NE3-16	NE3-17	NE3-18	NE3-19	NE3-20	NE3-21	NE3-22	NE3-23	NE3-24	NE3-25	NE3-26	NE3-27	NE3-28	NE3-29
Al_2O_3	11.32	11.63	11.88	12.21	12.87	13.56	14.37	14.05	14.71	17.21	17.63	18.23	17.64	18.13
B_2O_3	9.72	5.35	5.51	9.32	6.30	5.53	7.90	12.11	7.27	7.70	6.15	5.30	10.35	10.16
BaO	0.06	0.06	0.04	0.01	0.01	0.06	0.01	0.04	0.04	0.05	0.01	0.05	0.06	0.01
CaO	4.36	4.29	3.71	4.08	0.20	3.51	1.19	3.97	4.39	3.58	1.21	2.41	0.01	3.32
CdO	0.20	0.19	0.15	0.02	0.04	0.21	0.01	0.14	0.15	0.18	0.01	0.18	0.19	0.01
Ce_2O_3	0.29	0.15	0.21	0.03	0.07	0.30	0.01	0.20	0.22	0.25	0.02	0.25	0.27	0.01
Cr_2O_3	0.13	0.11	0.07	0.11	0.12	0.06	0.11	0.11	0.02	0.11	0.17	0.02	0.07	0.06
CuO	0.11	0.10	0.08	0.02	0.03	0.11	0.01	0.07	0.08	0.09	0.01	0.10	0.10	0.01
Fe_2O_3	6.78	9.45	5.43	7.47	4.92	5.38	9.45	5.47	5.65	5.39	7.16	6.34	5.10	7.73
La_2O_3	0.07	0.07	0.05	0.01	0.02	0.07	0.01	0.05	0.05	0.06	0.01	0.06	0.07	0.01
Li_2O	4.43	4.68	5.75	5.10	5.49	7.16	4.56	4.98	5.02	7.00	5.20	5.40	4.69	4.23
MgO	0.19	0.18	1.27	0.83	0.17	0.91	0.05	0.28	1.13	0.80	0.32	1.30	0.49	0.74
MnO	4.55	1.60	0.33	0.85	1.38	5.66	3.53	3.90	0.36	3.35	3.75	2.03	1.99	4.77
Na_2O	12.93	14.69	14.76	14.09	17.52	12.84	13.50	10.55	16.75	12.21	16.28	14.49	13.52	13.95
NiO	0.58	1.42	2.06	0.51	0.01	0.70	0.81	0.01	1.44	0.43	0.92	1.03	1.03	0.44
PbO	0.15	0.15	0.12	0.02	0.03	0.16	0.01	0.11	0.12	0.13	0.02	0.14	0.15	0.01
SiO_2	38.51	46.37	45.62	43.53	47.71	40.75	41.50	42.41	40.17	39.68	38.61	39.36	41.29	34.76
SO_4^{2-}	0.37	0.38	0.27	0.07	0.13	0.40	0.07	0.30	0.29	0.33	0.07	0.34	0.37	0.07
TiO_2	1.72	0.84	0.73	0.81	0.91	1.42	1.55	0.39	0.45	1.35	0.62	0.93	0.05	0.43
ZnO	0.11	0.11	0.08	0.01	0.02	0.11	0.01	0.07	0.08	0.09	0.01	0.10	0.10	0.01
ZrO_2	0.17	0.16	0.13	0.02	0.05	0.17	0.01	0.11	0.12	0.14	0.02	0.15	0.17	0.02

Table III summarizes the results of the normalized PCTs for the glasses of this study. One of the triplicate vessels from the ccc version of glass NE3-26 had a solution weight loss problem and thus was not included in the analysis. Any measurement below the detection limit of the analytical procedure was replaced by one half of the detection limit in subsequent analyses. The PCT results are listed by heat treatment for each glass in the table. Several of the study glasses had normalized leachate for boron (NL [B]) values that approached or were greater than the benchmark EA glass[11] NL [B] of 16.695 g/L. Glasses NE3-04ccc, NE3-11ccc, NE3-21ccc, NE3-24ccc, NE3-25ccc, and NE3-29ccc had elevated (greater than 1.0 g/L) NL [B] values, while glasses NE3-10ccc, NE3-13ccc, NE3-26ccc, and NE3-27ccc had NL [B] values significantly higher than that of the EA glass.

Table III. Normalized PCT Values by Glass ID and Compositional View

Glass ID	Heat Treatment	NL [B] (g/L)	NL [Li] (g/L)	NL [Na] (g/L)	NL [Si] (g/L)
NE3-01	ccc	1.07	1.32	1.63	0.56
NE3-01	quenched	1.04	1.16	1.64	0.57
NE3-02	ccc	0.64	0.69	0.70	0.45
NE3-02	quenched	0.60	0.62	0.75	0.44
NE3-03	ccc	2.03	1.91	1.31	0.46
NE3-03	quenched	0.81	0.85	0.87	0.37
NE3-04	ccc	4.78	3.74	2.15	1.29
NE3-04	quenched	0.67	0.62	1.25	0.56
NE3-05	ccc	0.92	0.81	0.82	0.43
NE3-05	quenched	1.01	0.81	0.86	0.44
NE3-06	ccc	2.05	1.62	0.87	0.45
NE3-06	quenched	0.62	0.78	0.53	0.42
NE3-07	ccc	3.40	2.86	1.42	0.50
NE3-07	quenched	0.69	0.78	0.61	0.35
NE3-08	ccc	1.36	1.62	1.14	0.51
NE3-08	quenched	0.62	0.71	0.87	0.38
NE3-09	ccc	0.43	0.55	0.46	0.19
NE3-09	quenched	0.32	0.42	0.46	0.19
NE3-10	ccc	39.52	21.27	11.00	1.45
NE3-10	quenched	0.68	0.70	0.70	0.41
NE3-11	ccc	6.75	3.22	2.24	0.87
NE3-11	quenched	0.91	0.92	0.84	0.62
NE3-12	ccc	1.98	1.48	0.74	0.49
NE3-12	quenched	0.57	0.64	0.50	0.42
NE3-13	ccc	89.32	18.15	34.22	3.43
NE3-13	quenched	0.89	0.68	1.17	0.51
NE3-14	ccc	3.25	3.31	0.45	0.90
NE3-14	quenched	1.17	1.02	1.05	0.45

Table III. Normalized PCT Values by Glass ID and Compositional View (continued)

NE3-15	ccc	1.07	0.78	1.11	0.57
NE3-15	quenched	0.85	0.73	1.24	0.57
NE3-16	ccc	0.90	0.97	0.92	0.42
NE3-16	quenched	0.95	0.95	0.99	0.43
NE3-17	ccc	0.68	0.81	1.01	0.38
NE3-17	quenched	1.16	0.81	1.19	0.41
NE3-18	ccc	0.75	0.87	0.99	0.39
NE3-18	quenched	0.69	0.84	1.18	0.42
NE3-19	ccc	0.55	0.65	0.68	0.33
NE3-19	quenched	0.79	0.67	0.76	0.34
NE3-20	ccc	0.68	0.75	1.03	0.55
NE3-20	quenched	0.70	0.69	1.18	0.56
NE3-21	ccc	5.25	3.60	2.12	0.91
NE3-21	quenched	1.38	1.07	1.14	0.54
NE3-22	ccc	0.74	0.78	0.68	0.47
NE3-22	quenched	0.97	0.67	0.67	0.43
NE3-23	ccc	0.53	0.58	0.49	0.26
NE3-23	quenched	0.57	0.58	0.49	0.26
NE3-24	ccc	5.79	5.09	2.58	1.07
NE3-24	quenched	0.67	0.67	1.01	0.36
NE3-25	ccc	5.16	5.17	1.16	1.10
NE3-25	quenched	0.84	0.85	0.91	0.42
NE3-26	ccc	85.04	15.03	26.36	2.50
NE3-26	quenched	0.76	0.75	1.19	0.55
NE3-27	ccc	33.98	8.55	7.80	0.69
NE3-27	quenched	0.75	0.69	0.88	0.42
NE3-28	ccc	2.39	1.90	1.03	0.51
NE3-28	quenched	0.64	0.65	0.60	0.43
NE3-29	ccc	6.46	5.64	1.78	0.26
NE3-29	quenched	1.05	0.82	0.94	0.38

Many of the study glasses showed a statistically significant difference between the ccc and the quenched versions for one or more of the PCT elements. The glasses that crystallized nepheline after the ccc heat treatment had a statistically significant reduction in durability compared to their quenched versions for several or all of the four elements (B, Li, Na and Si). Glasses NE3-21 and NE3-29 are exceptions to this statement. However, a review of Table III shows that the quenched versions of these glasses had relatively high normalized release values, as well as the ccc versions. So, while the difference in PCT response with heat treatment was not statistically different, the glasses do show poorer durabilities as a result of nepheline crystallization.

Another aspect of interest in regard to the durabilities of the study glasses is their predictability by the PCCS models that predict the PCT response of a glass using its composition.

However, care has to be taken in making these comparisons as the models only apply to homogeneous glasses; the models would not be expected to predict the durabilities of glasses containing nepheline. Therefore, the PCTs for those glasses whose ccc versions were found to contain nepheline were not included in these comparisons. Figure 1 provides a plot of the PCCS model for B that relates the logarithm of the normalized PCT value to a linear function of a free energy of hydration term (ΔG_p, kcal/100g glass) derived from both heat treatments of the remaining glasses.[17] Prediction limits (at a 95% confidence) for an individual PCT result are also plotted along with the linear fit. The EA and ARM results are indicated on the plot as well. Note that there are some points in this plot that fall above and below the confidence limits for the study glasses. None of these points are of particular concern to the outcome of this study as the measured NL [B] values for these glasses are all less than 2.0 g/L, regardless of heat treatment or compositional view, which is considerably lower than the EA benchmark value of 16.695 g/L.

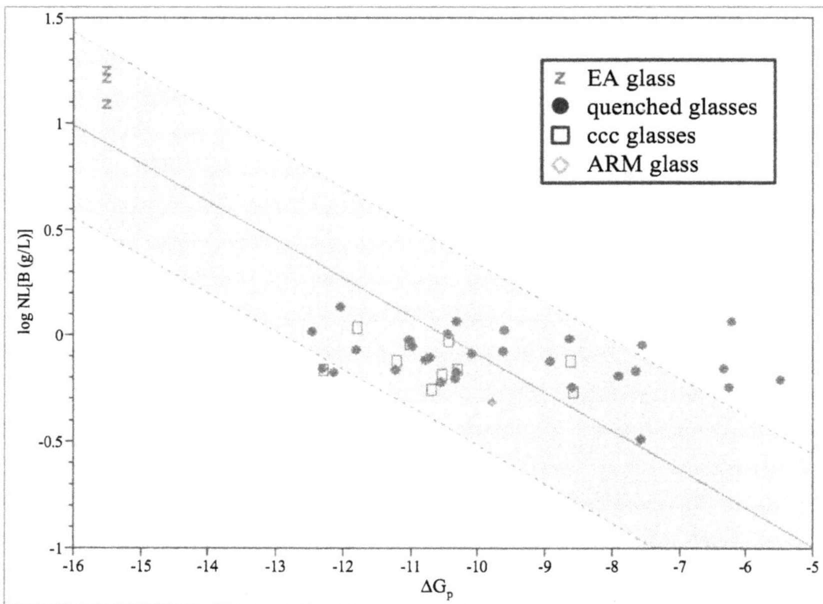

Figure 1. Measured NL [B] versus Predictions of the Model, with the 95% Confidence Interval for Individual PCTs.

A major objective of this study is reduction of conservatism in the nepheline discriminator in order to better predict the formation of nepheline and perhaps allow access to glass compositions at higher waste loadings. Two tools, the current nepheline discriminator[2] and the alternative nepheline discriminator,[12] were used in selecting the glasses that were fabricated and characterized. Figure 2 provides a series of plots for the two nepheline discriminator models to provide a graphical representation of these data. A review of the table and the plots show that the nepheline discriminator currently implemented in PCCS appears to be a conservative predictor of the formation of nepheline during slow cooling (ccc). None of the ccc glasses that were found to contain nepheline (the red circles in Figure 2) fall above the 0.62 line drawn by the current nepheline discriminator. There are, however, a small number of glasses that fall below the 0.62 line that did not crystallize nepheline after the ccc heat treatment (the black circles in Figure 2). This continues to point to some conservatism in the current nepheline discriminator model. Also apparent in Figure 2 is the inability of the alternative discriminator to predict the formation of nepheline in several of the study glasses after the ccc heat treatment. Several of the glasses that contained nepheline fall above the 0.50 line drawn by the alternative discriminator. It is possible that the addition of the data collected in this study to the process used to develop this alternative discriminator[12] may be useful in refining and improving the model.

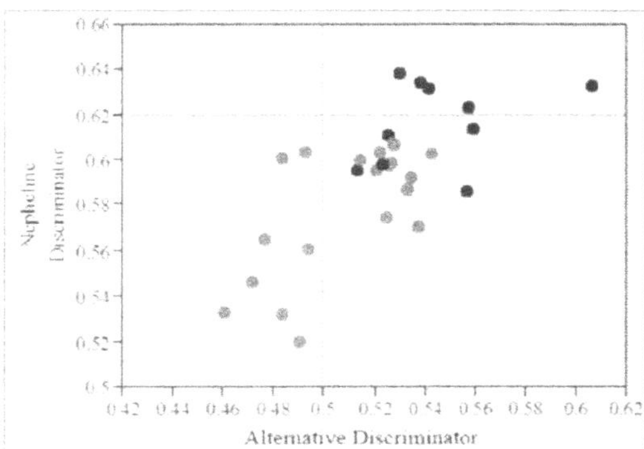

(• – nepheline present; • – no nepheline)

Figure 2. Plots of the Values for the Nepheline Discriminators for ccc Glasses by Compositional View.

Partitioning of the data using a statistical analysis software package[18] was completed as an additional attempt to identify any compositional trends that might indicate a preference for nepheline crystallization or a glass composition region where the current nepheline discriminator is conservative. The results did not provide the opportunity to draw specific conclusions. In general, those glasses included in this study with Al_2O_3 concentrations greater than 14.7 wt % were more prone to nepheline crystallization, although several glasses with lesser Al_2O_3 concentrations also formed nepheline. Any further trends were difficult to identify.

SUMMARY AND CONCLUSIONS

The study glasses generally met the target compositions with only minor issues. In particular, problems were identified with some of the oxides for glasses NE3-13 and NE3-17, although these problems had little impact on the outcome of the study. PCT results correlated well with the XRD analyses in that those glasses that were found to contain nepheline after the ccc heat treatment generally had normalized release values that were greater than their quenched counterparts on a statistically significant basis. Glasses that remained amorphous or contained spinels generally showed similar PCT responses for both their quenched and ccc versions. The glasses that did not contain nepheline had PCT responses that were generally predictable via the ΔG_p model, although a few of the measured values fell above or below the 95% confidence bands of the model. These results were of little concern as the measured NL [B] values for these glasses were all less than 2.0 g/L, regardless of heat treatment, which is considerably lower than the EA benchmark.

A main objective of this study was to identify any compositional region where conservatism in the current nepheline discriminator was preventing access to that region that would otherwise be acceptable for DWPF processing by the PCCS models. Four glasses (based on the measured compositions) were identified through this study that met that condition. However, a review of the individual compositions of these glasses revealed no clear trends that might indicate a driver for suppression of nepheline. Partitioning of the data using a statistical analysis software package gave results that did not provide the opportunity to draw specific conclusions. Another objective of this study was to evaluate an alternative nepheline discriminator model developed using theory of crystallization in mineral and glass melts. Unfortunately this new model, in its current state, was unsuccessful in predicting nepheline crystallization in the glass compositions studied here.

RECOMMENDATIONS

The results of this study show that the current nepheline discriminator as implemented at the DWPF continues to work well in predicting glass compositions that are prone to nepheline crystallization upon slow cooling. Some conservatism was identified in this study; four glasses with values below the 0.62 nepheline discriminator value did not contain nepheline or have statistically significant differences in PCT response after ccc. However, the experimental approach used was not successful in being able to identify any clear trends to explain this result. No clear compositional relationships were identified that would help locate a specific glass component or compositional region that contributed to conservatism in the current nepheline discriminator. The alternative nepheline discriminator that was investigated as part of this study did not correctly predict glass compositions that were prone to nepheline crystallization. It is recommended that the data collected in this study be incorporated into the new model for further refinement. A more general recommendation for future work is that studies aimed at identifying the factors responsible for suppression of nepheline crystallization use simplified glass compositions in order to develop a more fundamental understanding of how the structure of a borosilicate glass changes as aluminum and alkali concentrations are varied.

FOOTNOTES

[*]Jantzen, C. M. and J. M. Pareizs, *Journal of Nuclear Materials,* accepted for publication.

REFERENCES

[1]Li, H., P. Hrma, J. D. Vienna, M. Qian, Y. Su and D. E. Smith, "Effects of Al_2O_3, B_2O_3, Na_2O, and SiO_2 on Nepheline Formation in Borosilicate Glasses: Chemical and Physical Correlations", *J. Non-Crystalline Solids,* **331** 202-216 (2003).
[2]Edwards, T. B., D. K. Peeler and K. M. Fox, "The Nepheline Discriminator: Justification and DWPF PCCS Implementation Details," *U.S. Department of Energy Report WSRC-STI-2006-00014, Revision 0,* Washington Savannah River Company, Aiken, SC (2006).

[3]Fox, K. M., T. B. Edwards and D. K. Peeler, "Nepheline Formation Potential in Sludge Batch 4 (SB4) and Its Impact on Durability: Selecting Glasses for a Phase 3 Study," *U.S. Department of Energy Report WSRC-TR-2006-00053, Revision 0,* Washington Savannah River Company, Aiken, SC (2006).
[4]Fox, K. M., T. B. Edwards, D. K. Peeler, D. R. Best, I. A. Reamer and R. J. Workman, "Durability and Nepheline Crystallization Study for High Level Waste (HLW) Sludge Batch 4 (SB4) Glasses Formulated with Frit 503," *U.S. Department of Energy Report WSRC-STI-2006-00009, Revision 0,* Washington Savannah River Company, Aiken, SC (2006).
[5]Fox, K. M., D. K. Peeler, T. B. Edwards, D. R. Best, I. A. Reamer and R. J. Workman, "Nepheline Formation Study for Sludge Batch 4 (SB4): Phase 3 Experimental Results," *U.S. Department of Energy Report WSRC-TR-2006-00093, Revision 0,* Washington Savannah River Company, Aiken, SC (2006).
[6]Fox, K. M., T. B. Edwards, D. K. Peeler, D. R. Best, I. A. Reamer and R. J. Workman, "Nepheline Formation Study for Sludge Batch 4 (SB4): Phase 3 Experimental Results," *U.S. Department of Energy Report WSRC-TR-2006-00093, Revision 0,* Washington Savannah River Company, Aiken, SC (2006).
[7]Peeler, D. K., T. B. Edwards and T. H. Lorier, "Nepheline Formation Potential in Sludge Batch (SB4) Glasses," *U.S. Department of Energy Report WSRC-TR-2005-00153, Revision 0,* Westinghouse Savannah River Company, Aiken, SC (2005).
[8]Fox, K. M., J. D. Newell, T. B. Edwards, D. R. Best, I. A. Reamer and R. J. Workman, "Refinement of the Nepheline Discriminator: Results of a Phase I Study," *U.S. Department of Energy Report WSRC-STI-2007-00659, Revision 0,* Washington Savannah River Company, Aiken, SC (2008).
[9]Fox, K. M. and T. B. Edwards, "Refinement of the Nepheline Discriminator: Results of a Phase II Study," *U.S. Department of Energy Report SRNS-STI-2008-00099, Rev. 0,* Savannah River National Laboratory, Aiken, SC (2008).
[10]Edwards, T. B., K. G. Brown and R. L. Postles, "SME Acceptability Determination for DWPF Process Control," *U.S. Department of Energy Report WSRC-TR-95-00364, Revision 5,* Washington Savannah River Company, Aiken, SC (2006).
[11]Jantzen, C. M., N. E. Bibler, D. C. Beam, C. L. Crawford and M. A. Pickett, "Characterization of the Defense Waste Processing Facility (DWPF) Environmental Assessment (EA) Glass Standard Reference Material," *U.S. Department of Energy Report WSRC-TR-92-346, Revision 1,* Westinghouse Savannah River Company, Aiken, SC (1993).
[12]Fox, K. M. and T. B. Edwards, "Glass Compositions for the Nepheline Phase III Study," *U.S. Department of Energy Report SRNL-STI-2009-00375, Revision 0,* Savannah River National Laboratory, Aiken, SC (2009).
[13]SRNL, "Glass Melting," *U.S. Department of Energy Report Manual L29, ITS-0003,* Westinghouse Savannah River Company, Aiken, SC (2002).
[14]Marra, S. L. and C. M. Jantzen, "Characterization of Projected DWPF Glass Heat Treated to Simulate Canister Centerline Cooling," *U.S. Department of Energy Report WSRC-TR-92-142, Revision 1,* Westinghouse Savannah River Company, Aiken, SC (1993).
[15]ASTM, "Standard Test Methods for Determining Chemical Durability of Nuclear Waste Glasses: The Product Consistency Test (PCT)", *ASTM C-1285,* (2002).
[16]Fox, K. M. and T. B. Edwards, "Experimental Results of the Nepheline Phase III Study," *U.S. Department of Energy Report SRNL-STI-2009-00608, Revision 0,* Savannah River National Laboratory, Aiken, SC (2009).

[17]Jantzen, C. M., J. B. Picket, K. G. Brown, T. B. Edwards and D. C. Beam, "Process/Product Models for the Defense Waste Processing Facility (DWPF): Part I. Predicting Glass Durability from Composition Using a Thermodynamic Hydration Energy Reaction Model (THERMO)," *U.S. Department of Energy Report WSRC-TR-93-672, Revision 1,* Westinghouse Savannah River Company, Aiken, SC (1995).
[18]**JMP™, Ver. 6.0.3,** [Computer Software] SAS Institute Inc., Cary, NC (2006).

COLD CRUCIBLE VITRIFICATION OF SRS SB4 HLW SURROGATE AT HIGH WASTE LOADINGS

A.P. Kobelev, S.V. Stefanovsky, V.V. Lebedev, D.Y. Suntsov, M.A. Polkanov, O.A. Knyazev
SIA Radon
Moscow, Russia

J.C. Marra
Savannah River National Laboratory
Aiken, SC, USA

ABSTRACT

Savannah River Site (SRS) waste surrogate /additives mixtures at 60 wt.% and 70 wt.% SB4 waste loadings were fed into the SIA Radon bench-scale cold crucible as slurries with water contents of ~50 and ~70 wt.%. In total, ~569.1 kg of feed was vitrified and ~186.6 kg of glassy product was produced. Average slurry feed rate (capacity), glass pour rate (productivity) and specific melt productivity achieved were ~6.7 kg/hr, ~2.2 kg/hr, and ~1200 kg/(m^2×day), respectively. Melting ratio during processing of the slurry with water content of ~50 wt.% was lower (~14 kW×hr/kg) as compared with processing of the slurry with higher water content (~70 wt.%) – 31-34 kW×hr/kg, because additional heat expenses were spent on water evaporation for the high water content feed. The cold crucible operated for ~91 hr. After operation, the cold crucible did not have visible traces of corrosion and was suitable for further operation. The vitrified product obtained in the bench-scale cold crucible was poured into 12 canisters and two portions of the melt at ~60 and ~70 wt.% waste loading were poured into heat-insulated 1L carbon steel vessels followed by their slow cooling to room temperature. The products sampled from canisters 1 to 6 were composed of amorphous phase and spinel with traces of nepheline. The glassy products poured in canisters 7 to 12 were composed of glass, spinel and nepheline. Leach resistance, viscosity and electric resistivity parameters of the materials restrict maximized SB4 waste loading to ~55-60 wt.% for the SB4 feed when processing in the SIA Radon CCIM.

INTRODUCTION

Cold crucible inductive melting (CCIM) is an effective method for production of high-fusible materials.[1] Successful applications of the CCIM for production of various glasses, glass-ceramics, ceramics, and single crystals allowed recommending a cold crucible as an alternative to Joule heated ceramic melter (JHCM) currently used in vitrification of nuclear waste in USA, Russia and other countries.[2] The cold crucible has advantages over the JHCM such as higher temperature availability, higher specific productivity and longer lifetime. The vitrification of simulated Idaho National Laboratory (INL) sodium-bearing, Hanford C-106/AY-102, Savannah River Site (SRS) Sludge Batch 2 (SB2) and Sludge Batch 4 (SB4) wastes using the CCIM process has been successfully demonstrated.[3-6] In particular, it has been shown that ~50 wt.% SB4 waste loading may be achieved[6] whereas in the JHCM this value is limited to ~34 wt.%.[7]

Nepheline is known to be one of the troublesome components in nuclear waste glasses because its formation depletes the vitreous phase of alumina and silica and thus reduces chemical durability of the glass. The goal of this work was to determine a practical limit for SB4 waste loading in borosilicate glass that would not result in nepheline formation and a subsequent decrease in chemical durability of the glass waste form.

FEED PREPARATION AND EQUIPMENT

Feeds as slurries with water contents of ~50 and ~70 wt.% were prepared from SB4 waste surrogate (Table I) and Frit 503-R4 (8 wt.% Li_2O, 16 wt.% B_2O_3, 76 wt.% SiO_2) mixed in amounts to get 60 and 70 wt.% waste loadings in glass.

Table I. Targeted Concentration of Anion/Cation Content of SB4 Feed Simulant (Final Product).

Cation	wt% total solids	wt % calcined solids	Cation	wt% total solids	wt % calcined solids	Anion	Concentration, g/L
Al^{3+}	10.381	14.830	Mg^{2+}	1.287	1.839	SO_4^{2-}	1.386
Ba^{2+}	0.048	0.069	Mn^{2+}	3.447	4.924	PO_4^{3-}	0.284
Ca^{2+}	1.521	2.172	Na^+	10.679	15.256	CO_3^{2-}	7.449
Ce^{3+}	0.141	0.201	Ni^{2+}	1.004	1.434	SO_4^{2-}	1.319
Cr^{3+}	0.104	0.149	Pb^{2+}	0.273	0.390	NO_3^-	5.595
Cu^{2+}	0.031	0.045	Si	0.975	1.393	Cl^-	3.550
Fe^{3+}	15.602	22.289	Zn^{2+}	0.031	0.044	F^-	0.045
K^+	0.043	0.062	Zr^{4+}	0.052	0.074	I^-	0.085
						NO_2^-	13.762
						OH^-	5.158
Specific Gravity, g/mL			1.14	Total Organic Carbon (TOC)			<0.05
Total Solids, wt. %			20	Total Solids, g			276.0
Soluble Solids, wt. %			4.72%	Estimated Calcine Factor			0.7

The cold crucible test was performed at the Radon bench-scale facility (Figure 1, left) equipped with a 236 mm inner diameter cylindrical cold crucible (Figure 1, right).

The cold crucible (Figure 1, right) is a vessel of cylindrical shape with water-cooled walls and bottom. It is manufactured from stainless steel (Russian brand 12X18H10T) pipes. Sidewalls are fabricated from pipes 12 mm in diameter and 2 mm in thickness. Total crucible height is 521 mm, including the crucible itself – 500 mm, bottom – 16 mm, cover – 5 mm. The inner diameter is 236 mm which results in a melt surface area of 4.37 dm^2. The outside of the crucible walls is coated with protective zirconia-based putty. The cold crucible is equipped with a pouring unit consisting of a water-cooled tube-in-tube cylinder with a blocking rod (gate) to stop glass pouring (Figure 1, right).

Broken glass from a previous test at 50 wt.% waste loading and a portion of special high-sodium batch mixture were fed into the cold crucible to initiate melting followed by slurry feeding. At first, a slurry to produce the glass with 60 wt.% waste loading was fed and then a slurry to produce the glass with 70 wt.% waste loading was fed thus ensuring production of the material with variable waste loading from ~50 to ~70 wt.%. The melt was poured into 10 L canisters which were cooled in air.

Major CCIM process variables are given in Table II. In total, in the bench-scale cold crucible, ~569.1 kg of feed was vitrified and ~186.6 kg of glassy product was produced. Average slurry feed rate (capacity), glass pour rate (productivity) and specific melt productivity achieved were ~6.7 kg/hr, ~2.2 kg/hr, and ~1200 kg/(m^2×day), respectively. The highest values were ~9.7 kg/hr, ~3.6 kg/hr, and ~1940 kg/(m^2×day), respectively. Melting ratio during processing of the slurry with water content of ~50 wt.% was lower (~14 kW×hr/kg) as compared with processing of the slurry with higher water content (~70 wt.%) – 31-34 kW×hr/kg, because in the case with higher water content in the feed additional heat expenses were spent on water evaporation.

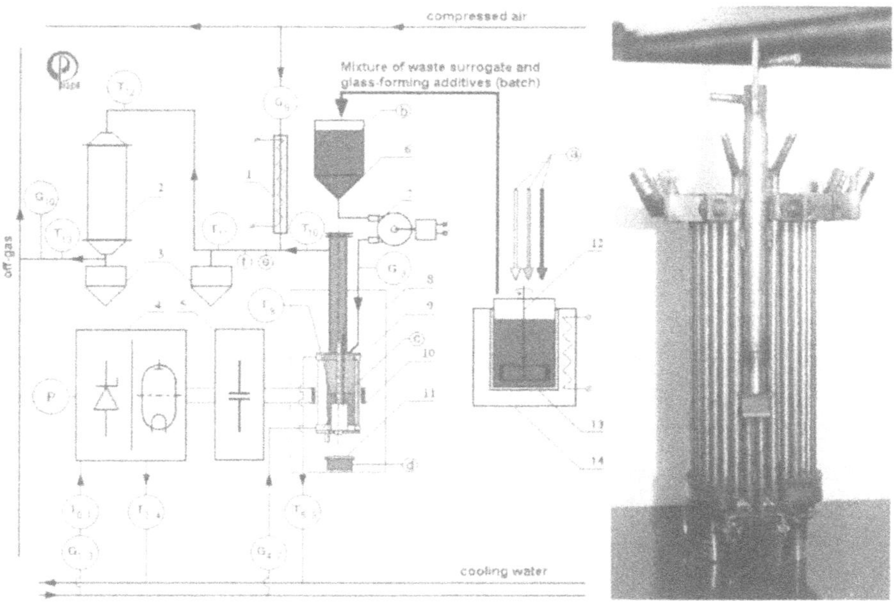

Figure 1. Flow sheet of the cold crucible based experimental plant (left) and cold crucible design (right).

Equipment: 1- electric heater of compressed air, 2- heat-exchanger, 3- condensate collectors, 4- high-frequency energizer, 5- battery of capacitors, 6- batch vessel, 7- feeder, 8- process box, 9- cold crucible, 10- inductor, 11- container for glass product, 12- stirrer, 13- slurry batch preparation vessel, 14- electric furnace. Sampling points: a- raw materials (source reagents), b- slurry, c- "yellow phase", d- glass, e- aerosols, f- off-gas.

The parameters to be controlled are: $T_{0...8}$ – cooling water temperature, T_9 – melt temperature, T_{10} – off-gas temperature at crucible outlet, $T_{11...13}$ – off-gas temperature at heat-exchanger inlet and outlet, $G_{1...7}$ – cooling water flow rate in the induction system, G_8 – slurry feeding rate, G_9 – compressed air flow rate, G_{10} – off-gas flow rate. The energy is supplied from a high frequency power supply, VCI-11-60/1.76 generator, with technical parameters as follows:

The dynamics of slurry feeding and glass production is illustrated in Figure 2. It is seen that the CCIM process was stable with rather minor variations in slurry feeding and glass production rates.

The materials in canisters were broken (Figure 3) and sampled for XRD analyses. The materials from canisters ## 3, 6, 9, and 12 at ~55, ~60, ~65 and ~70 wt.% waste loadings (target), respectively, were also examined by electron and optical microscopy and infrared spectroscopy. Moreover, in an effort to achieve cooling conditions similar to the canister centerline cooling (CCC) regime, portions of melts with ~60 wt.% and 70 wt.% waste loadings were poured into demountable 1 L carbon steel molds (Figure 4) which were placed within 10 L carbon steel containers. The spaces between the molds and inner walls of the containers were filled with foam-chamotte and sandstone. The containers were covered by asbestos sheets and coated with a layer of a foam-chamotte bricks. After cooling to room temperature the 1 L carbon steel molds were removed from the 10 L containers and dismantled.

Table II. Major Process Variables during the CCIM Test.

Parameter	Feed-1 (60 wt.% WL, 50 wt.% WC)	Feed-2 (60 wt.% WL, 70 wt.% WC)	Feed-3 (70 wt.% WL, 50 wt.% WC)	Feed-4 (70 wt.% WL, 70 wt.% WC)
Maximum slurry feed rate, kg/hr	9.65	6.75	9.42	7.56
Average slurry feed rate, kg/hr	8.15	6.31	7.05	6.59
Maximum glass pour rate, kg/hr	3.56	1.43	3.31	1.48
Average glass pour rate, kg/hr	3.11	1.36	2.50	1.39
Maximum specific slurry feed rate, kg/(m^2×hr)	~220	~153	~214	~172
Average specific slurry feed rate, kg/(m^2×hr)	~185	~143	~160	~150
Maximum specific glass pour rate, kg/(m^2×hr)	~81	~33	~75	~34
Average specific glass pour rate, kg/(m^2×hr)	~71	~31	~57	~32
Average vibration power, kW	45.4	46.3	47.6	46.4
Average melting ratio, kW×hr/kg	14.6	34.0	14.4	31.4
Melt surface temperature, °C	1100-1600	1100-1600	1100-1600	1100-1600

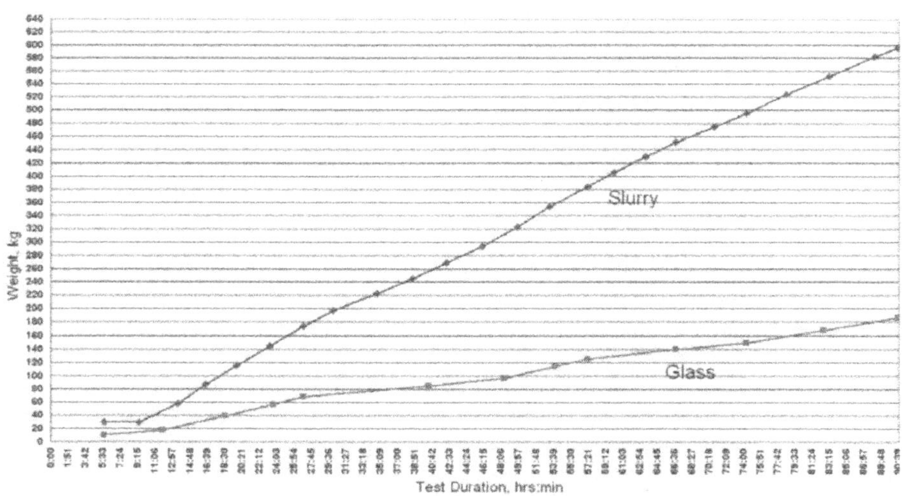

Figure 2. The Dynamics of Slurry Feeding and Glass Production during the CCIM Test.

The vitrified products were analyzed by X-ray fluorescence (XRF) spectroscopy using a PW-2400 Philips Analytical BV unit equipped with quantitative analytical Philips SuperQuantitative & IQ-2001 Software, inductive-coupled plasma - atomic emission spectroscopy (ICP-AES) using a Perkin-Elmer 403 spectrometer (at SRNL). The glassy materials were characterized with X-ray diffraction (XRD) using a DRON-4 diffractometer (Fe K – radiation, $\lambda = 1.936$ Å, scan range of 10 to 90 degrees 2-theta) optical microscopy using an OLYMPUS BX51 polarizing microscope, and scanning electron microscopy with energy dispersive system (SEM/EDS) using a JSM 5610 LV + JED-2300 analytical unit (voltage was 25 KeV, beam current was 1 nA, probe diameter was 1 to 3 μm, dwell time was 100 s; metals, oxides and fluorides were used as standards) and infrared (IR) spectroscopy (compaction of powdered glass in pellets with KBr). Chemical durability was determined by the Product Consistency Test (PCT) procedure.[8]

Figure 3. View of canisters with broken glass.

Figure 4. View of 1L Containers with Glassy Samples Cooled under Conditions Close to CCC (left) and Glassy Samples Removed from Containers (right).

GLASS CHARACTERIZATION

Target and actual chemical compositions of the glassy materials sampled from canisters are given in Table III. The materials are slightly depleted with alkali, boron, and sulfur oxides and, probably, chlorine which were partly lost during melting.

Table III. Target and Actual Chemical Compositions of the Materials Sampled from Canisters.

Oxides	Canister #3		Canister #6		Canister #9		Canister #12	
	Target	Actual	Target	Actual	Target	Actual	Target	Actual
Al_2O_3	14.93	17.35	16.08	19.95	17.42	20.15	18.76	21.2
B_2O_3	7.20	5.475	6.40	3.93	5.60	3.80	4.80	4.005
BaO	0.05	0.13	0.05	0.1	0.05	0.10	0.05	0.165
CaO	1.64	2.77	1.78	3.105	1.93	3.23	2.07	3.33
Cl	0.88	ND	0.97	ND	1.06	ND	1.14	ND
Cr_2O_3	0.12	0.11	0.13	0.11	0.14	0.14	0.15	0.125
CuO	0.03	0.065	0.03	0.04	0.04	0.07	0.04	0.085
F	0.01	ND	0.01	ND	0.01	ND	0.01	ND
Fe_2O_3	16.97	16.20	18.26	18.40	19.78	20.00	21.30	20.25
I	0.02	ND	0.02	ND	0.03	ND	0.03	ND
K_2O	0.05	0.095	0.05	0.09	0.06	0.08	0.06	0.095
Li_2O	3.60	3.035	3.20	2.67	2.80	2.57	2.40	2.43
MgO	1.65	1.89	1.79	2.09	1.94	2.22	2.09	2.38
MnO_2	3.38	3.83	3.63	4.14	3.94	5.20	4.24	4.875
Na_2O	11.41	11.6	12.71	11.85	13.77	11.15	14.83	11.80
NiO	0.97	0.865	1.04	0.925	1.13	1.63	1.22	1.025
P_2O_5	0.18	ND	0.20	ND	0.22	ND	0.23	ND
PbO	0.21	0.19	0.21	0.17	0.23	0.19	0.25	0.22
SO_3	0.85	0.1	1.22	0.075	1.33	0.10	1.43	0.075
SiO_2	35.80	34.4	32.13	31.15	28.46	29.45	24.79	28.15
TiO_2	-	0.04	-	0.03	-	0.03	-	0.03
ZnO	0.03	0.07	0.03	0.04	0.04	0.06	0.04	0.05
ZrO_2	0.06	0.095	0.06	0.105	0.07	0.12	0.07	0.1
Total	100.00	98.31	100.00	98.97	100.00	100.26	100.00	100.39
Waste loading	55	~55	60	~60	65	~65	70	~66

ND – not determined.

As seen from Table III, the highest actual waste loading achieved was ~66 wt.%. This means that the material with 70 wt.% waste loading cannot be produced in the cold crucible due to volatile losses of constituents at the very high process temperatures needed to process this material at this waste loading (up to 1600 °C – see Table II). XRD data (Figure 5, *1-6*) showed that the glassy materials from canisters #0 and # 2 to 6 contained spinel (magnetite/trevorite solid solution) as the only crystalline phase. The material from canister #1 contained trace nepheline. This was due to the effect of high-sodium batch mixture added to the broken glass during melter start-up. The glassy material in canister #6 had ~60 wt.% waste loading. This was the maximum value when nepheline was still not formed. Increase of waste loading by about 2 wt.% resulted in nepheline formation (Figure 5, 7). It was somewhat surprising that the nepheline content in the materials from canisters # 9 to 12 seemed to be lower than in canisters #7 and 8, but this can be explained by the larger volume of the glassy material in canisters #7 and 8 and, as a result, slower cooling rate.

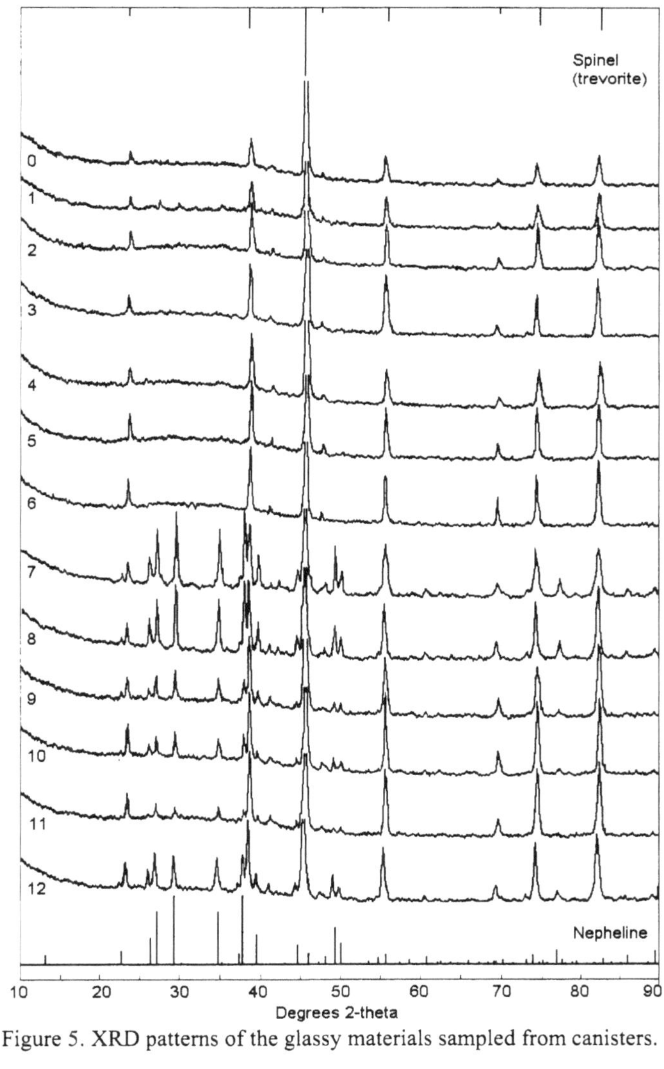

Figure 5. XRD patterns of the glassy materials sampled from canisters.

Analytical data obtained by different methods (SEM/EDS, optical microscopy, IR spectroscopy) are in a good agreement with the XRD data. SEM/EDS and optical microscopy data were described in a separate paper.[9] In this paper we briefly discuss major results.

SEM data (Figure 6, *1-4*) demonstrated an increase in the amount of crystalline content (degree of crystallinity – Table IV) in the materials from canister #3 (55 wt.% waste loading) to canister #9 (65 wt.% waste loading) and #12 (~66 w.% waste loading). At lower waste loadings (~55 wt.%), mainly individual spinel crystals in the homogeneous vitreous matrix were present (Figure 6, *1*). At higher waste loadings (60-66 wt.%) dendrite crystals were predominant (Figure 6, 2-4). In some areas, the degree of crystallinity may have reached ~35 vol.% in samples from canisters and up to 50 vol.% in samples annealed in heat-insulated molds.

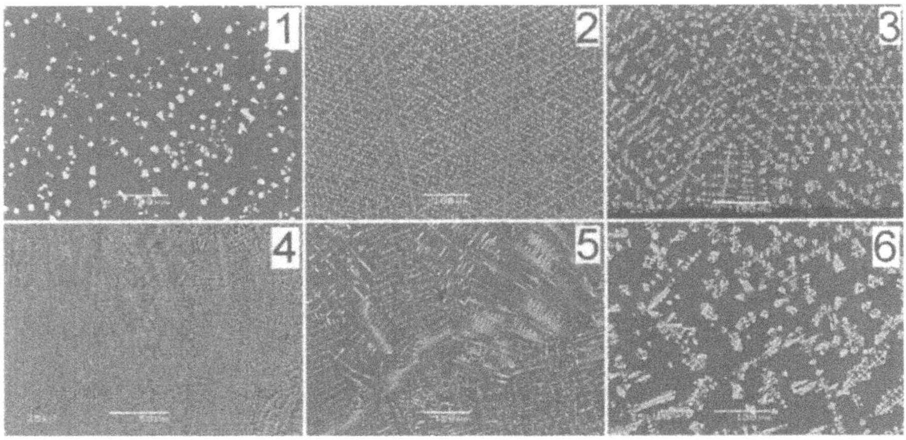

Figure 6. SEM images of the glassy materials sampled from canisters #3 (1), #6 (2), #9 (3), #12 (4) and heat-insulated molds with melts sampled during pouring in canisters #6 (5) and #12 (6).

Table IV. The Degree of Crystallinity Values in the Materials Sampled from Canisters and Heat-Insulated Molds.

Specimen	Average degree of crystallinity, vol.%	Spinel	Nepheline
Canister #3 (~55 wt.% WL)	6.5±1.0	6.5	-
Canister #6 (~60 wt.% WL)	12.0±2.0	12.0	-
Canister #9 (~65 wt.% WL)	13.2±2.7	~10.2	~3
Canister #12 (~66 wt.% WL)	20.0±4.0	~15	~5
Annealed, ~60 wt.% WL	11.0±3.0	~8	~3
Annealed, ~70 wt.% (target) or ~66 wt.% (actual) WL	22.4±3.6	~13.4	~9

Optical microscopy data confirmed SEM data, but provided some information about textural features due to interaction of the flowing glass melt with previously poured and solidified melt in the canisters. The bulk of the glassy materials had a layered texture with various degree of saturation of layers with spinel crystals.[9]

The glassy materials annealed into heat-insulated molds were composed of vitreous phase and dendrite spinel crystals rather uniformly distributed over the bulk (Figure 6, *5,6*). Annealing of the material with ~66 wt.% waste loading yielded larger spinel crystals. Moreover, fine elongated nepheline crystals could be distinguished in the major bulk (Figure 7). Nepheline formation was confirmed by XRD. While the total degree of crystallinity increased by ~2 times with the increase of waste loading from ~60 to ~66 w.%, nepheline content increased by ~3 times and became comparable with spinel content (Table IV).

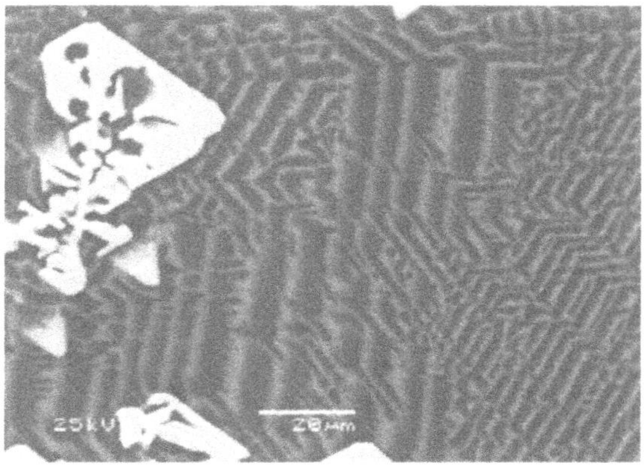

Figure 7. Nepheline in the glassy material with ~66 wt.% actual waste loading annealed in heat-insulated mold.

IR spectra of the glassy materials (Figure 8) were similar to the spectra of vitrified SB4 waste produced at bench-, full-scale units[10] and in a resistively heated laboratory furnace.[11] They consisted of the bands within the ranges of 3000-3700 cm^{-1} (stretching modes in molecules of structurally-bound and absorbed water and hydroxyl groups), 2700-3000 cm^{-1} (vibrations of hydrogen bonds), 1550-1800 cm^{-1} (straining modes in the water molecules), and below 1600 cm^{-1} (stretching and straining modes in the anionic motif of glass network and crystalline phases).[12] There was expected correlation between the intensities of the bands at 3000-3700 and 1550-1800 cm^{-1}.

Within the range of 1600-400 cm^{-1} corresponding to vibrations in the glass network the strongest band at 1200-800 cm^{-1} was due to asymmetric stretching modes in the silicon-oxygen units. Location of the maximum of this band at 985-990 cm^{-1} indicated relatively low degree of polymerization of the glass network whose base was short metasilicate chains formed by one- or two-fold bound [SiO_4] units. The bands due to symmetric stretching modes of the Si—O bonds in such units were positioned within the range of 700-800 cm^{-1} and had minor intensity. Straining modes in the SiO_4 units were responsible for the bands within the range of 400-550 cm^{-1}.[13] The bands with maxima at 1395-1410 cm^{-1} and 1260-1270 cm^{-1} were components of double degenerated asymmetric stretching ν_3 O—B—O mode and the bands at ~720 cm^{-1} and ~655 cm^{-1} were components of straining ν_3 O—B—O mode in boron-oxygen triangular units BO_3.[14]

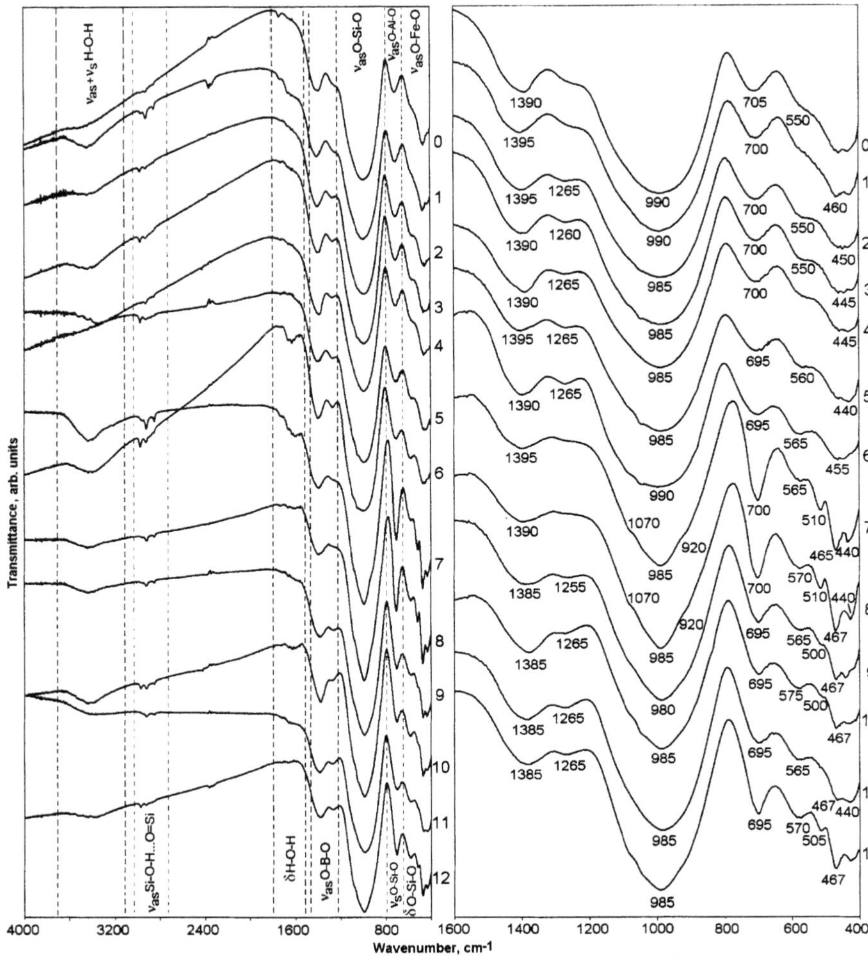

Figure 8. IR spectra of the glassy materials from canisters ## 0 to 12 within the range of 4000-400 cm^{-1} (left) and excerpts within the range of 1600-400 cm^{-1} (right).

Occurrence of ternary coordinated boron in the given glasses was in a good agreement with theoretical representation on the structure of complex alkali aluminoborosilicate glasses: that is, in glasses with high Al_2O_3 concentrations, Al is mainly four-coordinated whereas B remains three-coordinated.[15]

It should be noted, that Al—O vibrations in the AlO_4 units made contribution to the band at 650-800 cm^{-1}. Absorption within the range of 500-600 cm^{-1}, in particular the shoulder at 550-570 cm^{-1}, was probably due to vibrations of the Fe—O bonds in the FeO_4 units partitioned between the vitreous and spinel phases.[16]

At increased waste loading some changes in the IR spectra correlated with the XRD data (Figure 5) especially within the range of 1600-400 cm^{-1} (Figure 8, right). At ~55 wt.% waste loading (canister #3) up to ~60 wt.% waste loading (canister #6), the IR spectra demonstrated minor changes, mainly growth in intensity of the bands within the range of 500-600 cm^{-1}, where the shoulder gradually transformed to the clearly-apparent band with a maximum at 565 cm^{-1}. The maximum of the band at 650-800 cm^{-1} was shifted from 705 cm^{-1} to 695 cm^{-1}. These changes were due to an increase of spinel phase concentration in the glassy material.

The IR spectrum of the sample from canister #7 was appreciably different from that of the sample from canister #6. The strongest band at 1200-800 cm^{-1} became narrower and was split into three components: 1070, 985 and 920 cm^{-1}, the band with a maximum at 700 cm^{-1} became narrower and stronger, a new band at 510 cm^{-1} was formed, and components of the band 400-550 cm^{-1} appeared. These changes demonstrated removal of degeneration from stretching and straining modes due to decrease in symmetry of the structure-forming units. In the given case, these changes may be explained by nepheline crystallization. The bands in IR spectra due to nepheline were positioned at ~1085, 1045, ~1000, ~700, ~520, and ~480 cm^{-1}.[17] Here they can be identified as narrow bands at ~1070, ~1040, 985, 700, 510, and ~470 cm^{-1} observed in the spectra of the glassy materials from canisters ## 7, 8, 9, and 12 where, as follows from XRD patterns, nepheline content was the highest. In the spectrum of the material from canister #11 at rather low nepheline content the band 500-510 cm^{-1} was nearly absent and the strongest band 1200-800 cm^{-1} was nearly structureless.

The bands at 1385-1390 cm^{-1} and ~1265 cm^{-1} in all the samples had low intensity due to low B_2O_3 concentrations in the glassy materials and in this case their intensity was reduced with a decrease of boron content. In all the materials boron was present in a three-coordinated form.

Chemical durability testing demonstrated that increase of waste loading in the glassy materials resulted in an increase of normalized releases of B, Li and Na but a decrease of normalized release of Si (Table V). The normalized releases of B and Li increased strongly (by 4-5 times) from the material sampled from canister #12 as compared with the material from canister #9 although actual waste loading and nepheline content in both samples appeared similar. At the same time, Na release increased only by ~1.3 times and Si release even decreased. The latter was likely due to silica saturation of the leach solution..

Table V. Product Consistency Test Results for Maximum SB4 Waste Loading CCIM Testing

Glass ID	Normalized Release (g/L)			
	B	Li	Na	Si
Canister #3	0.81	0.72	0.57	0.27
Canister #6	0.77	0.89	0.65	0.25
Canister #9	0.95	0.88	0.69	0.23
Canister #12	4.32	3.30	0.90	0.20
Environmental Assessment Standard (EA)	16.69	9.34	12.81	3.99

Note, that the normalized releases of B and Li from the material sampled from canister #12 exceed those of Na by 4-5 times and Si by approximately 20 times. This was consistent with our suggestion[10] that vitreous phase contained lithium-borate groups or even chemically ordered areas with primarily lithium borate composition, yielding B and Li that leached congruently from this low durability phase. Significant fractions of Na and Si in this material entered nepheline, which is chemically durable itself.

It should be noted that the normalized releases of B, Li, Na, and Si from the material sampled from canister #12 were found to be lower by factors of ~4, ~3, ~14, and 8, respectively, than those from the Environmental Assessment (EA) glass that is used as the benchmark for HLW glass product acceptability.

CONCLUSION

Vitrification of SRS SB4 waste with high sodium, aluminum, and iron oxides content yielded glassy materials containing spinel and nepheline crystalline phases. The highest waste loading achieved was ~66 wt.%. Wet slurry feeding in the relatively small cold crucible (236 mm inner diameter) demonstrated rather low efficiency of the CCIM process: maximum slurry feed rate (capacity) and glass pour rate (productivity) did not exceed ~8 kg/hr and ~3 kg/hr, respectively, and specific glass pour rate (productivity) ~1750 kg/(m^2×d) during processing of slurry with ~50 wt.% water content. Melting ratio was ~14 kW×hr/kg of glass during processing of slurry with ~50 wt.% water content and increased to 31-34 kW×hr/kg of glass with increasing of water content in the slurry to ~70 wt.%. With the increase of waste loading in the materials their degree of crystallinity increased from ~7 vol.% at ~55 wt.% waste loading to ~20 vol.% (in some areas up to ~35 vol.%) at ~66 wt.% waste loading. The samples annealed in heat-insulated molds had higher degree of crystallinity (up to ~50 vol.% in some areas). The glassy materials had a tendency to chemical differentiation (elemental segregation??). The glass network was built from rather short metasilicate chains; boron was mainly in three-coordinated form and entered with lithium in a separate structural constituent. Chemical durability of the materials remained high even at high waste loadings and normalized releases of B, Li, Na, and Si were several (3 to 14) times lower than those from standard EA glass. Maximum SB4 waste loading recommended was estimated to be ~55-60 wt.%.

ACKNOWLEDGEMENTS

The work was performed under contract "CCIM Testing to Evaluate Processing Behavior of High Waste Loading Feeds" - 2008 & 2009 SCOPE (TASK 1). Subtask 3: CCIM Testing to Demonstrate Maximized Waste Loading with SB4 sponsored by Office of Environmental Management of the US DOE. Special thanks to Mr. K.D. Gerdes for funding.

Authors thank personnel of Experimental Works of Applied Radiochemical Technologies at SIA Radon who participated in the tests and personally thank Mr. S.Y. Shvetsov, Dr. A.G. Ptashkin, Dr. O.I. Stefanovskaya, Ms. M.S. Zen'kovskaya, as well as Mr. B.S. Nikonov, Dr. B.I. Omelianenko (IGEM RAS), and Mr. A.A. Akatov (Institute of Technology, St-Petersburg) for their help in sample analyses.

REFERENCES

[1] Y.S. Kuz'minov, E.E. Lomonova, V.V. Osiko, High Fusible Materials from Cold Crucible (Russ.), Moscow, Nauka (2004).
[2] S.A. Dmitriev, S.V. Stefanovsky, Management of Radioactive Waste (Russ.), Moscow, RCTU (2000).

[3] A.P. Kobelev, S.V. Stefanovsky, T.N. Lashchenova, V.N. Zakharenko, M.A. Polkanov, O.A. Knyasev, C.C. Herman, D.F. Bickford, E.W. Holtzscheiter, R.W. Goles, D. Gombert, Vitrification of INL and Hanford HLW Surrogates at the SIA Radon Bench-Scale Cold Crucible Unit, *ICEM '05: Proc. The 10th Int. Conf. on Environmental Remediation and Radioactive Waste Management*, Glasgow, Scotland, 4-8 September 2005. ID ICEM05-1222 (2005).
[4] A.P. Kobelev, S.V. Stefanovsky, O.A. Knyasev, T.N. Lashchenova, J.C.Marra, E.W. Holtzscheiter, C.C. Herman, Induction Heated Cold Crucible Melter Testing with Troublesome High Level Waste Components, *Proc. 107th Amer. Ceram. Soc. Meeting*. Baltimore, MD, April 10-13, 2005,159-169 (2005).
[5] A.P. Kobelev, S.V. Stefanovsky, O.A. Knyazev, T.N. Lashchenova, A.G. Ptashkin, M.A. Polkanov, E.W. Holtzscheiter, J.C. Marra, Results of a 50% Waste Loading Vitrification Test Using the Cold Crucible Melter for Savannah River Site, *Proc. Waste Management '06 Conf.* Tucson, AZ, February 26 – March 2, 2006. CD-ROM. ID 6127 (2006).
[6] S.V. Stefanovsky, A.P. Kobelev, V.V. Lebedev, M.A. Polkanov, A.G. Ptashkin, O.A. Knyazev, J.C. Marra, Vitrification of Savannah River Site SB4 Waste Surrogate in the Radon Bench-Scale Cold Crucible Unit, *Waste Management 2008 Conf.*, February 24 -28, 2008, Phoenix, AZ. ID 8035 (2008).
[7] C.C. Herman, Summary of Results for Expanded Macrobatch 3 Variability Study. Report WSRC-TR-2001-00511 (2001).
[8] Standard Test Method for Determining Chemical Durability of Nuclear Waste Glasses: The Product Consistency Test (PCT). ASTM Standard C1285-94. ASTM, Philadelphia, PA (1994).
[9] S.V. Stefanovsky, A.P. Kobelev, V.V. Lebedev, M.A. Polkanov, D.Y. Suntsov, J.C. Marra, The Effect of Waste Loading on the Characteristics of Borosilicate SRS SB4 Waste Glasses, *ICEM '09/DECOM '09: 12th International Conference on Environmental Remediation and Radioactive Waste Management*, October 11-15, 2009, Liverpool, UK (2009).
[10] A.A. Akatov, B.S. Nikonov, B.I. Omel'yanenko, S.V. Stefanovsky, J.C. Marra, Structure of Borosilicate Glassy Materials with High Concentrations of Sodium, Iron and Aluminum Oxides, *Glass Phys. Chem.* **35** (2009) 243-257.
[11] A.A. Akatov, B.S. Nikonov, B.I. Omelianenko, O.I. Stefanovskaya, S.V. Stefanovsky, D.Y. Suntsov, J.C. Marra, The Effect of Content of Iron-Aluminum High Level Waste Surrogate on Phase Composition and Structure of Glassy Materials for Their Immobilization, *Glass Phys. Chem.* (in press).
[12] K. Nakamoto, "Infrared Specra of Inorganic and Coordination Compounds", John Wiley & Sons, New York, London (1962).
[13] J. Wong, C.A. Angell, Glass Structure by Spectroscopy, Marcel Dekker, Inc. New York – Basel (1976).
[14] V.A. Kolesova, Vibrational Spectra and the Structure of Alkali Borate Glasses, *Glass Physics and Chemistry* (Russ.) **12** [1] (1986) 4-13.
[15] A.A. Appen, Chemistry of Glass (Russ.), Khimiya, Leningrad (1974).
[16] I.I. Plyusnina, Infrared Spectra of Minerals, MGU, Moscow (1977).
[17] I.I. Plyusnina, Infrared Spectra of Silicates, MGU, Moscow (1967).

AN EXTRACTION OF PLATINUM GROUP METALS AND MOLYBDENUM FROM MOLTEN BOROSILICATE GLASS USING Cu AND Cu_5Si

Kazuyoshi Uruga
EcoTopia Science Institute, Nagoya University
Furo-cho, Chikusa-ku, Nagoya, Japan

ABSTRACT

Copper silicide (Cu_5Si) was used for the extraction and removal of Pd and MoO_3 from molten borosilicate glass in an oxidizing environment. The oxidation rate of Cu in Cu_5Si up to 1273 K was very slow, less than 0.04% min^{-1}, in an oxidizing environment of O_2:Ar = 20:80 (v/v) and Cu was kept in the metallic state efficiently. Removal experiments of Pd and MoO_3 were performed by heating a mixture of glass cullet containing Cu_5Si and cullet containing Pd or MoO_3 at 1473 K for 4 h in the oxidizing environment. As for Pd, more than 80% of Pd was successfully extracted into Cu_5Si and removed from the glass as spherical Pd-Cu metal buttons. Though the removal of MoO_3 was also possible, the removal fraction was as low as 19%. Mixing additional Si in the glass as a reducing agent was effective for improving the removal fraction of both MoO_3 and Cu, and the maximum fraction removed were 68 and 95%, respectively.

INTRODUCTION

High-level radioactive raffinate generated from the reprocessing of spent nuclear fuel is vitrified with borosilicate glass for safe geologic disposal. However, some of the elements included in the raffinate, mainly platinum group metals (PGMs) and molybdenum, separate from the molten glass during the vitrification process because of their low solubility in the borosilicate glass. This separation may cause serious consequences such as an erosion of the glass furnace, clogging up of the drain nozzle of the furnace by PGM-glass mixed sludge and degradation of the chemical durability of the waste glass itself.

Extraction using fused metal is one of the effective methods to separate PGMs that are contained in glass or ceramics bodies. By heating ceramics with metal extractant, PGMs are extracted in the fused metal and can be partitioned from the molten ceramic sludge as a separable metal phase. This method was studied mainly to recover PGMs from exhausted automotive catalysts. Burnham *et al.* [1] recovered Pt and Pd from the catalysts using Fe or Cu as collecting metals in a plasma arc furnace. Benson *et al.* [2] discussed the recovery mechanism of PGMs from the catalyst using a gravity model described by Stokes' law. Studies for the application of this method toward the high-level waste were originally started at the Pacific Northwest Laboratory around early 1980s and later at Japan to collect PGMs as future resources. Jensen *et al.* [3] recovered PGMs from simulated high-level waste and evaluated the recovery performance using various metal oxides and reducing agents. Naito *et al.* [4,5] studied a recovery from simulated insoluble residue produced in the high-level raffinate using Pb, and evaluated the decontamination factor of Ce as a simulant of Pu.

We have studied and proposed this fused metal method for the removal of PGMs and Mo from molten glass as one of the options to solve the above vitrification problems.[6-9] Figure 1 shows the concept of our proposed removal process. The laboratory-scale crucible tests showed that fused Cu worked well as an extractant of metallic Pd, Ru and Mo, and we successfully removed these metals from the glass as metal buttons just by mixing Cu powder with the glass and melting them. Oxides such as RuO_2 and MoO_3 were also removed by addition of Si to the glass as a reducing agent of the oxides to the metals by following reactions:

$$RuO_2 + Si \rightarrow Ru + SiO_2 \quad (1)$$
$$MoO_3 + 1.5Si \rightarrow Mo + 1.5SiO_2 \quad (2)$$

The reason for using Si is that the SiO_2 produced by the reduction reaction dissolves in the molten glass as one of the major glass components without causing glass foaming. In this removal method, keeping Cu in the metallic state is essential since oxides such as CuO and Cu_2O dissolve in the molten glass and do not work as extractants. All of these tests were thus performed in an inert environment to keep Cu in the metallic form. However, if removal is possible in oxidizing environments, this method will become much applicable for practical use.

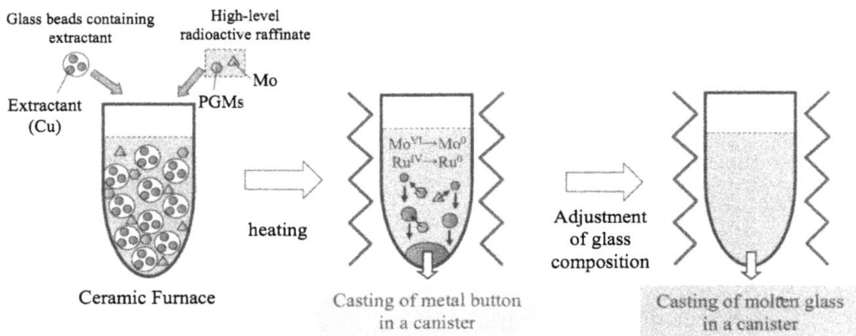

Figure 1. Removal concept of PGMs and Mo during the vitrification process of high-level radioactive raffinate

Extraction under oxidizing condition has been achieved by adding reducing agents together with copper oxides. A commercial plant for the recovery of PGMs from exhausted automotive catalysts uses CuO and coke as the extractant and reducing agent, respectively.[10] This method, however, requires relatively large amounts of reducing agents; at least more than an equivalent amount of the reducing agent. An increase in the amount of reducing agent added would entail the change of the glass composition and the glass viscosity when Si is used as the reducing agent.

In this study, copper silicide (Cu_5Si) was used as an extractant to achieve the removal of PGMs and Mo from molten glass under oxidizing conditions. The oxidation rate of Cu_5Si was determined by thermal gravimetry. Palladium was selected as a representative of the PGMs and the extraction efficiencies of Cu_5Si for both Pd and MoO_3 were experimentally determined. The behavior of Cu_5Si in molten glass was also examined by small-scale crucible tests.

EXPERIMENTAL

Glass Sample Preparation

The composition of the borosilicate glass used in this study was SiO_2 57%, B_2O_3 25%, Na_2O 18% by mass. The glass was prepared by melting the mixture of SiO_2 (reagents grade, Wako Pure Chemical Ind., Ltd., Japan), B_2O_3 (purity 99.9%, Aldrich, Japan) and Na_2CO_3 (purity 99.9%, Wako Pure Chemical Ind., Ltd., Japan) in a platinum crucible at 1473 K. After quenching the glass melts, the glass was ground using a ball mill (Pulverisette 7, FRITSCH, Germany). Two glass specimens containing Cu_5Si powder (purity >99.9%, Kojundo Chemical Lab. Co., Ltd., Japan) or Cu powder (purity >99.85%, Kanto Chemical Co., Inc., Japan) were prepared. Both powders were sieved to adjust the particle size to 20-45 μm. Several amounts of Cu_5Si or Cu powders were mixed with 5 g of the glass powder in alumina crucibles and heated in a tube-type furnace (ϕ65×1200 mm, KTF434N, Koyo Thermo System, Japan) at 1123 K for 15 minutes. The inside of the furnace tube was filled with Ar gas (purity 99.9999%) following a vacuuming under 20 Pa. Argon was introduced at a constant flow rate of 300 cm^3 min^{-1} during the melting to maintain the inert atmosphere. At 1123 K, the glass started melting and the mixture became a unit glass lump in which Cu_5Si or Cu powder was dispersed visually homogeneously. Glass specimens containing Pd (less than 200 μm in diameter, purity >99.9%, Rare Metallic Co., Ltd.) and containing MoO_3 (reagents grade, Kanto Chemical Co., Inc., Japan) were prepared separately. The amounts of Pd and MoO_3 were fixed at 100 and 150 mg per 5 g of the glass, respectively because the mass fractions of the PGM elements and Mo are both approximately 1% in vitrified waste glass from the reprocessing of fuel from light water reactors.[11] The Pd specimen was prepared in the same manner as the Cu_5Si specimen. The mixture of MoO_3 and borosilicate glass was put in a platinum crucible and heated at 1273 K for 2 h in air to dissolve MoO_3 in the glass homogeneously. All of the glass specimens containing Cu_5Si, Cu, Pd and MoO_3 were crushed into cullet of less than 2 mm in diameter and used for the removal experiments.

Removal Test of Pd and MoO_3

For the removal of Pd, the two types of glass cullet containing Cu_5Si and Pd were mixed together in an alumina crucible and melted at 1473 K for 4 hours in the tube-type furnace. In order to control the atmosphere for melting, the furnace was filled with O_2:Ar = 20:80 (v/v) mixed gas to get an oxidizing environment and the mixed gas was introduced with a constant flow rate of 300 cm^3 min^{-1} during the melting. For the removal of MoO_3, Si (less than 20 μm powder, purity 99%, Kojundo Chemical Lab. Co., Ltd., Japan) was added to the mixture of the glass cullets of Cu_5Si and MoO_3 as a reducing agent. Metal aggregates formed after melting were separated from the quenched glass by breaking the crucible with a hammer. The aggregates were dissolved in aqua regia solution. The solution and the insoluble residue produced in the solution were centrifugally separated (H-26F, Kokusan, Japan), and the concentrations of Pd, Mo, Cu and Si in the solution were then measured by ICP-AES (ICPS-7000, Shimadzu, Japan). The insoluble residue was fused with KOH and KNO_3 at 773 K, and then dissolved in HCl solution. The concentrations of Pd, Mo, Cu and Si in the solution were measured by ICP-AES. The results were evaluated by calculating the fraction;

$$\text{Removal fraction of } i, \% = \frac{[\text{Amount of } i \text{ in the metal buttons}]}{[\text{Total addition amount of } i]} \times 100 \quad (3)$$

where i is the element to be evaluated.

RESULTS AND DISCUSSION

Oxidation of Cu_5Si

The oxidation behavior of Cu_5Si was examined by thermal gravimetry and compared with the behavior of Cu. Figure 2 shows the change of the weight fraction when the temperature increased from room temperature to 1473 K with a rate of 2 K min^{-1}. The weight fraction of Cu increased from 473 K and all the Cu was changed to CuO when the temperature reached 973 K. On the other hand, the fraction of Cu_5Si was almost constant until the temperature reached 1273 K and the oxidation of Cu in Cu_5Si was successfully prevented. From the derivative of the fraction curve, the oxidation rate of Cu at 1273 K was calculated to be 0.04% of the initial moles of Cu per minute, in which we postulated that the increase of the weight fraction came from the oxidation of Cu to Cu_2O. At high temperatures, the surface of the Cu_5Si particles are covered by a passivated SiO_2 layer since the Si atom in Cu_5Si moves to the surface and is oxidized to SiO_2.[12-14] The inner Cu is protected by the SiO_2 layer from oxidation. The sudden increase of the weight fraction was observed at temperatures greater than 1273 K, which may be due to breakage of the surface SiO_2 layer and the subsequent oxidation of Cu. However, this sudden oxidation does not occur in the removal process since the Cu_5Si particle will be in the molten borosilicate glass at temperature greater than 1123 K, and the oxygen potential of the inside of the molten glass is much less than that of air.

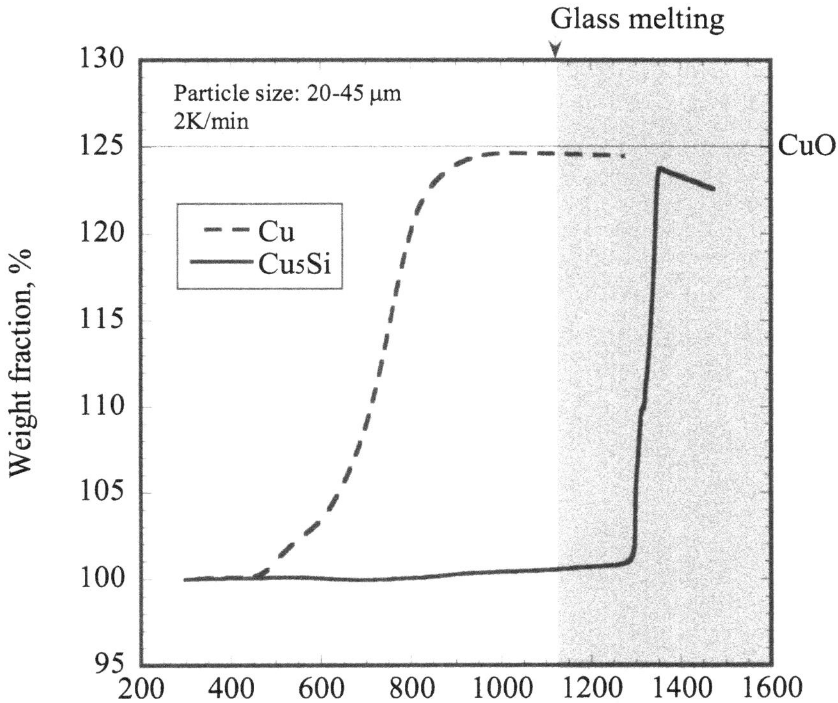

Figure 2. Change of the weight fraction of Cu_5Si and Cu in a temperature range from room temperature to 1473 K. The data was obtained in an atmosphere of O_2:Ar = 20:80 (v/v).

Formation of Metal Button

Before the removal experiments of Pd and MoO_3, we observed the behavior of Cu_5Si in the molten glass and determined the amounts of Cu and Si removed from the glass as metal buttons to clarify the advantages of using Cu_5Si as the extractant instead of Cu. All of the experiments except Run 1 were performed in binary Cu_5Si-glass or Cu-glass systems. The 5.544 g of glass cullet containing 0.544 g of Cu_5Si was mixed with 5 g of borosilicate glass and heated under a variety of conditions for 4 h. The heating conditions and removal fractions of Cu and Si in each run are shown in Table I. The effect of temperature on the formation of metal button was determined in Runs from 2 to 5. In Run 2, Cu_5Si powder settled on the bottom of the crucible and formed Cu_5Si-glass mixed sludge at 1173 K (Fig. 3(a)). No separable buttons or agglomerates of Cu_5Si formed anywhere in the glass. However, In Runs 3 and 4, spherical metal buttons formed in the bottom of the crucible (Fig. 3(b)) and almost all of the added Cu_5Si was successfully removed by increasing the heating temperature higher than 1273 K. The same experiment was performed using the glass cullet containing Cu particles. From Run 5, Cu powder assembled and floated on the glass surface but no spherical button formed even at 1343 K (Fig. 3(c)). The floating Cu was firmly attached to the glass. These behaviors arise from the differences of the melting point of Cu_5Si and Cu: the melting point of Cu_5Si is approximately 1123 K,[15] which is lower than that of the Cu, 1356 K.[15] Using Cu_5Si may thus have an advantage in decreasing the removal temperature. Heating only Cu_5Si powder at 1273 K (Run 1) resulted in no button formation. The results of Runs 1 and 3 show that molten glass has an important role for the button formation; the surface SiO_2 layer of Cu_5Si may dissolve in the molten glass and the agglomeration of Cu_5Si particles is enhanced.

Table I. Metal button formation and removal fraction of Cu in several conditions

Run	Addition amount, g				Temperature	Atmosphere	Removal fraction, %	
	Glass[a)]	Cu_5Si	Cu	Si	K		Cu	Si
1		0.544			1273	Ar	no button	
2	10	0.544			1173		no button	
3	10	0.544			1273	Ar	96	96
4	10	0.544			1343		96	96
5	10		0.500		1343		no button	
6	10	0.544					89	21
7	10		0.500				26	-
8	10		0.500	0.044			69	2
9	10		0.500	0.088	1473	O_2:Ar = 20:80	80	1
10	10		0.500	0.110			84	15
11	10		0.500	0.176			66	15
12	10		0.500	0.220			62	5

a) For Runs 2 to 5, $Na_2B_4O_7$ was used as a glass material instead of borosilicate glass in order to decrease the viscosity of the glass. Blank means no addition.

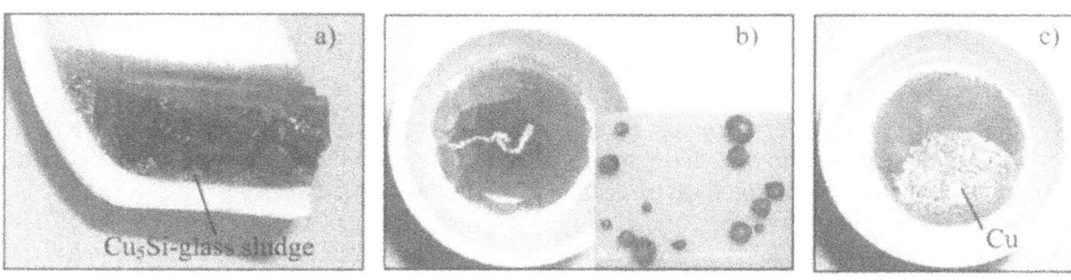

Figure 3. Photos of the glasses and buttons after the melting. a) Run 2, b) Run 4 and c) Run 5

Runs 6 to 12 compared the removal efficiency of Cu in an oxidizing environment between the use of the Cu_5Si cullet and the use of a mixture of the Cu cullet and Si. When Cu_5Si cullet was used, nearly 90% of Cu was removed from the glass. The close removal fraction of

84% was obtained from the use of Cu-Si mixture in Run 10. However the mass ratio of Si to Cu added in the glass was 0.22 which was larger than that in Cu_5Si, 0.088. Using Cu_5Si efficiently prevents Cu from oxidation. From Runs 11 and 12, metal-glass mixed sludge that was similar to Run 2 shown in Fig. 3(a) was observed, and the amounts of the metal buttons were less than in Run 10. Excessive addition of Si in the glass prevented the formation of Cu-containing buttons and had a negative effect on the Cu removal.

The results obtained from Table 1 indicate that the use of Cu_5Si is more effective for the removal of Cu itself rather than the use of Cu and Si mixture. Therefore, the removal experiments of Pd and MoO_3 in an oxidizing environment were performed using Cu_5Si in the next section.

Removal of Pd and Mo Using Cu_5Si

Table II shows the removal fractions of Pd and Mo in the inert or oxidizing environments when Cu_5Si was used as the extractant. From Run 13, Cu_5Si showed good extraction ability for Pd as well as Cu.[6] This indicates that the existence of Si in Cu results in no significant change in the Pd extraction. The removal fraction of Pd in the oxidizing environment was still more than 80% and Cu_5Si worked well as the extractant. The removal fraction of Cu was as same as that in Run 6 and the fraction was independent of the existence of Pd. From Runs 15 and 16, removal of MoO_3 with Cu_5Si was possible in both inert and oxidizing environments but the removal fraction decreased in the oxidizing environment. This was because in the oxidizing environment, the Si in Cu_5Si was used not only as the reducing agent of MoO_3 but also as the reducing agent of Cu and self oxidation while all the Si was available for the reducing agent of MoO_3 in the inert environment. Mixing additional Si in the glass was effective for improving removal fractions of both MoO_3 and Cu, and the fractions increased up to 68 and 95%, respectively in Run 18.

Table II. Removal fractions of Pd and MoO_3 obtained by using Cu_5Si as extractant

Run	Addition amount, g				Atmosphere	Removal fraction, %		
	Pd	MoO_3	Cu_5Si	Si		Pd or MoO_3	Cu	Si
13	0.100		0.544		Ar	98	97	-
14	0.100		0.544		O_2:Ar = 20:80	82	87	33
15		0.150	1.088		Ar	53	96	-
16		0.150	1.088		O_2:Ar = 20:80	19	88	5.6
17		0.150	1.088	0.088	O_2:Ar = 20:80	31	92	36
18		0.150	1.088	0.176	O_2:Ar = 20:80	68	95	51

- Not measured. Blank means no addition.

CONCLUSION

Removal of Pd and MoO_3 contained in molten glass in the oxidizing environment of O_2/Ar = 20% was performed using Cu_5Si as the extractant and Si as the reducing agent for MoO_3. Both Pd and MoO_3 were removed from the glass as metal buttons and the maximum removal fractions were 82 and 68%, respectively. Added Cu as Cu_5Si was also removed by nearly 90% and the removal fraction of Cu increased up to 95% when Si was added in the glass. Two advantages of the use of Cu_5Si instead of Cu were found: the formation of metal buttons are possible at lower temperatures, and that Cu is kept in the metallic state with smaller amounts of Si and a higher removal fraction of Cu is obtained in comparison with the use of Cu and Si mixtures.

ACKNOWLEDGEMENTS

This study was supported by the Grant-in-Aid program for the research fellow of Japan Society for the Promotion of Science.

REFERENCES

[1]R. F. Burnham, J. E. Harry, A. Gibbon, *et al.*, Plasma Arc Melters, *European Patent Application 0096 493 A2*, (1987).
[2]M. Benson, C. R. Bennett, J. E. Harry, *et al.*, The Recovery Mechanism of Platinum Group Metals from Catalytic Converters in Spent Automotive Exhaust Systems, *Resources, Conservation and Recycling*, **31**, 1 (2000).
[3]G. A. Jensen, A. M. Platt, G. B. Mellinger, *et al.*, Recovery of Noble-Metals from Fission-Products, *Nuclear Technology*, **65**, 305 (1984).
[4]K. Naito, T. Matsui, H. Nakahira, *et al.*, Recovery of Noble Metals from Insoluble Residue of Spent Fuel, *J. Nucl. Sci. Technol.*, **23**, 540 (1986).
[5]K. Naito, T. Matsui, H, Nakahira, *et al.*, Recovery and Mutural Separation of Noble Metals form the Simulated Insoluble Residue of Spent Fuel, *J. Nucl. Mater.*, **184**, 30 (1991).
[6]K. Uruga, K. Sawada, Y. Arita, Y. Enokida and I. Yamamoto, Removal of platinum group metals contained in molten glass using copper, *J. Nucl. Sci. Technol.*, **44**, 1024-1031 (2007).
[7]K. Uruga, K. Sawada, Y. Enokida and I. Yamamoto, Liquid metal extraction for removal of molybdenum from molten glass Containing Simulated Nuclear Waste Elements, *J. Nucl. Sci. Technol.*, **45**, 1063-1071 (2008).
[8]K. Uruga, K. Sawada, Y. Arita, Y. Enokida and I. Yamamoto, Influence of the size of collecting metal on removal of platinum group metals from molten glass, *J. Power and Energy Systems*, **2**, 198-205 (2008).
[9]K. Uruga, K. Sawada, Y. Arita, Y. Enokida and I. Yamamoto, Vitrification of high-level radioactive waste considering the behavior of platinum group metals, *Progress in Nuclear Energy*, **50**, 514-517 (2008).
[10]N. Esawa, H. Inoue, M. Takada and H. Masuda, Hakkinzoku Kinzoku Kaishuu Houhou, *Japanese Patent*, 4-317423 (1992) [in Japanese].
[11]T. Advocat, P. Jollivet, J.L. Crovisier, *et al.*, Long-term alteration mechanisms in water for SON68 radioactive borosilicate glass, *J. Nucl. Mater.*, **298**, 55 (2001).
[12]A. Cros, M. O. Aboelfotoh and K. N. Tu, Formation, oxidation, electronic, and electrical properties of copper silicides, *J. Appl. Phys.*, **67**, 3328-3336 (1990).
[13]T. C. Frank and J. L. Falconer, Surface compositions of copper-silicon alloys, *Applications of Surface Science*, **14**, 359-374 (1982).
[14]S. Hymes, K. S. Kumar and S. P. Murarka, Thermal stability of copper silicide passivation layers in copper-based multilevel interconnects, *J. Appl. Phys.*, **83**, 4507-4512 (1998).
[15]H. Okamoto, *Phase Diagrams for Binary Alloys*, Materials Park, Ohio, p 308 (2000).

MICROSTRUCTURE OF LASER-MELTED ZIRCONIUM CARBIDE CERAMICS

H.F. Jackson[1], D.D. Jayaseelan[1], W.J. Clegg[2], M.J. Reece[3], F. Inam[3], D. Manara[4], C. Perinetti Casoni[4], W.E. Lee[1]

[1]Department of Materials, Imperial College London, London, UK
[2]Department of Materials Science and Metallurgy, University of Cambridge, Cambridge, UK
[3]Nanoforce Technology Ltd., London, UK
[4]European Commission Joint Research Centre Institute for Transuranium Elements, Karlsruhe, Germany

ABSTRACT

Zirconium carbide (ZrC) is a candidate material for advanced nuclear fuels for high-temperature gas-cooled reactors, but its durability and performance under irradiation is still under investigation. In particular, the effect of carbon vacancies on thermodynamic, mechanical, and heat transport properties of ZrC must be better understood.

As part of a novel thermal analysis technique, melting of ceramics in the Zr-C system was investigated via pulsed laser heating. Ceramics were prepared by spark plasma sintering of commercial powders or powders derived from carbothermal reduction of zirconia. Laser pulses (400-3600 ms) heated samples above their liquidus temperatures (3200-3700 K) up to 4000 K.

Dendritic microstructure observed on laser-heated surfaces confirmed the formation of liquid. The laser-modified microstructure consisted of single-phase ZrC and lamellar ZrC+C eutectic, in proportions consistent with the equilibrium phase diagram. Non-equilibrium distribution of phases prevailed in subsurface regions of samples undergoing non-isothermal solidification. Porosity, generated by impurities and coalesced by material flow, existed throughout the heat-affected region.

INTRODUCTION

Zirconium carbide (ZrC), like other carbides of the transition metals of Groups IV, V, and VI, exhibits an unusual combination of properties that are appealing for refractory applications. These carbides combine the cohesive properties of covalently-bonded ceramics – high melting point, high strength and hardness – with the electronic properties of metals – high thermal and electrical conductivity, metallic slip systems[1,2]. A low thermal neutron capture cross-section for ZrC has led to significant interest in its implementation in high-temperature nuclear reactor fuels.

In particular, ZrC is targeted for use as a dense diffusion barrier layer in the ceramic composite tristructural isotropic (TRISO) coated particle fuel, which comprises the core of the very high-temperature gas-cooled reactor (VHTR), one next-generation reactor concept[3]. Fuel operating temperatures will be in excess of 1523 K and up to 1873 K, and the ZrC layer must maintain structural integrity and inertness while maximizing transfer of fission heat and neutrons.

A thorough understanding of the thermodynamic, mechanical, and heat transport properties of ZrC is limited by a paucity of reliable experimental data as a function of composition. The Zr-C phase diagram, most recently assessed in 1995 by Fernanded-Guillermet[4] (Figure 1), identifies zirconium monocarbide as the only intermediate compound. ZrC exists between 37.5-49.5 at% C ($ZrC_{0.6-0.98}$, temperature-dependent), its wide compositional range accommodated by carbon vacancies in its NaCl crystal structure ($Fm\overline{3}m$, space group 225).

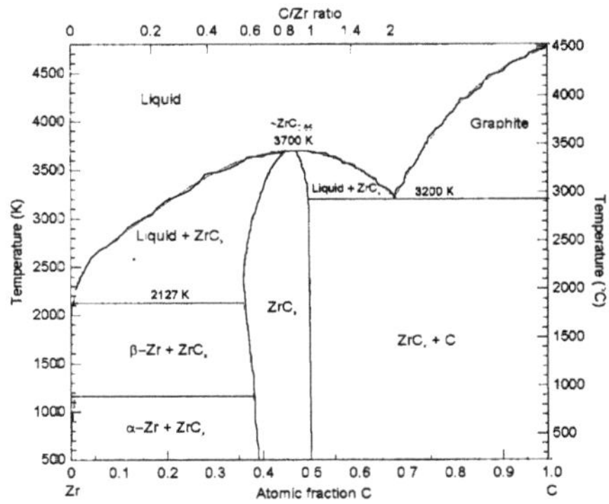

Figure 1. Zr-C phase diagram, after Fernandez-Guillermet[4].

This tolerance for structural vacancies has been suggested to impart a tolerance for radiation-induced defects[5,6]. However, durability, dimensional stability, and thermal property degradation of well-characterized ZrC under irradiation have not been extensively investigated.

The nature of chemical bonding in ZrC which stabilizes non-stoichiometry is not completely understood. The ZrC phase has a congruent melting point at 3700 K at the non-stoichiometric composition of $ZrC_{0.85}$ (46 at% C), and solidus is a strong function of composition, decreasing with increasing C content toward the ZrC-C eutectic at 3156 K and 49.5 at% C and with decreasing C content toward the Zr-ZrC eutectic at 2100 K and 37.5 at% C. Indentation hardness of ZrC decreases modestly with C/Zr ratio from 25 GPa for $ZrC_{0.98}$ to 18 GPa for $ZrC_{0.6}$, suggesting a reduction in overall bond strength as carbon is removed from the lattice[7]. ZrC exhibits a steep decline in room-temperature thermal conductivity[8] with departure from stoichiometry: 45 $W\,m^{-1}K^{-1}$ for $ZrC_{0.98}$ versus 10 $W\,m^{-1}K^{-1}$ for C/Zr ratio lower than 0.9. This indicates that conduction via phonon transport dominates for near-stoichiometric ZrC while electronic conduction dominates for highly non-stoichiometric ZrC.

Implications of non-stoichiometry on the thermodynamic stability of ZrC are of particular interest in this work. In a previous study[9], laser-induced melting of commercial, near-stoichiometric ZrC was investigated. The present work aims to extend this technique to other compositions in the Zr-C system. ZrC_x ceramics of various compositions were subjected to heat treatment and thermal analysis using a novel pulsed laser heating system, and microstructural evolution due to melting was characterized.

The laser melting technique offers several benefits over traditional methods for studying high temperature phase transitions in refractory materials, but introduces its own complications. Extremely high temperatures are achieved rapidly (tens of ms), with laser pulse profile controllable to optimize heating and cooling rates. Brief exposure to high temperatures and containment under inert atmosphere suppress sample vaporization and oxidation.

Earlier investigators[10] found that blackbody cavities filled with liquid or specimens collapsed between the solidus and liquidus temperatures. Liquidus transitions reported were more accurate for isothermal transitions, and the cooling curve was not recorded. By melting a surface rather than a blackbody cavity, these limitations are removed, but knowledge of the material's emissivity in the temperature range of interest is required.

Partial melting of the sample permits containerless heating and post-melting analysis of the largely intact, uncontaminated sample. However, localized melting of the surface leaves the unmelted bulk as a heat sink, obscuring latent heat exchanges on melting.

Global thermodynamic equilibrium is compromised at such rapid heating and cooling rates. High thermal gradients contribute to undercooling phenomena, complicating thermal analysis on freezing. Diffusion is sluggish in refractory covalent compounds like ZrC, and solute segregation and composition gradients are likely during rapid solidification.

Bearing in mind the associated caveats, thermal analysis derived from laser melting is promising for characterizing extremely high temperature phase transitions where other techniques are problematic. Areas of the phase diagram with few data points, such as liquidus transitions at hypoeutectic ZrC+C compositions, may be assessed.

EXPERIMENTAL PROCEDURE

Ceramic Preparation

Powders used as raw materials were zirconium carbide (ABCR, Germany), zirconium oxide (Sigma Aldrich, UK), and acetylene black (ABCR, Germany).

Compositions in the ZrC single phase field were prepared by carbothermal reduction of ZrO_2 by acetylene black according to the reaction equation

$$ZrO_2 + (3-y)C \rightarrow ZrC_{1-y} + 2CO_{(g)} \quad (1)$$

Reactant powders were mixed in proportions to produce a range of carbides having C/Zr ratios of 0.66 to 1.1. Reactant powders were homogenized on a rolling ball mill for 12 h in an ethanol slurry with ZrO_2 media. Carbothermal reduction synthesis was carried out in a vertical graphite tube furnace (Astro, Thermal Technology, Inc., Santa Rosa, CA, USA) under Ar atmosphere with a 2 h dwell at 1973 K (+15 K min^{-1}, -30 K min^{-1}). Temperature was monitored by W-Re thermocouple.

Compositions in the ZrC+C two-phase field were prepared by mixing commercial ZrC and acetylene black powders in proportions to produce overall C/Zr ratios of 1.04 to 2.6. Homogenization of powder mixtures was conducted as above.

Powders were densified by spark plasma sintering (model HP D 25/1, FCT Systeme, Rauenstein, Germany). Powder (6-7 g) was packed in a graphite die (20 or 40 mm diameter inner diameter, 10 mm diameter wall thickness) lined with graphite foil and pre-compacted in a laboratory press. A carbon foam blanket surrounded the die during sintering to reduce heat losses. Temperature above 673 K was monitored by optical pyrometer into a hollow in the top punch, focused 0.05 cm above the sample, differing from sample temperature by not more than 25 K. Temperature control between room temperature and 673 K was by thermocouple. Sintering was carried out under vacuum (typically 0.05 mbar for a 20 mm diameter die, 5.5 mbar achieved for a 40 mm diameter die). Samples were sintered at 2273 K (+200 K min^{-1} from 673-1273 K, +50 K min^{-1} from 1273-1373 K as load was applied, +100 K min^{-1} from 1373-2273 K) under a uniaxial pressure of 50 MPa (while temperature was 1273 K or greater). Piston displacement was monitored during sintering to characterize specimen densification; the high-temperature dwell was terminated manually after piston displacement rate decreased to <0.01 mm min^{-1}. Sample dwell times were typically 6-8 min. Cooling rate was -200 K min^{-1} from 2273-1273 K, -100 K min^{-1} from 1273-673 K, and water cooling to room temperature after a further

10 min. The entire sintering program was completed in <1 h.

Microstructural analysis of powders and as-sintered cross-sections was performed by scanning electron microscopy (SEM, LEO Gemini 1525, Carl Zeiss, Cambridge, UK; JSM-840 and JSM-5610, JEOL, Tokyo, Japan).

Carbon content of powders and crushed sintered specimens was determined by combustion (FlashEA 1112, ThermoFisher, Waltham, MA) and oxygen and nitrogen content by inert gas fusion (ONH-2000, Eltra, Neuss, Germany); Zr plus Hf was taken to comprise the balance. These analyses were combined to yield the C/Zr ratio.

Phase identification was made by X-ray diffraction (XRD, PW 1729 Generator, PW 1710 Diffractometer Control, Philips, Amsterdam, The Netherlands) operating at 40 kV and 40 mA, using Ni-filtered CuKα radiation. Samples were scanned between 2θ angles of 20-100°, with a scanning step of 0.04° and 2 s per step. Peak positions were calibrated with an external Si standard. The International Centre for Diffraction Data (ICDD) Powder Diffraction File (PDF) was consulted for standard peaks of possible phases. Systematic error in lattice parameter versus reflection angle was approximated using the Nelson-Riley function[11] and lattice parameters calculated for at least eight reflections for each specimen. Lattice parameter extrapolated to $2\theta = 180°$ is reported.

Laser Melting Apparatus

The laser melting apparatus has been described previously by Manara *et al.*[12], with its application to the melting of commercial $ZrC_{0.96}$ reported by Jackson *et al*[9]. Prismatic specimens 3-5 mm on a side were sectioned from sintered pellets and held in a graphite fixture within a pressure vessel fitted with a sapphire window. Experiments were carried out under 0.2 MPa argon. The sample surface was heated by a 4.5 kW Nd:YAG CW laser (HLD4506, TRUMPF, Schramberg, Germany), programmed to deliver a single pulse with a user-defined peak power, ramp time, and hold time. Heating and cooling rates as well as peak temperature were varied via the choice of pulse profile. The laser pulse was conveyed to the sample surface via optical fibers, lenses having 8 mm diameter focal spot, and mirrors. Laser optics were aligned with the sample center using a low-power He-Ne pilot laser coaxial with the heating laser. Samples were quenched via rapid shut-off of the laser.

Sample surface temperature was measured with a fast pyrometer ($\lambda = 644$ nm), with analog signals transferred to a Nicolet Pro 44C transient digitizer (Madison, WI). The pyrometer diaphragm (0.4 mm) was focused on the center of the pilot laser spot, to minimize the dependence of the measured temperature on radial thermal gradients. Pyrometer voltage signals were converted to brightness temperature by calibration against standard lamps to 2500 K. Calibration linearity to 3400 K was verified using a graphite blackbody.

Brightness temperature at wavelength λ was converted to true temperature according to Wien's approximation to Planck's law. Effective total transmittance of the optical path between the sample and pyrometer as measured by probe laser and spectrophotometer was 0.854. A value of 0.6 was adopted as the best approximation for high-temperature emissivity of ZrC, in accordance with typical values from the literature (for instance, Zapadaeva *et al.*[13]), though scattered and conflicting data introduce uncertainty into the reported temperatures. The experimental uncertainty of the temperature measurement has been estimated as ±1.5% for ZrC at 3200 K.

A key experimental aim was thermal analysis, and all samples were heated to temperatures above their liquidus temperatures and up to 4000 K using suitable laser pulse profiles. Samples were usually heated multiple times on a surface; a substantial depth of unmelted material permitted further experiments on the reverse surface.

Post-Melting Characterization

As-melted surfaces and ceramographic cross-sections were characterized by SEM and XRD. Thin sections for transmission electron microscopy (TEM, 2000FX, JEOL, Tokyo, Japan) were prepared from melted cross-sections by Ga focused ion beam milling (FIB200-SIMS, FEI Company, Hillsboro, OR, USA). Sections 6 x 9 μm were thinned to electron transparency (90-100 nm) before extraction by *ex situ* liftout. Selected area electron diffraction (SAD) and energy dispersive X-ray spectroscopy (EDX) were performed in conjunction with TEM.

RESULTS AND DISCUSSION

Powders and As-sintered Ceramics

Properties of the commercial powders as reported by the manufacturer and as characterized in this work are reported in Table I. Commercially available ZrC had non-negligible O and N content. ZrC supplied by ABCR contained 1.9 wt% (0.6 at%) Hf as analyzed by the manufacturer, but Hf analysis for other Zr compounds was not reported. As such, compositions have not been corrected for Hf. A level as high as 1.9 wt% Hf would increase atomic fraction C by approximately 0.5% and carbon-to-metal ratio by 1%.

Table I. Properties of commercial powders

	Mean particle size (μm)	wt% C	wt% O	wt% N	wt% Hf	C/Zr ratio	*a* (Å)
ZrC (ABCR)	0.5-10[a], 6.8 ± 2.1[b]	11.2[d]	0.4[d]	0.2[d]	1.9[c]	0.96	4.693
ZrC (Sigma Aldrich)	0.08-2[a] 3.56[c]	11.4[d]	0.5[d]	0.003[d]	-	0.98	4.699
ZrO_2 (Sigma Aldrich)	0.15-0.45[a] 0.6 ± 2[b]	-	-	-	<2[c]	-	-
Acetylene black (ABCR)	0.04-0.11[a]	99.99[c]	-	-	-	-	-

[a] By SEM observation
[b] By laser diffraction particle size analysis
[c] Manufacturer analysis
[d] By combustion or inert gas fusion

Composition and lattice parameter of as-sintered ceramics are summarized in Table II (compositions throughout refer to sintered samples). Samples produced by carbothermal reduction had high O content, indicating that reduction did not proceed to completion. Recovery of C, N, and O from Zr by elemental analysis is hindered by its refractory nature, and ZrC is expected to behave similarly.

Table II. Composition and lattice parameter *a* of sintered ceramics

	wt% C	wt% O	at% C	C/Zr ratio	*a* (Å)
ZrC_x by carbothermal reduction of ZrO_2	7.9	n.d.	39.0	0.66	4.688
	11.9	0.9	49.5	1.04	4.696
ZrC (Sigma Aldrich) + C powder mixture	16.2	n.d.	65.8	1.92	4.697
	19.4	n.d.	72.3	2.61	4.697
Commercial ZrC (ABCR)	11.2	0.3	48.6	0.98	4.692

Reaction product powders were consistent with ZrC by XRD (Figure 2). Here, square root of intensity rather than intensity was plotted to amplify low-intensity peaks, such as the monoclinic ZrO_2 peaks observed in one sample. While a small amount of ZrO_2 did not react, the original ZrO_2-C mixtures were largely converted to the fcc ZrC phase.

ZrC has a considerable solid solubility for impurity elements O and N, which may be introduced during processing and readily occupy sites on the C sublattice. ZrC_x ceramics produced by carbothermal reduction more accurately consist of an oxycarbonitride phase with the same structure as ZrC. The lattice parameter of an oxycarbonitride is lower than that of a carbide of the same carbon content, with the effect amplified at lower carbon content[14] O and N impurities are likely to reduce melting temperature, but at this concentration the effect on phase equilibria is unknown, as the thermodynamics of the Zr-C-O-N system have not been determined.

Hf impurities mix with Zr on the metallic sublattice, forming the ternary (Hf,Zr)C carbide. Substitution of Hf for Zr would slightly reduce lattice parameter and raise melting temperature[15].

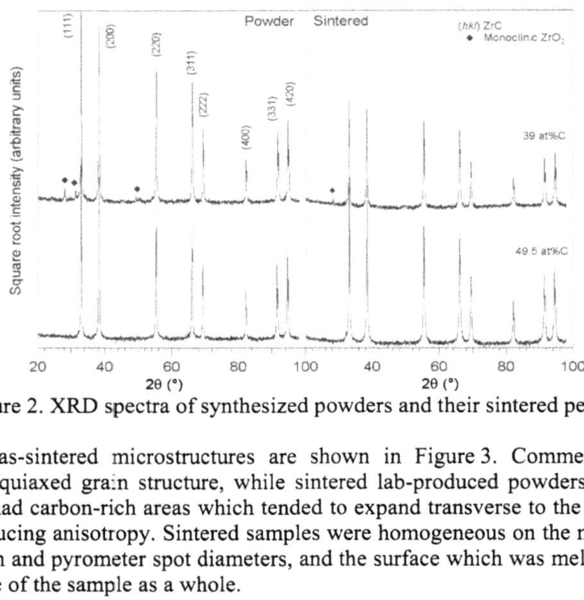

Figure 2. XRD spectra of synthesized powders and their sintered pellets.

Typical as-sintered microstructures are shown in Figure 3. Commercial ZrC had a homogeneous, equiaxed grain structure, while sintered lab-produced powders and ZrC-carbon black mixtures had carbon-rich areas which tended to expand transverse to the uniaxial pressing direction, introducing anisotropy. Sintered samples were homogeneous on the macroscopic scale of the laser beam and pyrometer spot diameters, and the surface which was melted is assumed to be representative of the sample as a whole.

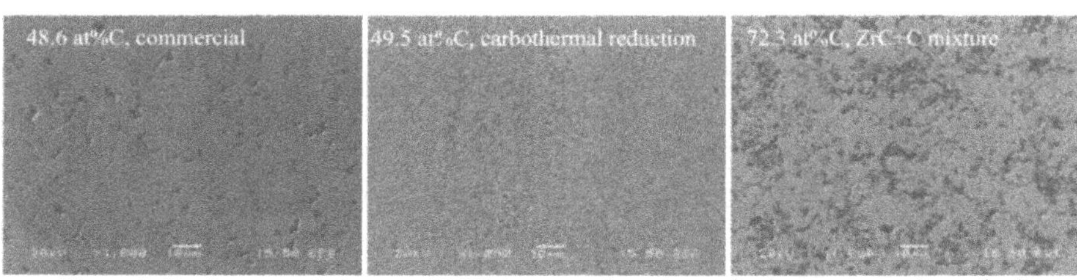

Figure 3. Backscattered electron (BSE) images of as-sintered microstructure (unetched).

Ceramography of Melted Ceramics

Initial experiments using a 3 mm diameter laser beam produced a 1 mm diameter melted spot[9]; the majority of experiments used an 8 mm diameter beam, which heated the entire sample surface. The wider beam and lower laser radiation intensity also lessened the severity of sample breakage due to thermal shock. Heat treatment history of the samples is summarized in Table III.

Table III. Laser pulses applied to ZrC_x samples

	No. of pulses	Peak power (W)	Ramp time (ms)	Hold time (ms)	Figure
39 at% C, carbothermal reduction	2	1800-2200	400	250-600	4,6,9
48.6 at% C, commercial	4	2500-3800	150	250	5,6,7,8,9
49.5 at% C, carbothermal reduction	3	2000-2200	400	250	4,5,6
65.8 at% C, ZrC+C mixture	1	1800	400	250	6,10
72.3 at% C, ZrC+C mixture	1	990	780	2820	6,9
same, 2nd sample	3	1800-2475	400	250	10

Dendritic structures were revealed on the melted surface, with dendrites incompletely fused at the center of the melt pool (Figure 4a). Small radial thermal gradients could contribute to faster quenching in the center of the surface, and if the melt pool solidified radially inward, as if from the walls of a crucible, then shrinkage in the center could account for this observation.

The dendrites in cross-section revealed the nature of the interdendritic liquid, which depended on the alloy composition (Figure 4b,c). For compositions in the two-phase ZrC+C field, a eutectic was observed as quenched in the interdendritic spaces. No interdendritic eutectic was found in samples within the single phase ZrC phase field.

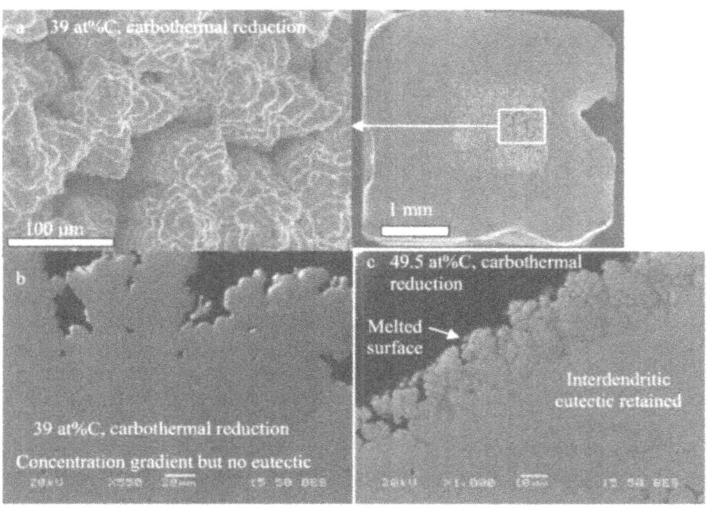

Figure 4. (a) Secondary electron images of unfused dendrites at melt center. (b,c) BSE images of surface dendrites in cross-section.

On a macroscopic scale, cross-sections through the melted surfaces revealed a layered structure consisting of a dense surface layer atop an intermediate layer containing pores and metastable phases (Figure 5). The interior microstructure was that of as-sintered material.

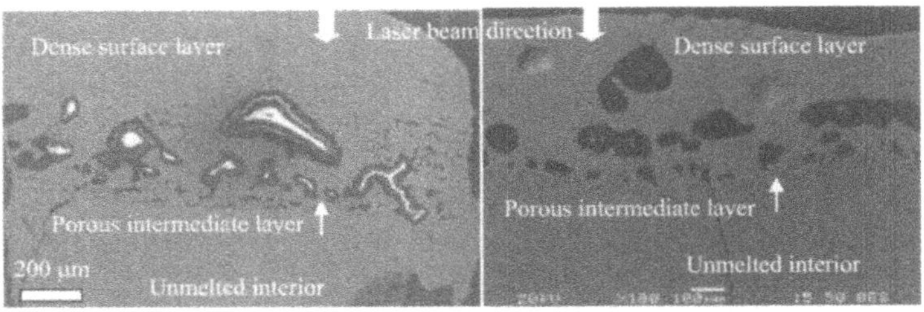

(a) 48.6 at%C, commercial (b) 49.5 at%C, carbothermal reduction

Figure 5. Macroscopic layered structure observed in cross-section.

The microstructure of the surface layer varied with composition consistent with the equilibrium phase diagram (Figure 6a-e). The depth of this surface layer varied with the nature and quantity of the laser pulses applied, but ranged from 100-300 μm.

The intermediate layer usually showed different microstructure or phase distribution than the surface layer (Figure 6f-j). In compositions undergoing non-isothermal freezing through a liquid+ZrC 2-phase field, metastable phases were observed. For example, in commercial ZrC (48 at% C), the surface layer was homogeneous single-phase ZrC, while the intermediate layer consisted of primary ZrC in a matrix of eutectic. In hypoeutectic ZrC (49.5 and 65.8 at% C), both the surface and intermediate layers consisted of primary ZrC and eutectic, but the proportion of eutectic was greater in the intermediate layer. A fully eutectic microstructure existed at the surface of an alloy with 72.3 at% C, while primary ZrC grains existed as a metastable phase at the interface between melted and unmelted material. In contrast, no ZrC+C eutectic was detected in an alloy with 39 at% C, as this composition is to the left of the congruent melting composition.

The equilibrium surface microstructure contrasted with that of the intermediate layer, which was clearly heat-affected but did not reach equilibrium. Repeated melting exposed the material nearer the surface to higher temperatures for longer, while material at intermediate depths may not have reached the liquidus temperature. Solute segregation during rapid solidification also encouraged formation of metastable eutectic. As solid ZrC nucleated, the interdendritic liquid became enriched in C; the last liquid to solidify had a sufficiently high C concentration to solidify as eutectic. Repeated melting improved the distribution of solute and reduced the severity of segregation during solidification, homogenizing the surface layer.

The XRD pattern[9] of the surface layer (48.6 at% C, commercial) was that of the fcc ZrC structure with $a = 4.698 \pm 0.001$ Å. The small quantity of melted material and its dendritic microstructure contributed to a low intensity, textured pattern. The proximity of unmelted material also influenced the analysis.

Selected area electron diffraction (SAD) patterns were taken at various zone axes of a TEM section containing material solely from the melted surface (Figure 7). The surface layer crystallized in the fcc crystal structure ($a = 4.7$-4.8 Å). The SAD lattice parameter is within 2% of the X-ray lattice parameter of the melted sample.

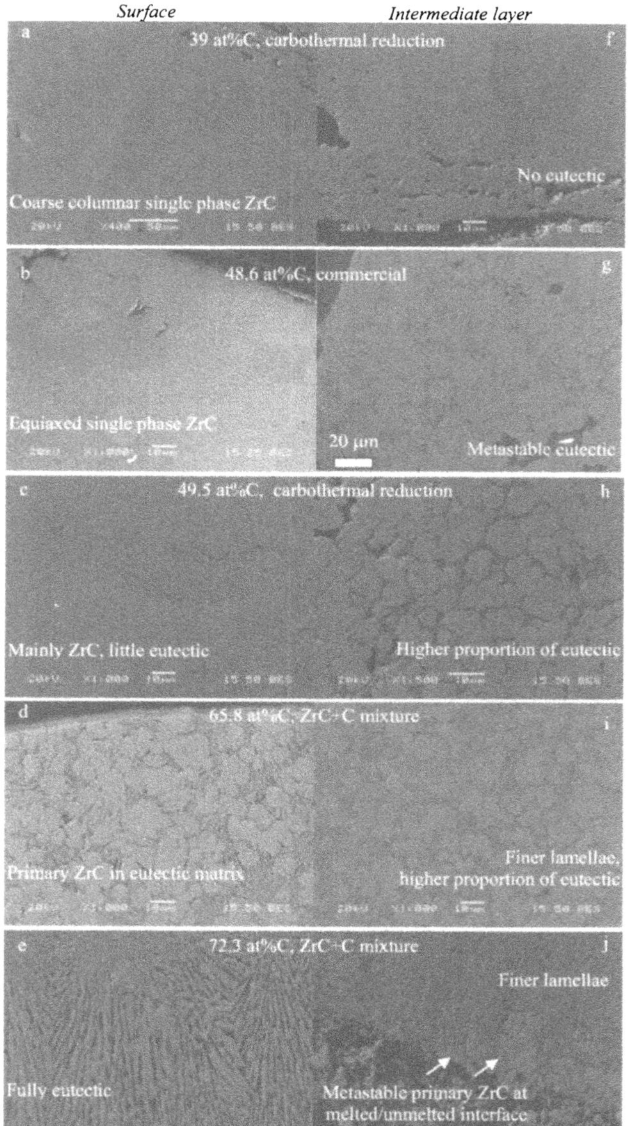

Figure 6. BSE images of surface and intermediate layer microstructure.

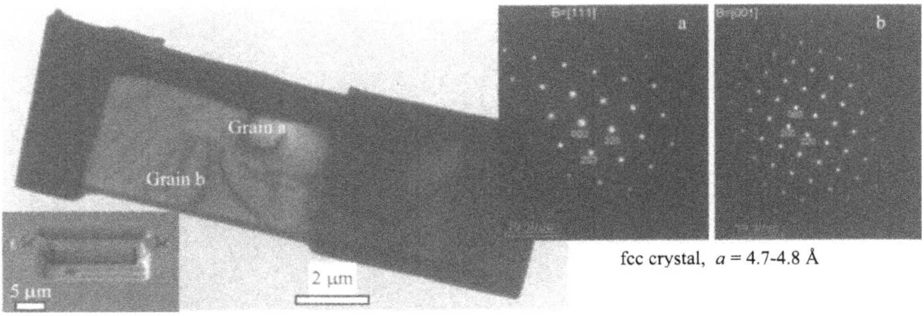

Figure 7. TEM section from surface layer of melted ZrC (48.6 at% C, commercial). SAD patterns from two grains are consistent with fcc ZrC.

A TEM section from the intermediate layer containing metastable eutectic was analyzed by SAD and EDX (Figure 8). The dark phase was Zr-rich by EDX (Figure 8c) with a SAD pattern (Figure 8f) consistent with that of ZrC in the surface layer (Figure 7b). The light phase was C-rich, and its SAD pattern (Figure 8e) had rings with linearly oriented higher-intensity arcs, indicating a textured polycrystalline material. The interplanar spacing of the arcs was 3.3-3.4 Å, corresponding to that of graphite basal planes. The ring pattern could not be attributed to the carbon support film on the TEM grid, which was amorphous or non-textured with a different d-spacing. The spot pattern occurring with the rings in Figure 8e was due to neighboring ZrC grains. FIB milling resulted in minor Ga ion implantation and slight bending of the sections.

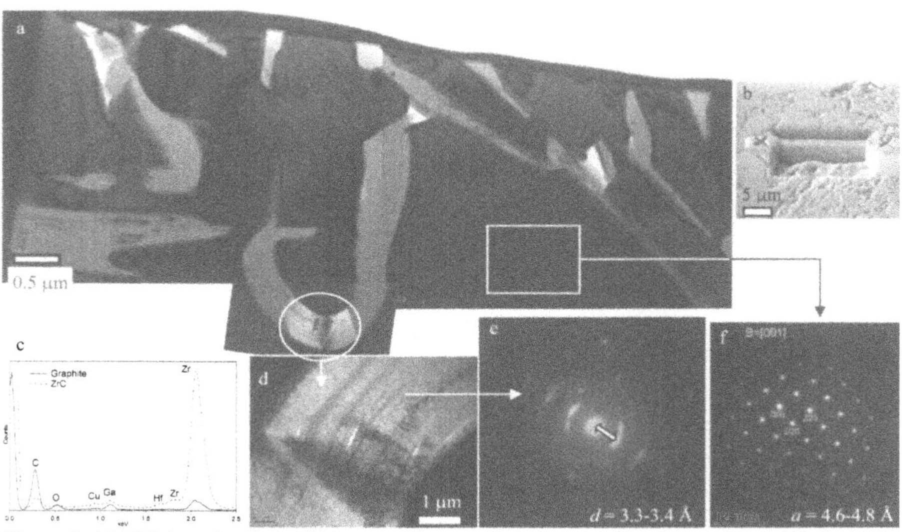

Figure 8. (a) TEM section from (b) metastable eutectic in intermediate layer of ZrC (48.6 at% C, commercial). Light phase (d) is graphitic by (c) EDX and (e) SAD, dark phase is ZrC (f).

Porosity and bubbles were prevalent in melted samples (Figure 9). Presumably, macroscopic (>100 μm) gas bubbles escaping through the melt (Figure 9a) originated in the intermediate layer as smaller (10-20 μm) pores (Figure 5b) or shrinkage porosity/cavities at grain boundaries (Figure 9b).

Remnant porosity in the sintered structure is a potential source of voids, enhanced by high-temperature material flow or separation of grain boundaries and dissimilar phases due to thermal shock (Figure 9c). Another source is CO gas generated by further carbothermal reduction of oxycarbonitride phases or unreacted ZrO_2 induced by laser heating. Both commercial and laboratory-synthesized samples exhibited this phenomenon.

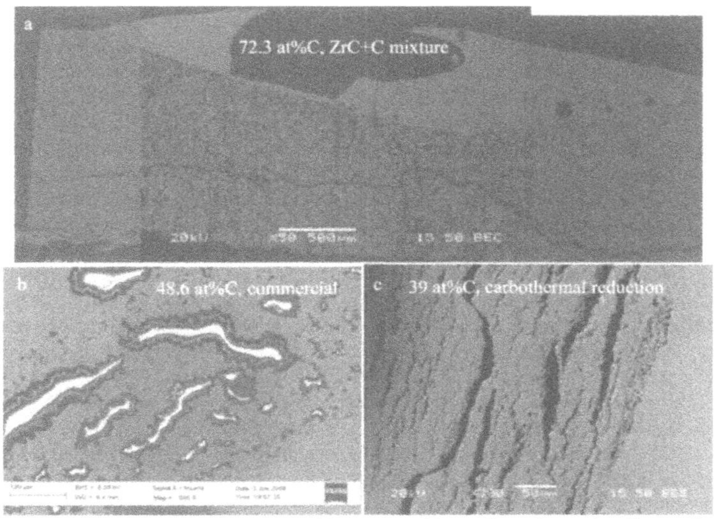

Figure 9. Porosity, bubbles, and voids occurred in various forms in all samples.

On melting of hypoeutectic samples containing 66 and 72 at% C, the liquid coalesced into a droplet or stalagmite, leaving uncovered areas of solid, rather than wetting the entire surface (Figure 10). The high surface tension in liquid ZrC+C eutectic is evident.

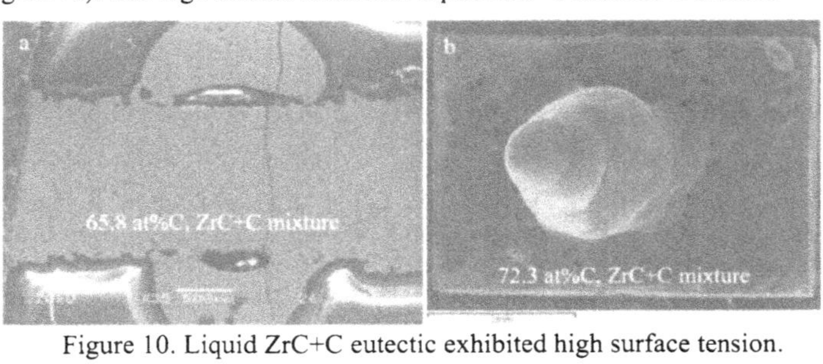

Figure 10. Liquid ZrC+C eutectic exhibited high surface tension.

CONCLUSIONS

Ceramics in the Zr-C system were heated above their liquidus temperatures using pulsed laser heating. The aim of these experiments was to validate a laser melting technique for thermal analysis of ZrC. Microstructural analysis of compositions in the ZrC_x and ZrC+C phase fields was undertaken to aid the interpretation of thermal analysis data.

First, microstructural analysis provided increased confidence in the capability of laser heating to melt ZrC. The laser power densities tested were sufficient to heat ZrC to peak temperatures of 2500-4000 K, permitting measurement of eutectic, solidus, and liquidus transitions in the Zr-C system. Dendritic solidification structures confirmed melting occurred, and elemental analysis and diffraction identified solidified phases as ZrC or a lamellar ZrC-graphite eutectic, with phase distribution consistent with the equilibrium phase diagram. After 1-4 laser pulses, each with a timescale of 100-1000 ms, a melted layer several hundred microns in depth was produced. A thermal gradient with depth was inferred, based on the observation of coarsening of eutectic lamellae at the surface, as well as the formation of metastable subsurface phases which were less prevalent or absent nearer the surface. Repeated or prolonged heating aided homogenization of the samples and the attainment of thermodynamic equilibrium.

Second, microstructural analysis may be used to corroborate phase transitions identified in thermogram data and offers insight into the thermodynamic and kinetic processes occurring during melting and solidification. Characterization of a thermal arrest as a eutectic transition is supported by the observation of eutectic microstructure in the recovered sample. This is particularly relevant to the observation of eutectic in samples for which this is a non-equilibrium phase, in which case, solute segregation during freezing is implicated.

Third, microstructural evolution during pulsed laser heating was characterized. ZrC is mechanically robust after repeated and extended exposure to temperatures above 2000 K with heating and cooling rates on the order of 1000-10000 K s^{-1}. Samples were prone to cracking, particularly those with multiple phases and anisotropic grains, but samples did not shatter or disintegrate. Heated in an inert atmosphere, ZrC showed no evidence of oxidation, though vaporization may be relevant at the temperatures involved. Sample processing history – dissolved gases in raw materials, remnant porosity from sintering processes, and the degree of homogenization of powder mixtures – contributed to the porosity generated in melted samples.

The present work indicates that pulsed laser heating is viable for producing equilibrium microstructures in ZrC as a prerequisite for precision measurement of phase transitions. However, raw material purity and sample processing history must be accounted for when interpreting microstructural and thermodynamic behavior. As the focus of this research has been the melting of ZrC and recording of thermogram data, a systematic study focusing on laser heating parameters such as power, pulse length, cooling rate, and number of pulses would also be of interest. Such a study could more clearly illustrate the relationship between laser surface processing and microstructural evolution in ZrC, such as melted depth, grain morphology, homogeneity, and solute segregation.

ACKNOWLEDGMENTS

The authors wish to thank Dr. Rudy Konings, Dr. Konstantinos Boboridis, and Mr. Franck De Bruycker of the Institute for Transuranium Elements for fruitful discussions, assistance with laser melting experiments, and development of metrological techniques. Carried out under collaboration agreement no. 30966 between Imperial College London and the European Commission Joint Research Centre Institute for Transuranium Elements (ITU). This work was sponsored by the Towards a Sustainable Energy Economy (TSEC) programme Keeping the Nuclear Option Open (KNOO) under UK Engineering and Physical Sciences Research Council grant no. EP/C549465/1. This work was also supported by the Pottery Mechanics Institute and the Worshipful Company of Armourers and Brasiers.

REFERENCES

[1]L. E. Toth, *Transition Metal Carbides and Nitrides*, Academic Press, New York (1971).

[2]E. K. Storms, *The Refractory Carbides, Refractory Materials: A Series of Monographs*, Vol. 2. ed., J. Margrave. Academic Press, New York (1967).

[3]U.S. DOE Nuclear Energy Research Advisory Committee and the Generation IV International Forum. A Technology Roadmap for Generation IV Nuclear Energy Systems, 2002. Available at http://www.gen-4.org/ (accessed 23 January 2009).

[4]A. Fernandez-Guillermet, Analysis of Thermochemical Properties and Phase Stability in the Zirconium–Carbon System, *J. Alloys Compd.*, **217**, 69–89 (1995).

[5]A. I. Gusev, A. A. Rempel, and G. P. Shveikin, Radiation Damage Resistance of Materials and Nonstoichiometry, *Dokl. Phys. Chem.*, **357**, 373–76 (1997).

[6]D. Gosset, M. Dolle, D. Simeone, G. Baldinozzi, and L. Thome, Structural Behaviour of Nearly Stoichiometric ZrC Under Ion Irradiation, *Nucl. Instrum. Methods Phys. Res., Sect. B*, **266**, 2801–05 (2008).

[7]L. Ramqvist, Variation of Lattice Parameter and Hardness with Carbon Content of Group 4B Metal Carbides, *Jernkontoret. Ann.*, **152**, 517-23 (1968).

[8]E.K. Storms and P. Wagner, Thermal Conductivity of Sub-Stoichiometric ZrC and NbC, *High Temp. Sci.*, **5**, 454-62 (1973).

[9]H.F. Jackson, D.D. Jayaseelan, W.E. Lee, M.J. Reece, F. Inam, D. Manara, C. Perinetti Casoni, F. De Bruycker, and K. Boboridis, Laser Melting of Spark Plasma-Sintered Zirconium Carbide: Thermophysical Properties of a Generation IV Very High-Temperature Reactor Material, *Int. J. Appl. Ceram. Technol. EarlyView*, Published online: 15 August 2009, doi:10.1111/j.1744-7402.2009.02434.x.

[10]E. Rudy, D. P. Harmon, and C. E. Brukl. Ternary Phase Equilibria in Transition Metal Boron-Carbon-Silicon Systems, Part 1. Related Binary Systems, Volume 2. Ti–C and Zr–C System, Report no. AFML-TR-65-2, Aerojet-General Corp., Sacramento, CA, 1965.

[11]B. D. Cullity, *Elements of X-Ray Diffraction*, 2nd edition, Addison-Wesley, Reading, MA, (1978).

[12]D. Manara, M. Sheindlin, W. Heinz, and C. Ronchi, New Techniques for High-Temperature Melting Measurements in Volatile Refractory Materials Via Laser Surface Heating, *Rev. Sci. Instrum.*, **79**, 113901 (1–11) (2008).

[13]T. E. Zapadaeva, V. A. Petrov, and V. V. Sokolov, Emissivity of Stoichiometric Zirconium and Titanium Carbides at High Temperatures, *High Temp.*, **19**, 313–20 (1981).

[14]V.A. Mitrokhin, R.A. Lyutikov, and R.S. Yurkova, Change in the Lattice Constant of Zirconium Carbide in the Region of Homogeneity, *Inorg. Mater.*, **11**, 978-80 (1975).

[15]E. Rudy, Ternary Phase Equilibria in Transition Metal-Boron–Carbon–Silicon Systems, Part 5. Compendium of Phase Diagram Data, Report no. AFML–TR–65-2, Aerojet-General Corp., Sacramento, CA, (1969).

ON THE MECHANISM OF RADIATION DAMAGE IN ZIRCON BY HIGH-ENERGY ELECTRONS

N. Jiang
Department of Physics, Arizona State University, Tempe AZ 85287-1504

ABSTRACT

Radiation effect in zircon by high-energy electron irradiation in transmission electron microscopy (TEM) was studied by means of quantitative electron energy-loss spectroscopy (EELS). The decomposition of zircon under electron irradiation was measured using O K, Si L_{23}, and Zr M_{45}-edge EELS. The immediate transfer from zircon to Zr_xSi_y alloys due to the preferential sputtering of O species was discovered by analyzing the change in the fine structures of Si L_{23}-edeg EELS.

INTRODUCTION

Radiation effects in zircon have been increasingly attracting much attention with the development of fission reactors and concern of nuclear waste disposal [1]. The mechanisms for radiation damage processes have thus been extensively investigated [2]. In general, the primary damage process in zircon by energetic particle is displacement of one or more atom species by transfer of kinetic energy to atom species (knock-on damage) [3]. One of the essential instruments for experimental studies of radiation effects is transmission electron microscopy (TEM), usually operating between 100 and 400 keV. Although the kinetic energies of electrons in TEM are less than the usual maximal kinetic energy of β particles (around 1 MeV) emitted during radioactive decay, they are definitely high enough to displace some atom species either ballistically or radiolytically. Additional problem on top is that the specimen volume in TEM observations is very small, and thus the surface to volume ration is relatively large. Therefore, electron irradiation induced damages in specimen in TEM are crucially important for the understanding of radiation processes by radioactive decays in real materials. Unfortunately, the possible damage by electron irradiation in TEM has not been paid much attention. There were only a few studies involving electron irradiation induced damage, and it was considered the damage process by electron beam was similar to that by other energetic particles [4, 5]. In our previous study [6], however, we discovered the different damage process in zircon by electron irradiation in TEM, in which the damage was induced by the preferential sputtering of O resulting formation of Zr_xSi_y nanocrystalline alloys.

In this study, we further discuss the mechanism of the damage process by electron beam in TEM. The detailed composition changes during the damage were measured using quantitative electron energy-loss spectroscopy (EELS) technique. The quantitative analysis of the formation of Zr_xSi_y alloys was also presented here.

EXPERIMENTAL

Commercially synthesized zircon $ZrSiO_4$ crystals were used in this study (Aldrich Chemical Company, Inc.). TEM specimens were prepared by grinding the samples into powders in acetone, and picking them up using a Cu grid covered with a lacy carbon thin film. The specimens were then immediately transferred into a JEOL-2010F TEM with a field-emission gun operating at 200keV. Electron energy-loss spectra (EELS) were acquired using a parallel-detection electron energy-loss spectrometer (Gatan Enfina). The full width at half maximum (FWHM) of the zero-loss peak was about 1.0 eV. The EELS entrance aperture was 2.0 mm. The dispersion of the spectrometer was 0.2eV/channel in order to obtain both Si L_{23} and Zr M_{45}-edge, along with the zero-loss peak. Although the O K-edge EELS spectra were acquired separately, all the spectra, including Si L_{23}, Zr M_{45} and O K-edge, were recorded using the same integrating time, and from the same area under the same electron beam intensity. In other words, EELS spectra were acquired during the beam damage process, and thus can be described as an *in situ* observation. For quantitative analysis, all the edge spectra were deconvoluted to remove plural scattering contribution using Fourier-ratio method. The integration windows of all edges are 40eV starting from the edge onset energy. The data were processed using Digital Micrograph software.

RESULTS AND DISCUSSION

In the previous study, we have discovered that the damage process in zircon under high-energy electron irradiation in TEM is due to the preferential sputtering of oxygen from surface [6]. The conclusion was obtained by comparing the O K-edge EELS intensities between the fully damaged area with an undamaged fresh area with similar values of t/λ. However, the inelastic scattering mean-free-path, λ, may have different values in undamaged and damaged areas. Therefore, we collected a series of O K-edge EELS spectra during the beam damage process. In order to compare with the changes in Si and Zr composition, both Si L_{23} and Zr M_{45}-edge EELS spectra were also alternatively acquired during the same damage process. The integrated intensities of O K-edge EELS during the damage process are plotted in figure 1a, while the corresponding integrated intensities of Si L_{23} and Zr M_{45} EELS are given in figure 1b.

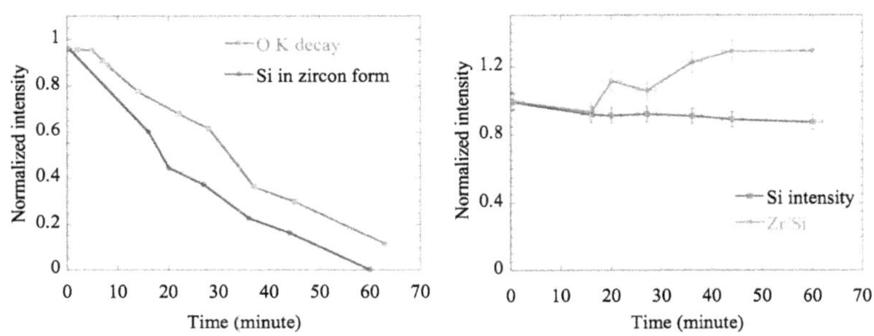

(a) (b)

Figure 1 (a) Comparison of the integrated intensities of O K-edge EELS (blue) with the portion of Si L_{23}-edge intensities associated with the zircon structure; (b) Comparison of the integrated intensities of Si L_{23}-edge EELS with the intensity ratios of Zr M_{45} to Si L_{23}-edge. Note: all the data were normalized to their corresponding first values.

As shown in figure 1a, the O intensity drops "linearly" with the increase of electron irradiation (or increase of electron dose). After 60 minutes of irradiation, less than one fifth of O left in the irradiated area. Although the Si intensity (figure 1b) also decreases with the increase of electron irradiation, the rate is much smaller than that of O. After 60 minutes of irradiation, there are still more than four fifth of Si left in the irradiated area. Obviously, the damage process is due to the preferential sputtering of O and the electron irradiation results in decomposition in zircon.

It is also noticed that within the first 20 minutes of irradiation, the change of Zr intensity (figure 1b) is consistent with that of the Si, but the relative amount of Zr is dramatically increased after 20 minutes. This phenomenon is probably induced by experimental error in the background subtraction of Zr M_{45}-edge. As shown in figure 2a, Zr M_{45}-edge sits on the tail of Si L_{23}-edge and is very close to the Si L_1-edge. It is even worse that the fine structure of Si L_1-edge and the intensity of Si L_{23}-edge tail are changing during the damage process. It is seen that a small peak indicated by an arrow disappears in the damaged spectra. In addition, the onset of Zr M_{45}-edge shifts about 3.6eV to lower energy in the damaged spectrum. All these factors can seriously affect the accuracy of background fitting for the Zr M_{45}-edge. Nevertheless, it is definite that the Zr intensity does not drop as much as the O K-edge. Therefore, it is conclusive that the preferential sputtering of O is responsible for the damage of zircon induced by electron irradiation in TEM.

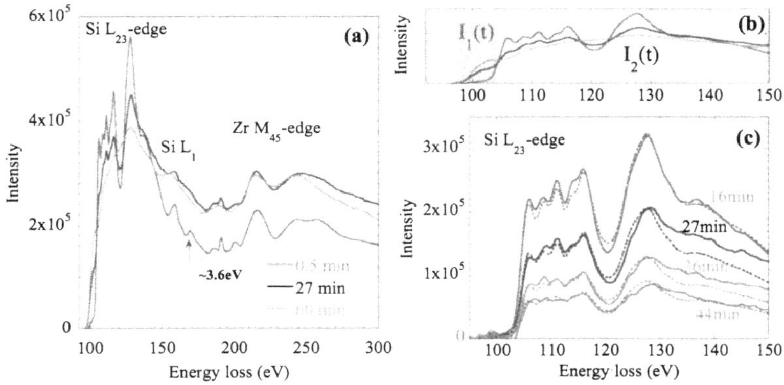

Figure 2 (a) Comparison of Si L_{23} and Zr M_{45}-edge EELS at different stage of damage; (b) enlarged spectra in (a) to show the detailed changes in the Si L_{23}-edge; (c) comparison of the weighted sum of undamaged and damaged spectra with the Si L_{23}-edge in the undamaged sample (dotted lines).

The detailed changes in Si L_{23}-edge EELS can also be observed in figure 2a: the pre-edge intensity increases with the increase of irradiation, while the fine structures of zircon gradually disappear. According to our previous study [6], the increase of pre-edge intensity is due to the increase of damaged regions, while the remaining regions are more or less still maintained zircon structure. Therefore, the intensity of the Si L_{23}-edge can be approximately considered as the sum of the volume fractions of damaged and undamaged areas within the electron irradiated region. To quantify the damage rate, we can utilize the property that the pre-edge intensity (shaded region in figure 2b) is only from the damaged area. We noticed that the damage was completed after about 60 minutes of irradiation; further irradiation did not make change in the fine structures of Si L_{23}-edge except for a slightly decrease in intensity. Therefore, we can use spectrum acquired after 60 minute as a reference for the damage. The total integration of Si L_{23}-edge consists of two portion: $I(t) = I_1(t) + I_2(t)$, as shown in figure 2b. In the undamaged sample, $I_1(initial)$ should be equal to zero. During the damage process, $I_1(t)$ is only from the damaged regions, while $I_2(t)$ is from both damaged and undamaged region. In the completely damaged sample, both $I_1(final)$ and $I_2(final)$ are exclusively from the damaged regions. Therefore, we can obtain a ratio in the fully damaged sample,

$$R = I_2(final) / I_1(final) \qquad (1)$$

This ratio is a constant if the sample becomes fully damaged, i.e. the fine structures in the Si L_{23}-edge do not change any more, and the plural scattering has been removed from the spectrum. Once we obtain this constant ratio *R*, we can decompose the intensity $I_2(t)$ into damaged and undamaged portions by

$$I_2^{undamaged}(t) = I_2(t) - R \cdot I_1(t) \quad (2)$$

The plot of $I_2^{undamaged}(t)$ is compared with the decay curve of O K-edge in figure 1a. It is seen that the decrease rate of undamaged region estimated from Si L_{23}-edge is very consistent with that observed in the O decay. This consistency further supports our previous conclusion that the sputtering of O does not uniformly occur within the illuminated area, but rather within small sporadic regions. With the increase of electron irradiation, these isolated regions gradually cover the whole area of the electron beam.

To justify the above decomposition procedure, we compare the original spectrum in the undamaged sample with the various decomposed spectra (figure 2c), in which the contributions from the damaged regions were removed. It is seen that the overall shapes of these two types of spectra fit very well each other.

As we knew, three types of mechanisms can involve in the knock-on damage in materials by energetic particles: Frenkel-pair production, vacancy-enhanced displacement, and surface sputtering. Usually, the displacement energy for Frenkel-pair creation is fairly high, and thus this mechanism may not play an important role in the damage process in zircon by electron irradiation. On contrast, both the vacancy-enhanced displacement and surface sputtering are low-energy processes. Therefore, they may be all responsible for the damage. A quantitative model considering both processes has been developed by Medlin and Howitt [7]. In this model, the specimen is envisaged as a sequence of layers, each the thickness of one displacement distance or an atomic spacing. The sputtering occurs only if the surface atoms exist in the layer. In other words, the sputtering rate in this layer is proportional to the number of surface atoms in the layer. The remaining atoms in this layer can also be knocked into vacancies in the adjacent layers. Similarly some of the vacancies which have been formed in this layer will be filled by atoms displaced from the proceeding layer. The net rate of loss from any layer is a balance between the rates at which material enters and exits that layer. Therefore, a differential equation can be obtained for each layer, and a set of equations for the foil can be solved numerically if the vacancy-enhanced displacement cross-section and surface sputtering cross-section are known. Later, this model has been adopted to multicomponent systems and applied to interpret the damage in intermetallics [8]. It has also been used to interpret the damage in semiconductor [9].

Using this quantitative model, the decay curve of certain species can be predicted by playing around with two cross-sections (the vacancy-enhanced displacement cross-section and surface sputtering cross-section), which are generally unknown from experimental point of view. In the end, the cross-sections can be obtained by fitting the decay curve. However, this model cannot apply to the damage process in zircon.

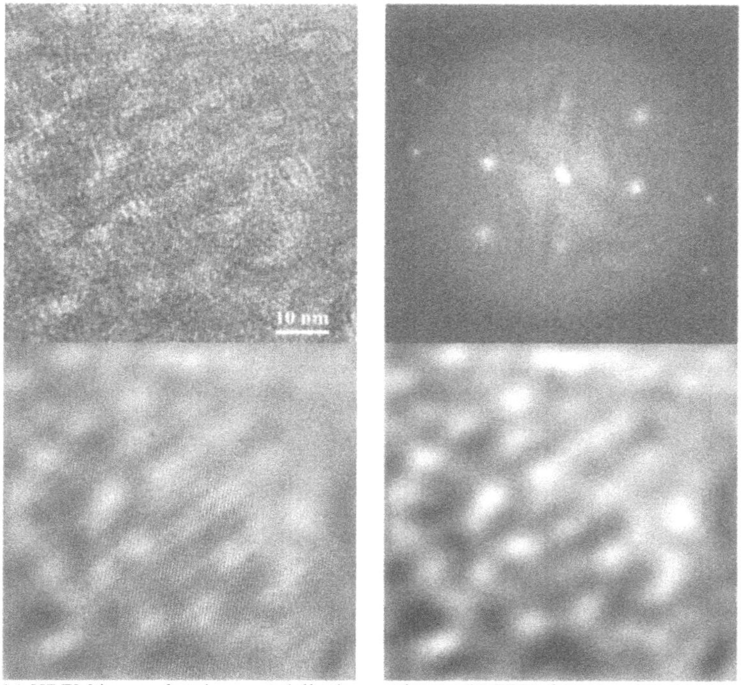

Figure 3 (a) HREM image showing a partially damaged area; (b) FFT of (a) showing the orientation of the area; (c) inverse FFT image produced by including all the spots in (b); inverse FFT image only using the central spot in (b).

As shown in figure 3, during the damage process, the illuminated area mainly kept zircon structure. The damage occurred randomly and sporadically. The random patches can be clearly seen in the FFT of HREM image in figure 3c and 3d. In other words, the decay of O was not uniformly in the whole illuminated area, but occurred mainly in the patch areas. This indicates that the damage rate obtained from the O K-edge intensity integrated in the whole illuminated area may largely underestimate the real damage rate in the patches, or the local damage rate, which is governed by the damage mechanism. Another difficulty to apply this model is immediate formation of Zr_xSi_y compounds in the damage area. This suggests that the zircon lattice collapse in the patches. In other words, the vacancies produced by removal of O may largely be reduced, and therefore, any possible model must consider the vacancy concentration as a variable. Ignoring this effect may again underestimate the cross-sections.

CONCLUSION

The decomposition of zircon under electron irradiation in TEM has been measured using quantitative EELS of O K, Si L_{23} and Zr M_{45}-edge. The formation of Zr_xSi_y following the removal of O has also been analyzed by decomposing Si L_{23}-edge into references in original and fully damaged zircon. Considering that the damage is not uniformed in the irradiated area and the collapse of zircon lattice is an instant event after removal of O, the current available damage model [7] cannot be applied to zircon, at least under electron irradiation in TEM.

ACKNOWLEDGMENT

This work is supported by NSF DMR0603993. The use of facilities within the Center for Solid State Science at ASU is also acknowledged.

REFERENCES

[1] R. C. Ewing, W. Lutze, and W. J. Weber, Zircon: A host-phase for the disposal of weapons plutonium J. Mater. Res. 10, 243 (1995).

[2] R. C. Ewing, A. Meldrum, L. M. Wang, and S. X. Wang, "Radiation-induced amorphization", in Transformation Processes in Minerals, edited by S. A. T. Redfern and M. A. Carpenter, Rev. Mineral Geochem. 39, 319 (2000).

[3] L. M. Wang and R.C. Ewing, Detailed in situ study of ion beam-induced amorphization of zircon, Nucl. Instru. Method Phys. Res. B 65, 324 (1992).

[4] A. Meldrum, L. A. Boatner, and R. C. Ewing J. Mater. Res. 12, 1816 (1997).

[5] R. Devanathan, W. J. Weber, and L. A. Boatner, "Response of zircon to electron and Ne+ irradiation", in Phase Transformations and Systems Driven Far From Equilibrium, edited by M. Atzmon, P. Bellon, and R. Trivedi, Mater. Res. Soc. Symp. Proc., vol 481, (Warrendale, Pennsylvania 1998), pp. 419-424.

[6] N. Jiang and J. C. H. Spence, Radiation damage in zircon by high-energy electron beams, J. Appl. Phys., 105, 123517 (2009).

[7] D. L. Medlin and D. G. Howitt, The role of sputtering and displacement damage in the electron SCRIBE process, Phil. Mag. Lett., 64, 133-141 (1991).

[8] D. A. Muller and J. Silcox, Radiation damage of $Ni_3Al$by 100keV electrons, Phil. Mag. A, 71, 1375-1387 (1995).

[9] K. A. Mkhoyan and J. Silcox, Electron-beam-induced damage in wurtzite InN, Appl. Phys. Lett., 82, 859-861 (2003).

ANELASTICITY IN AUSTENITIC STEELS

Ashwin Rao, P. John Bouchard, Michael E. Fitzpatrick
Materials Engineering, The Open University, Walton Hall, Milton Keynes MK7 6AA, U.K

ABSTRACT

Load-unload creep testing of AISI Type 316H austenitic steel has been undertaken to simulate the effect of power plant start-up and shut-down loading and study the influence on material integrity. The study was undertaken at two temperatures (550°C and 650°C) with applied loading removed for varying times during the course of testing. Anelasticity (recoverable strain) was observed during the unload period, saturating as the number of load-unload cycles increased. In addition, the load-off time was found to significantly influence the overall test rupture time, enhancing load-on life by 2-3 times when compared with the response of a monotonic creep rupture test. Transmission electron microscopy examination revealed changes in dislocation densities during the recovery process. It is postulated that this may be due to the interactions of dislocations with precipitates. The resultant dislocation motion appears to induce negative strain when compared to strains induced during the load-on cycle.

INTRODUCTION

Austenitic steels are widely used in the nuclear power plant industry to construct various components, many of which operate at temperatures where creep plays an important role in influencing the material lifetime. The basis for obtaining data for high temperature design is monotonic, uniaxial creep rupture testing. Also, the assessment procedures used for maintenance and repairs are based on such data. In reality, power plants are frequently subject to load changes during their operation. Fluctuations in load can have adverse effects on the material behaviour. During plant shut downs, the material experiences exposure to loads which are very low. When the plant is started up again, the immediate reload behaviour and also the subsequent material behaviour after initial loading will be significantly different to monotonic creep rupture tests. Thus the design codes used must be robust enough to cater for such situations.

Time-dependent creep recovery of a material subject to load removal has been reported extensively [1-4]. Anelasticity, negative creep, and creep recovery are all terms that have been used to describe this occurrence. Anelasticity has been reported fairly extensively and has been observed in a range of single phase materials and alloys, and also in precipitation hardened materials such as stainless steels[4], aluminium[5] and Ni-Cr based alloys[6]. Different theories have been put forth for understanding anelastic behaviour, and these include dislocation bowing[7] and the influence of internal stresses[8].

This paper describes a part of a project being carried to understand the anelasticity in austenitic steels. AISI 316H grade steels were used for this study. Uniaxial creep rupture and load relaxation tests have been undertaken on AISI 316H grade steels to study the influence of nuclear power plant start-up and shut-down loading on material life integrity.

EXPERIMENTAL PROCEDURES

Material Details

The material used for this study was AISI316H grade austenitic steel, having a chemical composition as shown in Table.1

	C	Mn	Si	P	S	Cr	Ni	Mo
Chemical Composition (%)	0.078	0.63	0.29	0.32	0.015	18.5	10.17	2.12

Table 1. Chemical Composition of As-Received Material

The material was received in bar form with a diameter of 12.5 mm. The bars were water quenched following heat treatment at 1100°C for 90 minutes, and 'dog-bone' tensile specimens prepared with a gauge length of 30 mm and a gauge diameter of 6 mm. Initial monotonic tensile tests showed that the proof strength at 0.2% strain was 225 MPa at room temperature and 110 MPa at 650°C. The as-treated microstructure is shown in Fig.1. The material had a fairly uniform microstructure and the grain size was estimated at 60 μm using the line intercept method.

Temperature(°C)	0.2% Proof Stress (MPa)	1% Proof Stress (MPa)
25°C	224	285
550°C	127	157
650°C	110	127

Table 2. Tensile Properties at Room and Elevated Temperatures

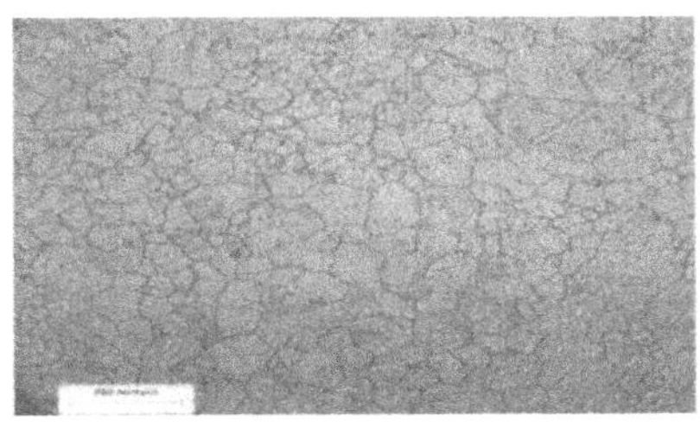

Fig. 1 Microstructure of As-received Material. Specimen was electrolytically etched using 60% Nitric Acid.

Equipment Details

The specimens were tested using lever arm creep machines with ratios of 10:1. The temperature chosen for this study was 650°C. The tests conducted were load controlled. The stress was controlled within ± 1MPa of the target load. Strain was measured using linear variable displacement transducers (LVDTs) which had a resolution of 1 μm. Strain monitoring was done using an in-house data acquisition system. Temperature was monitored using three N-type thermocouples which were placed along the gauge length. The specimen temperature was maintained at ±2°C of the required temperature.

RESULTS

Baseline Creep Testing

The temperature chosen for this study was 650°C and the stress applied was 180 MPa. Initially, creep rupture tests were conducted at the above conditions to ascertain the baseline data. The data for the creep rupture experiments are consistent with other sources[17].

	Rupture Time (hours)	Rupture Strain (%)	Secondary Creep Rate (/hour)
Test 1	562	6.9	3.26e-5
Test 2	480	8.5	4.46e-5

Table 3. Creep Rupture Data for 650°C 180MPa

Start up-Shut Down Testing

As stated previously, tests were carried out to mimic some aspects of material behaviour during plant start-ups and shut downs. Start-ups can be both hot (removal of load only) and cold (removal of both stress and temperature). A series of tests were conducted where loads were removed for 48 hours, for 4 hours, and for 5 seconds. In each case the stress was applied for 48 hours so that the material reached secondary creep before any unloading was done. Displacements were measured once every minute for 6 hours and then once every 10 minutes. All tests were taken to failure unless indicated. The strain curves are shown in Fig. 4.

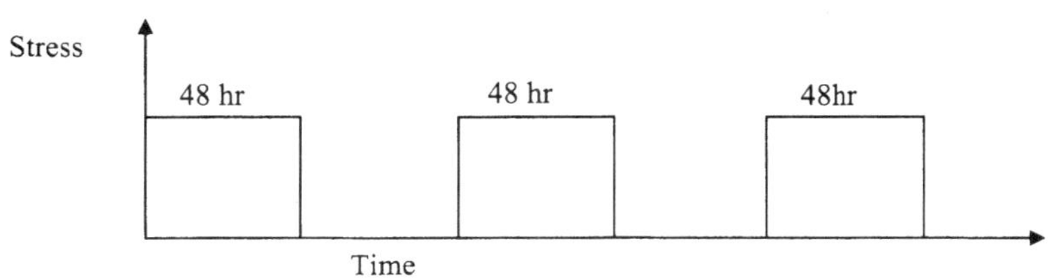

Fig.3 Stress Profiles- Unloads were done after loading period of 48 hours.

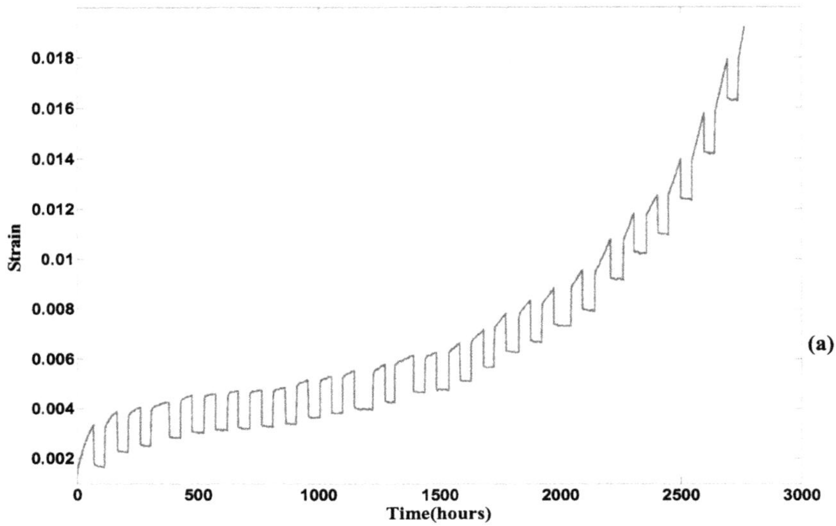
0.018
0.016
0.014
0.012
0.01
0.008
0.006
0.004
0.002
Strain
0
500
1000
1500
2000
2500
3000
Time(hours)
(a)

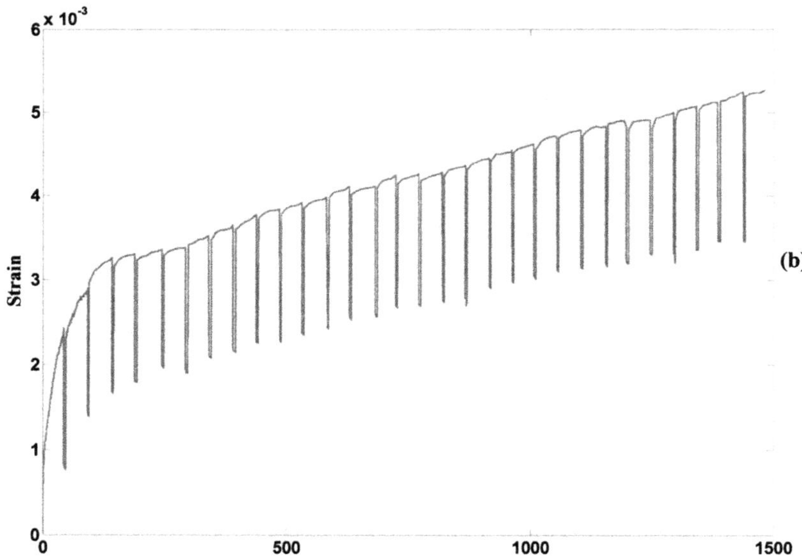
x 10^-3
6
5
4
3
2
1
0
Strain
0
500
1000
1500
(b)

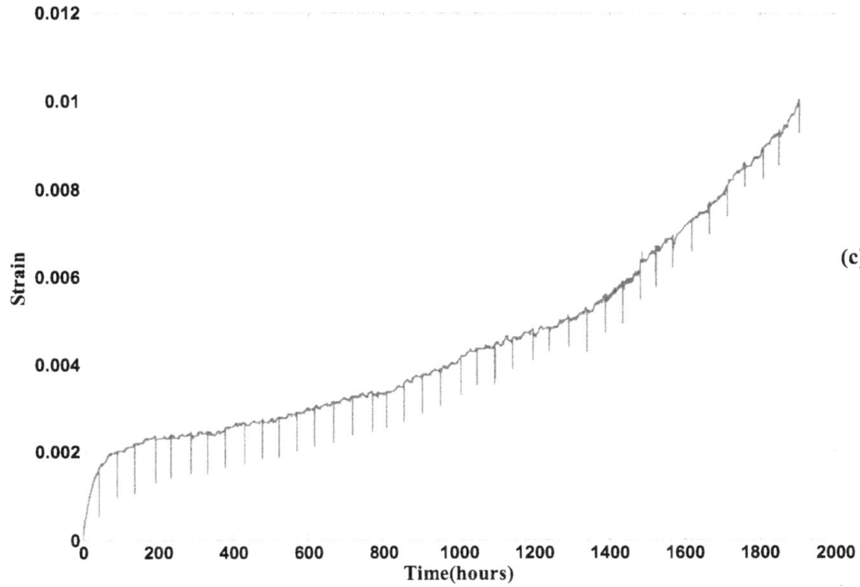

Fig.4 Strain Curves of Tests which were unloaded for (a) 48 hours(b) 4 hours(c) 5 seconds. Test (b) was not taken to rupture

Plastic strain recovery was observed during the unload phase. This recovery has been termed 'anelasticity'. A schematic of the phenomenon is shown in Fig 5. It was also observed that the majority of the strain was recovered relatively soon after unloading (<4 hours). Also, the magnitude of anelastic strain decreased as the number of load-unload cycles increased (Fig.6).

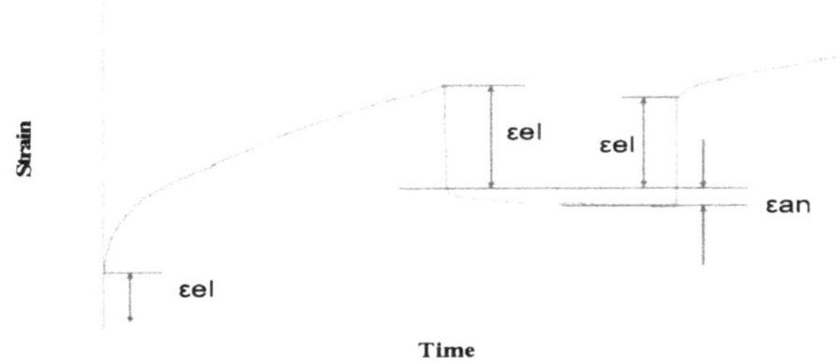

Fig 5. A Single Unload Cycle

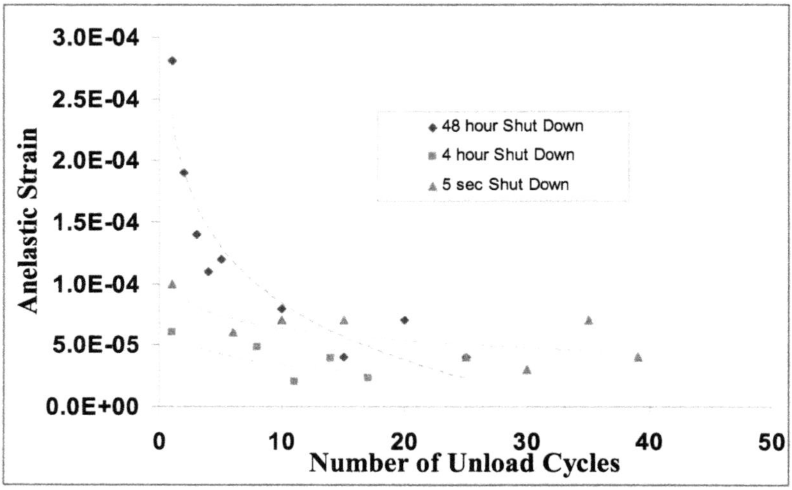

Fig.6 Saturation of Anelasticity with number of load-unload cycles.

The load-unload conditions had a significant effect on the material behaviour. Large deviations were observed when the data was compared to uniaxial creep rupture data. The on-load rupture life was about 2-3 times the uniaxial creep rupture life. The failure strain, on the other hand, reduced by about 50-60% when compared to constant stress results. The results obtained are consistently outside the scatter observed in the creep data.

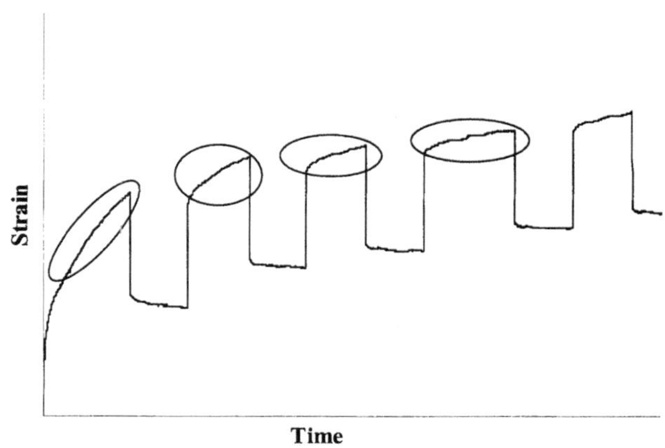

Fig 7. Calculating Lifetimes

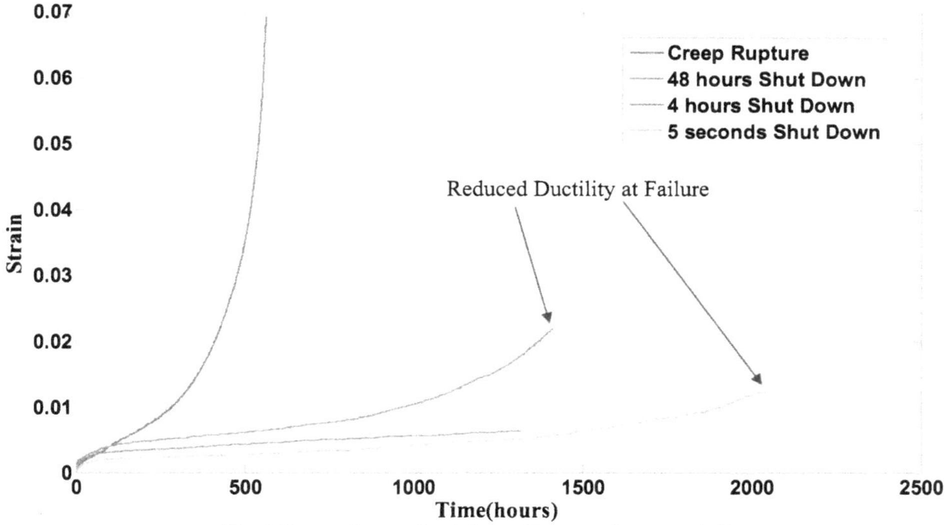

Fig.8 Computing on-load times Shows an Increase in Rupture Life but Reduced Ductility

DISCUSSION

In all the tests conducted to-date, anelasticity has been a dominant factor influencing the overall cyclic creep behaviour of the material. Most theories which have been proposed to explain this phenomenon are based on long-range internal stresses[8] or on the interactions between precipitates and dislocations[15].

Secondary creep can generally be described by an Arrhenius-type equation:

$$\varepsilon = A\sigma^n \exp\left(-Q/_{RT}\right) \tag{1}$$

Where n is the stress exponent, σ is the stress applied, Q is the activation energy, T the absolute temperature, R the gas constant and A is a material constant. Much higher stress exponent values have been found[10,11] when equation (1) is used to describe precipitate strengthened materials. In some cases a higher value of Q has also been found. Precipitates hinder the motion of dislocations, thereby inhibiting the recovery process[12]. Anelastic creep behaviour can also be described by using back stress models:

$$\varepsilon = A(\sigma - \sigma_b)^n \exp\left(-Q/_{RT}\right) \quad (2)$$

Where σ_b is the back stress in the material. Back stress can originate from a number of sources such as long range internal stresses[8] and dislocation interactions. Average values for dislocation numbers and velocities are usually considered, but Mclean[13] has criticised the above approach stating average densities cannot be used as only mobile dislocations contribute to deformation.

Dislocation densities and precipitate densities will affect the deformation behaviour of any material. Dislocations (both mobile and immobile) can form networks or loops which can impede local deformation. Precipitate densities and distributions affect the precipitation strengthening behaviour. In this study, TEM examination showed dislocation loops were mainly observed at temperatures below 650°C (Fig. 9,10). This is probably due to the higher energies available at higher temperatures.

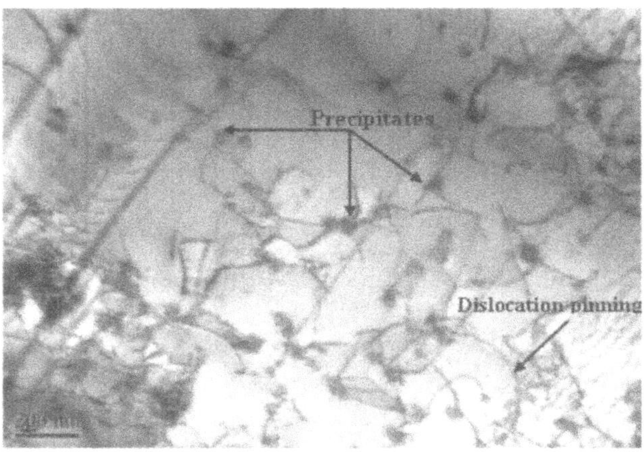

Fig 9. TEM image showing the interaction of dislocations with precipitates. The specimen used here is a creep rupture specimen. The conditions used were 650°C and 180MPa. The image was taken at 200KeV. Dislocation bending can be clearly seen

Load-unload tests conducted at 550°C showed lower levels of anelastic strains. From this, it was inferred that the mechanism(s) responsible for anelasticity had a strong thermal dependence. Similar results were also obtained by [15].

Plotting strain rate against time is a useful way to identify changes in the internal mechanisms of the material: see Fig. 4(a). There is a significant difference in the recovery strain rates between the initial (<1 hour) and later stages of the unload. The slope of the strain rate changes after the initial period, suggesting a change or saturation in the mechanism. Thus, it is possible that more than one mechanism may be prevalent during the unload. Similar studies[14,6] also propose the existence of two mechanisms during unload.

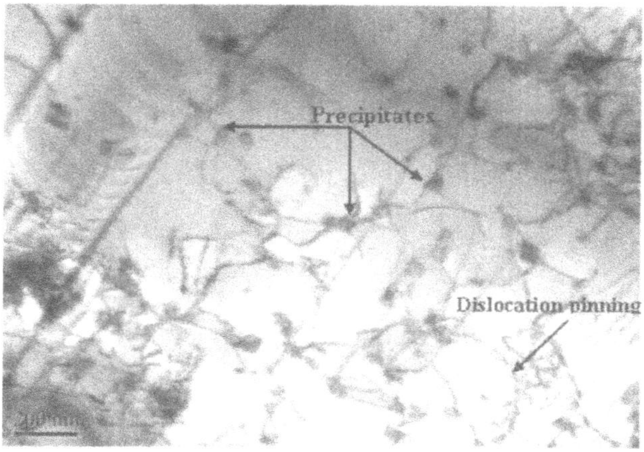

Fig 9. TEM image showing the interaction of dislocations with precipitates. The specimen used here is a creep rupture specimen. The conditions used were 650°C and 180MPa. The image was taken at 200KeV. Dislocation bending can be clearly seen

Microscopic examination was conducted by transmission electron microscopy (TEM) on specimens which were stopped at different unload times. The results showed a distinct decrease in the size of dislocations for specimens during the initial stages of the unload. Decrease in dislocation link sizes can occur when they interact with carbides. Comparing the first and second unload cycles of the test shown in Fig 4(a) suggests that there may be a dependence of strain on the precipitation behaviour during unload. Studies[16] carried out previously have shown that precipitation behaviour can be affected by strain.

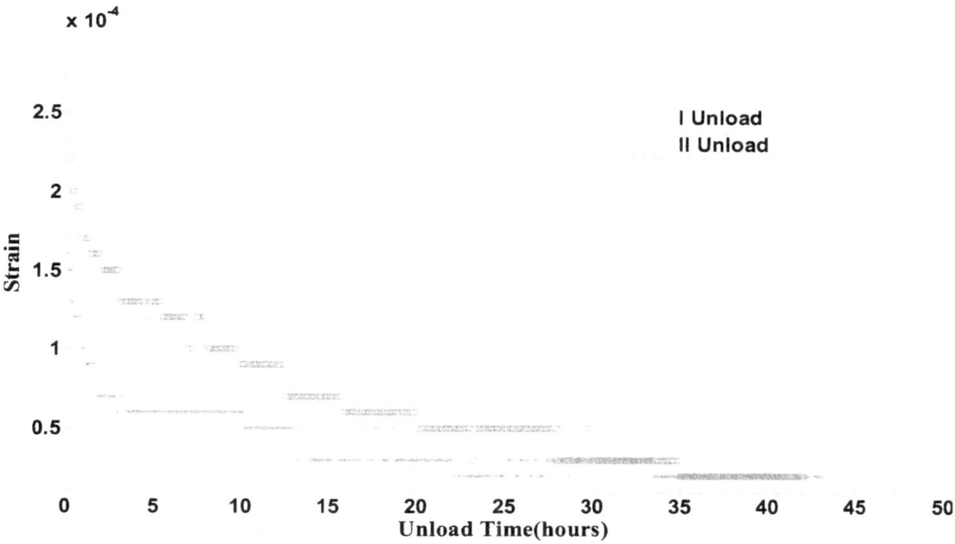

Fig. 12Comparison of I and II Unload for Experiment Shown in Fig. 4(a)

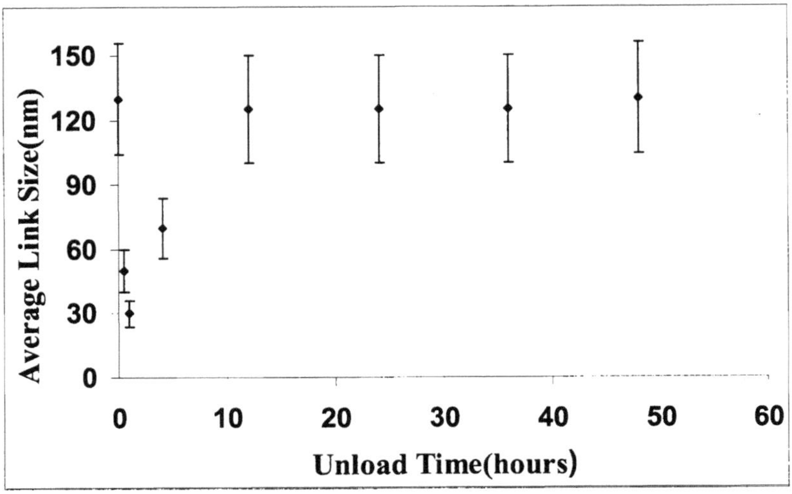

Fig 13. Change in Dislocation Density during the Unload. This may Indicate a Change in the Precipitate Densities.

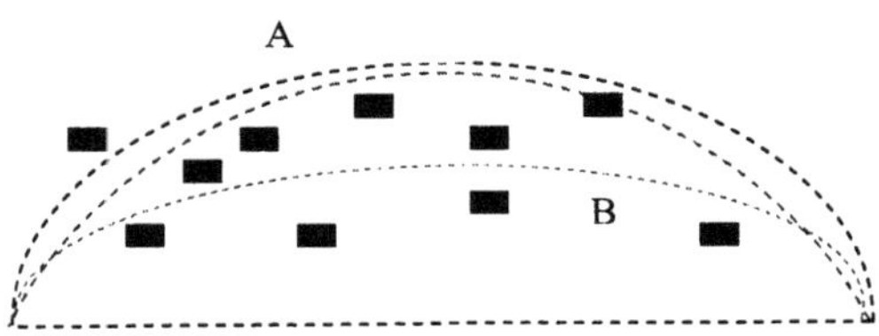

Fig14. Dislocation Unbowing. Depending on precipitate density, the amount a dislocation can bend will vary. The first stage might correspond to movement from A to B and the second stage may motion from B to a stable state.

Fig 14 shows the dislocation unbowing after load relaxations. It has been proposed that the first stage of the unbowing may be the dislocation moving from 'A' to 'B', and the second stage would be from B onwards. The unbowing behaviour would depend on the distribution of precipitates. If precipitates are encountered, then they have to be cut or circumvented in some way.

CONCLUSIONS

- Creep Behaviour of AISI316H steel has been studied. The creep behaviour seen under conditions of load-unload cycling is significantly different to the monotonic creep rupture behaviour.

- Tests which involved load-unload cycles were conducted to simulate material behaviour in power plants. Anelasticity was observed in all the tests. The tests showed an increase in the overall creep life but a decrease in the rupture ductility.

- TEM examination suggests strongly that dislocation interactions with precipitates were largely the origin of anelasticity. A decrease in dislocation sizes was observed during the initial stages of the unload. Also, comparison of different unload cycles suggested a strain influence on the precipitate densities. Further studies are being performed to understand this.

ACKNOWLEDGEMENTS

The work has been carried out as a part of the TSEC program KNOO and the authors are grateful to the EPSRC for funding under grant EP/C549465/1. The authors would also like to thank Prof. Mike Fitzpatrick for useful discussions and a thorough review of the paper.

REFERENCES

[1]A.E Johnson, Proc. Inst. Mech. Eng. 145 (1941) 210-220.
[2] J.D. Lubahn, Trans. ASM 45(1953) 787-838.
[3] A.E. Johnson, J. Henderson, B.Kanh, Metallurgica 63 (1963) 173-177
[4] D.G. Morris, Journal of Mat Sci 13 (1978) 1849-1854.
[5] L.O.Bueno, R.L.Bell Mat Sci and Eng A 410-411(2005) 72-78.
[6] V.Lupinc, F.Gabrielli, Mater Sci Eng. 37 (1979) 143-149
[7] M.F Ashby, Physics of Strength and Plasticity. M.I.T Press, Cambridge (1969)
[8] K.Sawada, K.Kimura, F.Abe, Mechanical Response of 9%Cr Heat Resistant Martensitic Steels to Abrupt Stress Loading at High Temperature, Mater. Sci. Eng. A358(2003), 52-58.
[9] B.J.Lazan Proc.Amer.Soc.Test.Mater.49 (1949) 757.
[10] K.R.Williams, I.R McLaughlin, Mater.Sci. 5, 1063 (1970).
[11] I.R McLaughlin, Metal Sci. 8, 247(1974)
[12] R.Lagneborg, J.Mater.Sci 3, 596(1968).
[13] D. Mclean, Can.J.Phys 45, 973(1967).
[14] T.B.Gibbons, V.Lupinc, D.Mclean, Met Sci,J.9(1975)437-444.
[15]. D.G.Morris, D.R.Harries, Jour of Mat. Sci. 14(1979) 2625-2636.
[16]. M.Hattestrand, H.O Andren, Acta Mater. 49(2001) 2123-2128.
[17] NIMS Data Sheets on the Elevated Properties of 18Cr-12Ni-Mo Stainless Steel Bars for General Application(SUS 316-B)

MOLTEN SALTS FOR NUCLEAR COGENERATION

Luke Olson, James Ambrosek, Guoping Cao, Kumar Sridharan, Mark Anderson, Todd Allen
Department of Engineering Physics
1500 Engineering Drive
University of Wisconsin
Madison, WI 53706

ABSTRACT

The utilization of process heat from a high temperature nuclear reactor for hydrogen production and chemical industry support could be a significant step towards increasing national energy independence and reducing greenhouse gases. For this co-generation, thermochemical process plants should be co-located with a nuclear power plant, but from a safety standpoint they must be separated by a reasonable distance. Molten salts have the potential to be excellent heat transport fluids for the heat exchange loop. Fluoride salts are specifically being considered for these heat transfer applications. However, materials corrosion has been identified to be a problem in molten fluoride salts, particularly at high temperatures. Corrosion tests have been performed to determine the compatibility of a few candidate high temperature materials for the heat transfer loop in a molten alkali fluoride eutectic salt, LiF-NaF-KF: 46.5-11.5-42 mol %, also known as FLiNaK. Generally, higher Cr content alloys experienced higher corrosion due to the propensity of Cr to selectively dealloy and dissolve into this fluoride salt. Additionally, data from three important experiments in the literature dealing with forced convective heat transfer in flowing molten FLiNaK were compiled and re-evaluated. For turbulent flow, in experiments conducted in systems constructed of Inconel alloys, it was determined that FLiNaK salt behaves as a normal fluid and can be modeled using the Dittus-Boelter (DB) correlation within ± 15% accuracy. The DB correlation can thus be used for preliminary calculations in design of the heat transfer equipment for molten FLiNaK.

1. INTRODUCTION

Molten salts have generated interest for use as heat transfer fluids in solar, nuclear, advanced oil recovery, biomass gasification, and several other applications where high temperatures and low pressures are desired.[1-5] Process heat from high temperature nuclear reactors is presently being considered for providing energy to chemical industries and desalination plants, as well as hydrogen production from water by either HTE (high temperature electrolysis) or a thermo-chemical process (sulfur-iodide (SI) process). It is advantageous to transfer heat at the highest possible temperatures for these applications. To further increase the efficiency of the plant, it is necessary for the process heat from the nuclear reactor to be transported efficiently to the associated industrial facilities.

It has been estimated that the hot side of a helium-cooled high temperature reactor for energy conversion could be in the 750°C to 900°C range, although the exact temperature has yet to be decided.[5] These high temperatures limit the available heat transfer fluid options. Molten salts provide a viable option as heat transport fluids at high temperatures because of their high boiling point, low vapor pressure, large specific heat and thermal conductivity, chemical stability, and high density at low pressures. These properties result in low pumping power and small pipe size requirements that would translate to lower capital and operational costs.[2, 6] However, these cost benefits must be balanced against materials corrosion issues in molten salts, and therefore the choice of materials for various components of the molten heat exchange system are of paramount importance. With this background, the present study examines corrosion of a few candidate alloys and revaluates heat transfer data from previous literature.

This study was undertaken to evaluate the stability of some candidate alloys against corrosion in FLiNaK salt at 850°C, while developing a fundamental understanding of the mechanisms of corrosion in these alloys in this molten salt environment.[7] Materials corrosion has been recognized for some time as an issue in high temperature molten salts, and particularly in the molten fluoride salts. In most high temperature industrial environments, materials derive their corrosion resistance from the formation of a tenacious surface film of an oxide of Cr, Al, or Si; elements that are intentionally added to the base alloy for this purpose.[8, 9] In molten fluoride salts these passive oxide films are chemically unstable, and corrosion occurs largely by the thermodynamically driven dissolution of alloying elements into the molten salt environment.[10, 11] An understanding of the propensity of molten fluoride salts to preferentially dissolve specific elemental constituents from an alloy can be achieved from phase diagrams, acid–base effects, and the redox potential of the alloy-salt system.[1]

In molten fluoride salts, the most active alloying element is selectively attacked by dissolution.[1] A National Aeronautic and Space Administration (NASA) study determined that the tendency for common alloying constituents to corrode in molten fluoride salts increased in the following order: Ni, Co, Fe, Cr, Al, with Al being the most prone to dissolution.[12] In Fe-Ni-Cr based alloys, the primary constituent that is most prone to dissolution is Cr, and in general, Cr content of an alloy correlates with weight-loss.[7, 12] The attack on Cr is all the more significant at carbide forming temperatures (~650°C-1200°C) due to the tendency of Cr-carbides to form at the grain boundaries, that lead to localized attack resulting in voiding along the grain boundary network.[13] Pure Ni is largely immune to dissolution attack in molten fluoride salts, but Ni in its elemental form lacks the high temperature strength to be used as a material for structural components.

It is desirable to use conventional superalloys that have relatively high Cr and C contents because of their well established high temperature creep rupture strength and resistance to air oxidation on the outside of a pipe. The alloys investigated in the present study include Ni-based high temperature alloys, Hastelloy-N, Hastelloy-X, Haynes-230, and Inconel-617, an Fe–Ni-based alloy Incoloy-800H, and Ni-201 which is a nearly pure Ni alloy with minor alloying additions.

2. EXPERIMENTAL

FLiNaK salt, purified by sparging with a H_2/HF gas mixture, was procured from Electrochemical Systems, Inc. (ECS), in Oak Ridge, TN, USA. Table I lists the compositions of the alloys evaluated in this study. The alloys were selected based on their high temperature strength, code certification status, and their history of prior usage in high temperature applications. An exception was Ni-201, which was selected to experimentally test the high corrosion resistance of Ni in molten fluoride salts.

Table I. Nominal compositions (weight %) of the alloys tested in the present study rounded to the nearest tenth of a percent for all constituents except for carbon which is rounded to the nearest hundredth of a percent.

Alloy	Cr	Mo	W	Al	Ti	Fe	C	Co	Ni	Mn
Haynes-230	22.5	1.2	14.1	0.3		1.8	0.1	0.3	Bal.	0.5
Inconel-617	22.1	9.6		1.1	0.4	1.1	0.08	12.4	52.9	0.1
Hastelloy-N	6.3	16.1	0.1			4.0	0.03	0.2	Bal.	0.5
Hastelloy-X	21.3	8.8	0.4	0.1		19.3	0.06	1.4	Bal.	0.5
Incoloy-800H	20.4			0.5	0.6	45.3	0.07		31.6	0.8
Ni-201						0.1	0.02		99.4	0.2

Corrosion tests were performed in graphite capsules because of the relative inertness of graphite to molten fluoride salts. Purified graphite, grade AXZ-5Q1 was obtained from POCO Graphite, Inc. This POCO graphite had an average particle size of 5μm and an average pore size of 0.7 μm.[14] Previous studies have shown that small pore sizes are desirable from the standpoint of minimizing salt infiltration of the graphite.[15] The graphite was machined into crucibles, crucible lids, holding rods, and fixturing screws. Prior to corrosion testing, the graphite crucibles and components were cleaned ultrasonically in methanol and water, to remove contaminants from the machining process, before being heated to 1000°C for 8 h to remove residual water and oxygen. After the graphite bake-out, the alloy samples were placed in their corresponding graphite crucibles. Three samples of a given test alloy were fixtured to the central graphite rod using graphite screws. Therefore, all alloys were tested in triplicate for increased confidence in corrosion results, with one crucible being dedicated to each of the alloys. To further limit contamination, the graphite crucible was in turn completely sealed in a 304 stainless steel capsule. The static corrosion tests were performed at 850°C for 500 h. The salts were kept in an Ar environment at all times to avoid environmental contamination. For the same reason, welding for the stainless steel air boundary was also performed inside an Ar atmosphere glove box after the salts had been introduced into the crucible.

It should be noted here that studies by the authors' group has shown that graphite accelerates corrosion because it is cathodic with respect to the test samples and can form a Cr carbide with Cr from solution. This acceleration effect allows for an evaluation of relative corrosion performance of the various candidate alloys in relatively short exposure times, but also raises questions about the use of molten salts in systems jointly constructed of graphite and steel.[7]

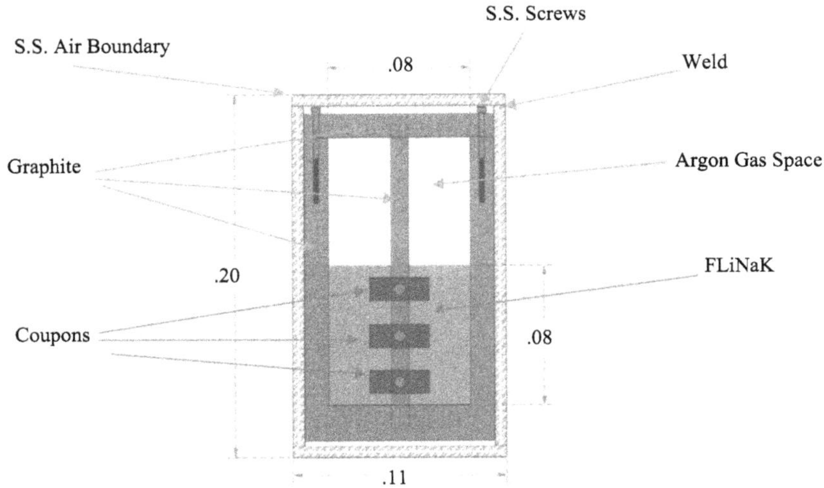

Figure 1. Schematic illustration of the capsules for the static corrosion testing of alloys in molten FLiNaK salt at 850°C for 500 hours. Dimensions in meters.

The final welds of the inlet holes used for introducing FLiNaK were performed in a furnace inside of the glove box at 600°C, immediately after being filled with FLiNaK, to lower the capsule's internal pressure at 850°C. Figure 1 shows a schematic illustration of the corrosion test capsule, designed to maximize safety and maintain salt purity, and Figure 2 (a) shows a photograph of the graphite crucible, fixturing rod, and screws. After the final weld to seal the corrosion capsule, the capsules were placed in an argon atmosphere, box furnace, where they were heated up to 850°C and maintained at this temperature for 500 h.

After the 500 h exposure, the furnace temperature was lowered to 500°C and the capsules were inverted to allow for the FLiNaK salt to drain away from the alloy samples prior to further cooling. After the capsules reached room temperature, the corrosion test capsules were cut open using an automated CNC mill, exposing the samples (Figure 2 (b)). The retrieved samples were cleaned of residual FLiNaK and than evaluated using weight change measurements and scanning electron microscopy.

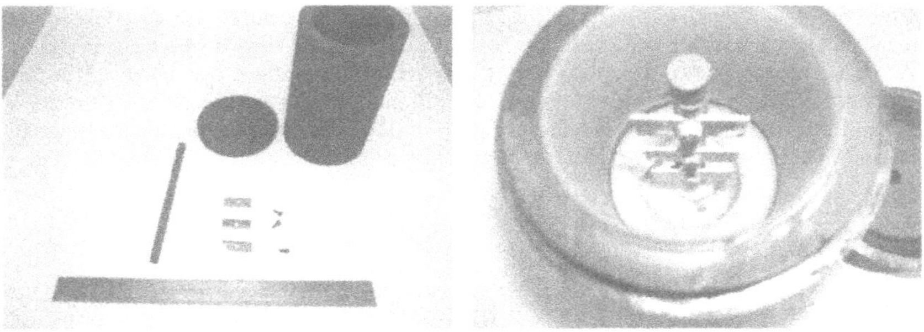

(a) (b)

Figure 2. Photographs of the corrosion capsule, (a) the graphite crucible, fixturing rod, and screws with samples, and (b) corrosion capsule after corrosion testing and after cutting the bottom of the capsule, showing alloy test coupons.

3. RESULTS AND DISCUSSION

The results of weight change measurements of the alloys after corrosion testing in molten FLiNaK at 850°C for 500 h is shown in Figure 3. For the four Ni-based alloys, Hastelloy-N, Hastelloy-X, Inconel-617, and Haynes-230, the weight loss due to corrosion was noted to increase with the Cr content of the alloy. Ni-201, a predominantly Ni-containing alloy with no Cr, was virtually immune to attack. The large variations in weight-loss between the Ni-base alloys and the Fe-Ni based alloy, all having Cr levels of about 20%, demonstrate the complex roles alloying elements such as Mo, W, Fe, Co, and carbon play in corrosion susceptibility. For example, it may be noted that Hastelloy-X and Inconel-617 have similar chemical compositions, but Inconel-617 demonstrated nearly twice the weight-loss per area as Hastelloy-X. The major difference between these two alloys is that the Fe in Hastelloy-X is to a large extent replaced by Co in Inconel-617.

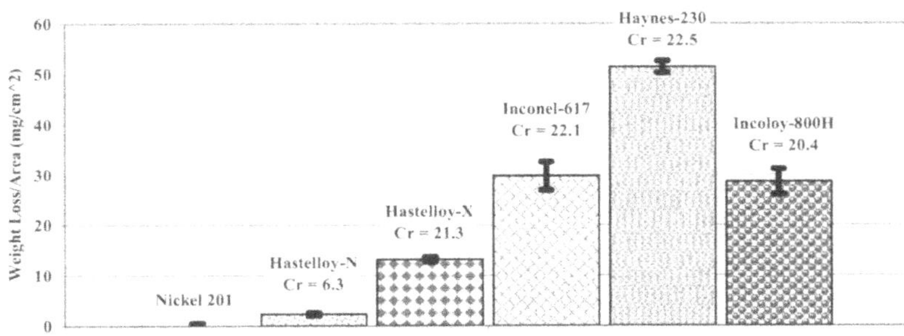

Figure 3. Weight loss due to corrosion for various alloys after exposure to FLiNaK at 850°C for 500 h in graphite capsules.

Generally, the alloys with high Cr and high carbon concentrations showed higher corrosion. The corrosion attack for Hastelloy-N (6.3% Cr) was the lowest of all the Fe-Ni-Cr alloys. As shown in Figure 4, Cr depletion up to depths of about 50μm was observed in this alloy. EDS x-ray mapping performed on cross-sectional images of this alloy showed a reasonably uniform depletion of Cr and the presence of Mo-enriched precipitates at the grain boundaries, confirming that Mo is generally resistant to dissolution in molten FLiNaK salt.

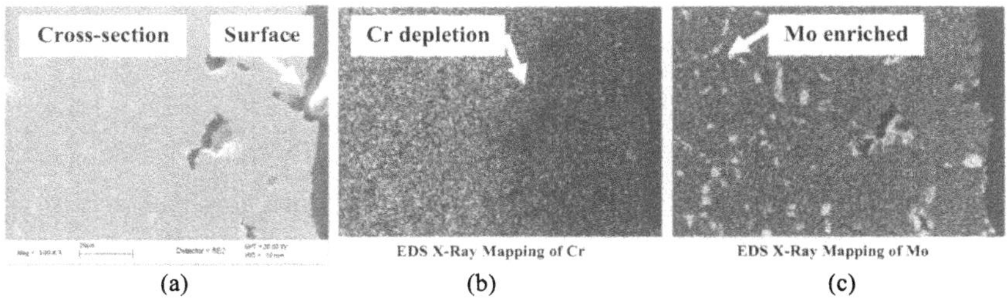

Figure 4. (a) SEM of Hastelloy-N cross-section with EDS, and x-ray maps of (b) Cr, and (c) Mo.

Figure 5 shows the microstructure from Haynes-230, which is prototypical of the chemical attack in molten salt. Here the depletion of Cr at the surface resulted in a highly Ni-enriched surface film. Haynes-230, which contains about 14% W, exhibited an abundance of W-rich precipitates, predominantly in the grain boundary regions. The refractory rich precipitates consisted primarily of the refractory metal and Ni, with low concentrations of other elements. For Haynes-230, the W-rich precipitates appeared resistant to dissolution, demonstrating that W, like Mo is resistant to attack from molten fluoride salt.

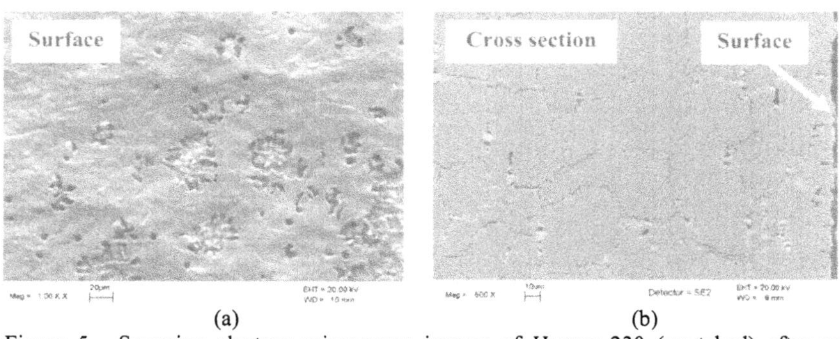

Figure 5. Scanning electron microscopy images of Haynes-230 (unetched) after corrosion testing in FLiNaK at 850°C for 500 h. (a) surface plan view and (b) cross-sectional view.

Hastelloy-X (21.3% Cr) exhibited grain boundary attack up to depths of at least 300 μm below the surface. Significant depletion of Cr was noted to have occurred throughout the grains near the surface, but this depletion was markedly more severe along the grain boundaries. As in the case of Hastelloy-N, Mo-rich precipitates at the grain boundaries were observed to remain intact after corrosion tests (Figure 6 (b)).

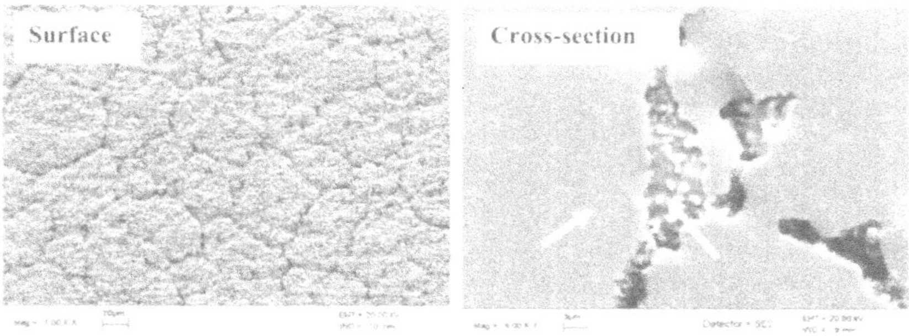

(a) (b)

Figure 6. Scanning electron micrographs showing of Hastelloy X after corrosion tests (a) plan view showing intergranular and general surface attack of the surface of Hastelloy-X, and (b) cross-section view with arrows indicating Mo enriched precipitates.

Inconel-617 (22.1% Cr) was uniformly depleted in Cr up to depths of about 100 μm from the surface, and experienced dramatic grain boundary corrosion throughout the thickness of the sample. Similar attack was observed for Haynes-230 (22.5% Cr) in that it appeared that the intergranular attack appeared to result in connected voiding. Additionally, the surfaces of Haynes-230 and the Inconel-617 exhibited a Ni-enriched surface film (Figure 7 (a)). Figure 7 (b) shows the cross-sectional image of Inconel-617 after the corrosion tests showing attack at the grain boundaries. EDS x-ray mapping detected Na in the grain boundary regions of the Inconel-617 alloy, indicating that molten salt had infiltrated into the alloy along the grain boundary void region (Figure 7 (c)).

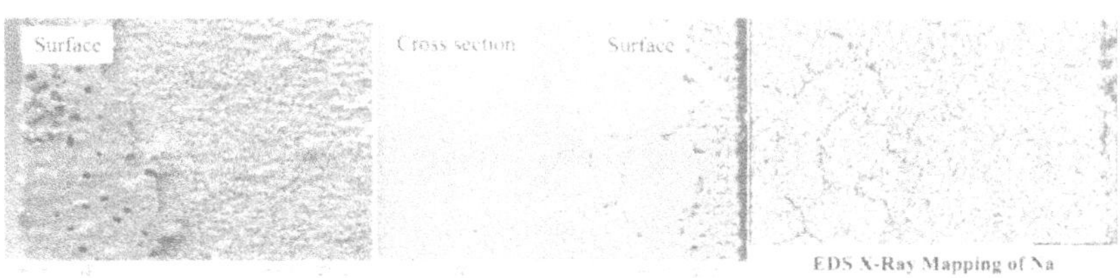

(a) (b) (c)

Figure 7. Scanning electron micrograph (a) showing partial Ni rich film on the surface of Inconel-617, (b) cross-sectional view showing grain boundary attack, and (c) EDS x-ray map showing Na concentrated at grain boundaries indicating FLiNaK infiltration into alloy along grain boundary.

The Cr dealloying for the Fe-Ni based Incoloy-800H (20.4% Cr) appeared to result in pitting of the surface. An SEM micrograph of the pitted surface of Incoloy-800H samples is shown in Figure 8 (a). As in the case of pervious alloys, severe depletion of Cr was observed particularly along the grain boundaries and near the surface and this effect was more pronounced for smaller grains compared to the larger grains. EDS analysis shows the uniformly Cr depleted region extends into the surface about 150 μm.

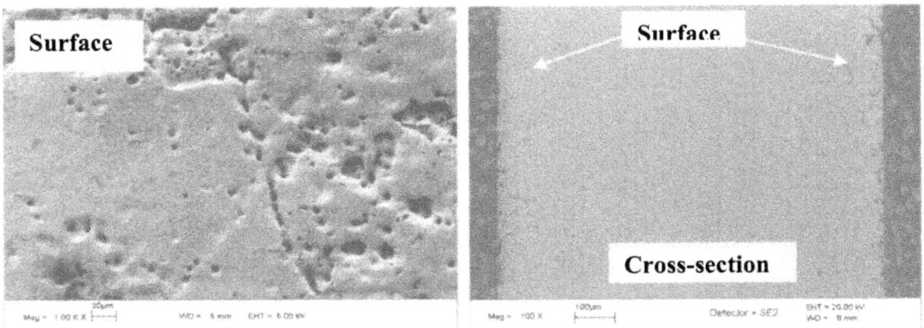

(a) (b)

Figure 8. Scanning electron micrograph (a) showing pits on the surface of Incoloy-800H after corrosion tests and (b) cross-sectional view showing grain boundary attack.

In general, for alloys with a given Cr content, higher carbon content of the alloy resulted in more severe intergranular attack due the precipitation of Cr-carbide phases at the grain boundaries. Barring minor attack at the grain boundaries, Ni-201 (Figure 9) exhibited very little corrosion attack due to the strong resistance of Ni to dissolution in molten fluoride salts. It should be noted that a Ni-plated or a developed Ni surface will not be protective over long time periods due to the high diffusion rate of Cr through Ni at the temperatures of interest used in this study.

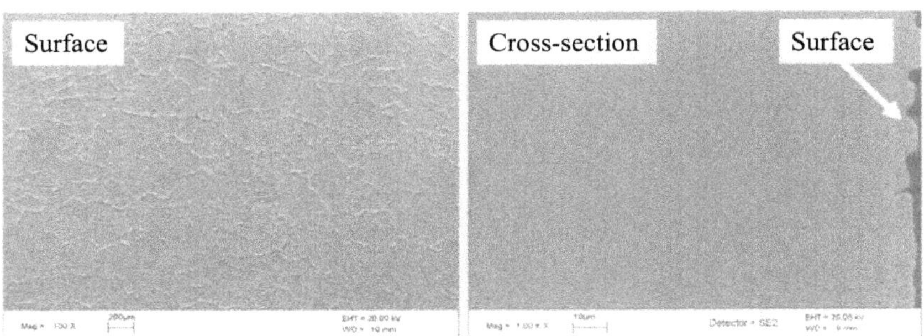

(a) (b)

Figure 9. SEM images of Ni-201 after corrosion tests (unetched). (a) surface plan view, (b) cross-sectional view. This alloy was largely immune to corrosion attack.

4. HEAT TRANSFER PROPERTIES

The thermo-physical properties of a salt are the first metric with which to compare salts for heat transfer applications. Many salts were considered initially for study as potential heat transfer fluids. Table II shows important properties of some of these salts. In the nuclear-cogeneration plant, it is necessary for the salts to be stable to temperatures in excess of about 800°C. This eliminates nitrate and carbonate based salts due to thermal decomposition of these salts at lower temperatures.

Table II. Comparison of salt properties for salts of interest in the present work and water. (Source[5])

Salt (composition in mol %)	Formula weight (g/mol)	Melting point (°C)	Boiling Point (°C)	vapor pressure 900°C (mm Hg)	density 700 °C (g/cm³)	volumetric heat capacity $\rho*Cp$, 700 °C (J/cm³-°C)	Viscosity μ, 700 °C (cP)	thermal conductivity k, 700 °C (W/m-K)
LiF-NaF-KF (46.5-11.5-42)	41.3	454	1570ᵃ	~ 0.7	2.02	3.81	2.9	0.92
LiCl-KCl (59.5-40.5)	55.5	355	~1400ᵃ	5.8	1.52	1.82	1.15	0.42
$KCl-MgCl_2$ (68-32)	81.4	426	>1418	< 2.0	1.66	1.92	1.40	0.40
$NaF-NaBF_4$ (8-92)	104.4	385	694	9500	1.75	2.64	0.90	0.40
$NaNO_3$-K NO_3 (64-36)	91.44	227	600ᵇ	--	1.84 @ 400°C	2.76 @ 400°C	1.80 @ 400°C	0.55 @ 400°C
Water (H_2O)	18	0	100	--	1.0 @ 20°C	1.84 @ 20°C	1.0 @ 20°C	0.6 @ 20°C

[a]Estimated by extrapolation of lower-temperature data (~1100°C) or assumption of ideal mixture behavior.

[b]This is the decomposition temperature of the $NaNO_3$-KNO_3 mixture.

Interest in the use of molten salts as a heat transfer media in the nuclear field gained traction with the successful use of uranium fuel bearing salts in the aircraft reactor experiment (ARE) which used NaF-ZrF_4-UF_4 (53-41-6 mol%) salt for the development of a nuclear powered aircraft.[16] This was followed by the molten salt reactor experiments (MSRE) at ORNL which utilized LiF-BeF_2-ZrF_4-UF_4 (65-30-5-0.1 mol%) for fuel salt and FLiBe ($2LiF$-BeF_2) for the secondary coolant.[17] Non-fuel bearing salts have also been investigated as heat transfer agents in several solar thermal systems ($NaNO_3$-KNO_3) as well as by ORNL (NaF-$NaBF_4$) as a secondary coolant for the molten salt reactor.[18] However, the decomposition of $NaNO_3$-KNO_3 (T_D~600C) is well below the desired temperatures for desirable for co-generation, and NaF-$NaBF_4$ exhibits undesirably high vapor pressure at the same temperatures.

With the goal of using molten FLiNaK salt as a heat transport fluid, the heat transfer properties of this salt from literature were examined. In the literature only three of four experiments contained enough information for re-evaluation. These experiments included the work by; Hoffman conducted at Oak Ridge National Laboratory, ORNL, in 1954,[19] Grele at The National Advisory Committee for Aeronautics, NACA, in 1953,[20] Vries'ma of The Delft University in 1979,[21] and finally Ignat'ev in 1983.[22] The description provided in the Ignat'ev work proved to contain insufficient information for re-evaluation, as was performed with the other three data sets.

The documented experimental data from each of these three experiments measuring convective heat transfer experiments was re-evaluated using consistent and more accurate and up-to-date fluid properties for the salt.[23] For turbulent flow, in experiments conducted in chambers constructed of Inconel alloys, it was determined that FLiNaK salt behaves as a 'normal' fluid and can be modeled using the Dittus-Boelter (DB) Correlation within ± 15% accuracy as shown in Figure 10. The DB correlation can thus be used for preliminary calculations in predicting the heat transfer of salts.

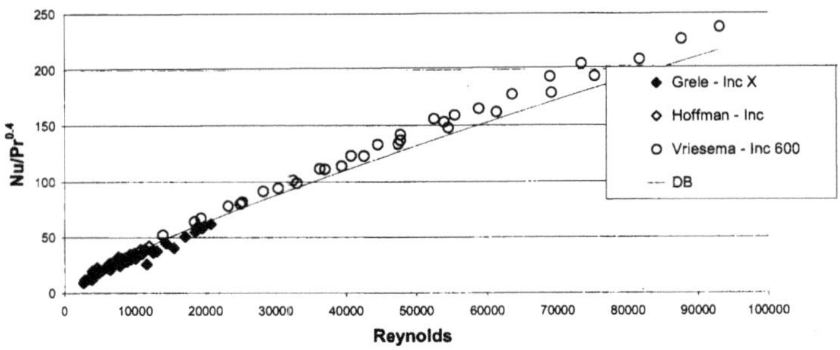

Figure 10. The experimental data of Hoffman, Grele, and Vriesema in Inconel based tubes was re-analyzed. The data for each of the three experiments agrees within ±15% of the DB Correlation.

The benefits of these molten salts following the DB correlation, i.e. normal fluid, is that the heat transfer of molten salts is physically similar to that of other normal fluids. This allows for use of other correlations that were developed for these normal fluids as long as the correct physical properties of these salts are used in the correlations.

5. CONCLUSION

A number of potential high temperature alloys and alloy Ni-201 were evaluated for corrosion in molten FLiNaK salt at 850°C for exposure durations of 500 h, using graphite crucibles encapsulated in stainless steel containments to ensure salt purity and for safety. In the Cr-containing alloys the weight-loss generally increased with Cr and carbon contents, and it was found that refractory rich phases of Mo and W were resistant to corrosion attack. In alloys with high Cr contents (~20%), voiding at the grain boundaries was observed. The Ni-201 alloy, which is predominantly Ni with minor alloying additions, was predictably resistant to corrosion. The enhanced corrosion rate induced by the graphite crucible and components provided an accelerated environment for evaluating the relative corrosion performance of various candidate alloys. Molten salts behave like normal fluids and the turbulent convective heat transfer coefficient can be predicted closely by the Dittus-Boelter correlation. This normal fluid behavior of molten salts allows for use of more unique correlations developed using other normal fluids for the design of heat transfer equipment utilizing molten salts in a co-generation plant.

6. ACKNOWLEDGEMENTS

This work was supported by DOE grants No. DOE-FC07-05ID14675 and DOE-FC07-07ID14826. This research utilized NSF-supported shared facilities at the University of Wisconsin.

This report was prepared as an account of work sponsored by an agency of the United States Government. Neither the United States Government nor any agency thereof, nor any of their employees, makes any warranty, express or implied, or assumes any legal liability or responsibility for the accuracy, completeness, or usefulness of any information, apparatus, product, or process disclosed, or represents that its use would not infringe privately owned rights. Reference herein to any specific commercial product, process, or service by trade name, trademark, manufacturer, or otherwise does not necessarily constitute or imply its endorsement, recommendation, or favoring by the United States Government or any agency thereof. The views and opinions of the authors expressed herein do not necessarily state or reflect those of the United States Government or any agency thereof.

7. REFERENCES

[1]D. Williams, L. Toth, and K. Clarno, Assessment of Candidate Molten Salt Coolants for the Advanced High-Temperature Reactor (AHTR), Oak Ridge National Laboratory, Oak Ridge, Tennessee, ORNL/TM-2006/12, 1-69, (2006).
[2]C. Davis, C. Oh, R. Barner, S. Sherman, and D. Wilson, Thermal-Hydraulic Analyses of Heat Transfer Fluid Requirements and Characteristics for Coupling a Hydrogen Production Plant to a High-Temperature Nuclear Reactor, INL/EXT-05-00453, Idaho National Laboratory, Idaho Falls, ID (2005).
[3]C. W. Forsberg, P. F. Peterson, and H. Zhao, High-Temperature Liquid-Fluoride-Salt Closed-Brayton-Cycle Solar Tower Powers, *J. Sol. Energy Eng.*, **129**, 141 (2007).
[4]K. Cornwell, The Thermal Conductivity of Molten Salts, *J. Phys. D: Appl. Phys.*, **4**, 441-445, (1971).
[5]D. F. Williams, Assessment of Candidate Molten Salt Coolants for the NGNP/NHI Heat Transfer Loop, ORNL/TM-2006/69, Oak Ridge National Laboratory, Oak Ridge, TN (2006).
[6]C. W. Forsberg, Goals, Requirements, and Design Implications for the Advanced High-Temperature Reactor, *14th International Conference on Nuclear Energy*, Paper No. ICONE 14-89305, (2006).
[7]L. Olson, J. Ambrosek, K. Sridharan, M. Anderson, and T. Allen, Materials corrosion in molten Lif-NaF-KF salt, *Journal of Fluorine Chemistry*, **130** (1), 67-73, (2009).
[8]R. Hurst, R. Lyon, and C. Nicholls, Ed., *Progress in Nuclear Energy, Series IV: Technology, Engineering and Safety*, **2**,164-179, (1960).
[9]J. Steinmetz, P. Steinmetz, and A. Huntz, Corrosion processes related to superalloys used in gas turbines, *Solid State Phenomena*, **21 & 22**, 223-276, (1992).
[10]D. Williams, D. Wilson, J. Keiser, L. Toth, and J. Caja, Research on Molten Fluorides as High Temperature Heat Transfer Agents, in *Global 2003, Session 2A: Coolant/Material Interactions in Advanced Reactor Systems, Embedded Topical Within 2003 American Nuclear Society Winter Meeting*, 1-12, (2003).
[11]J. Koger, in: *ASM Handbook: Corrosion: Fundamentals, Testing, and Protection*, Vol. 13A, S. Cramer, B. Covino, Ed., OH, ASM International, 124–128, (2003).
[12]A. Misra, and J. Whittenberger, Fluoride Salts and Container Materials for Thermal Energy Storage Applications in the Temperature Range 973 to 1400 K, National Aeronautics and Space Administration, NASA Lewis Technical Memorandum 89913, (1987).
[13]M. Streicher, Effect of composition and structure on crevice, intergranular, and stress corrosion of some wrought Ni-Cr-Mo alloys, *Corrosion-NACE*, **32** (3), 79–93, (1976).
[14]Poco Graphite Inc. AXZ-5Q, 2009, [Online], available: http://www.poco.com/tabid/92/Default.aspx [Accessed: 4/26/2006].

[15]H. McKoy, R. Beatty, W. Cook, R. Gehlbach, C. Kennedy, J. Koger, A. Litman, C. Sessions, and J. Weir, *Nuclear Applications & Technology*, **8**, 156-169, (1970).
[16]J. Lane, *Fluid Fuel Reactors*, Reading, Mass, Addison-Wesley Pub. Co., xxii-979, (1958).
[17]P. Haubenreich, and J. Engel, *Experience with the molten salt reactor experiment, Nuclear Applications and Technology*, **8**,118-136, (1970).
[18]T. Mancini, *Advantages of using molten salt, Sandia National Laboratory: National Solar Thermal Test Facility*, October 2006, [Online], Available: http://www.sandia.gov/Renewable_Energy/solarthermal/NSTTF/salt.htm [Accessed 4/24/2007].
[19]H. Hoffman, and J. Lones, Fused Salt Heat Transfer Part II: Forced Convection Heat Transfer in Circular Tubes Containing NaF-KF-LiF Eutectic, ORNL 1977, Oak Ridge National Laboratory, Oak Ridge, TN, (1955)
[20]M. Grele, and L. Gedeon, Forced-Convection Heat-Transfer Characteristics of Molten FLiNaK Flowing in an Inconel X System, NACA RM E53L18, National Advisory Committee for Aeronautics, Cleveland, OH, (1954).
[21]I. Vriesema, Aspects of Molten Fluorides as Heat Transfer Agents for Power Generation, WTHD No. 112, Delft University of Technology, Netherlands.
[22]V. Ignat'ev, S. Keronovskii, A. Surenkov, O. Shcherbanyuk, S. Manchkha, and Y. Smirnov, Heat Exchange During the Flow of a Melt of LiF-NaF-KF Fluoride Salts in a Circular Tube, *Sov. At. Energy*, **57** (2), 123, (1984).
[23]J. Ambrosek, M. Anderson, K. Sridharan, and T. Allen, Current Status of Knowledge of the Fluoride Salt (FliNaK) Heat Transfer System, *Nuclear Technology*, **165**, 166-173, (2009).

Green Engineering and Environmental Stewardship

DEVELOPMENT OF LOW-COST FUNCTIONAL GEOPOLYMERIC MATERIALS

Mazen Alshaaer
Deanship of Academic Research, University of Jordan, 11942 Amman, Jordan

Faten Slaty and Hani Khoury
Department of Geology, Faculty of Science, University of Jordan, 11942 Amman, Jordan

Hubert Rahier
Research Group of Physical Chemistry and Polymer Science (FYSC), Vrije Universiteit Brussel (VUB), Pleinlaan 2, 1050 Brussels, Belgium

Jan Wastiels
Department of Mechanics of Materials and Constructions (MEMC), Vrije Universiteit Brussel (VUB), Pleinlaan 2, 1050 Brussels, Belgium

ABSTRACT

Some important research on the development, and applications of Low-cost functional geopolymeric materials was conducted.. This paper focuses on geopolymerization technique intended for the stabilization of natural geomaterial samples collected from various deposits in Jordan. The obtained materials are environmentally friendly and need only a low energy amount to be produced. Observations confirmed that natural geomaterials, e.g. kaolinite, satisfy the criteria to be used as precursors of inexpensive and durable construction materials. Specimens with compressive strength of 41 MPa under dry conditions and 23 MPa under water saturated conditions were obtained. Also these materials maintained or improved their mechanical performance after heating up to 600 °C. Therefore, they can be used for applications where fire safety is important. It was observed that adding functional fillers such as zeolitic tuff increases the adsorption capacity of the end geopolymer products toward heavy metals such as cadmium, copper and lead. Such products also can be used for water storage and in waste water treatment plants.

INTRODUCTION

Geopolymer technology has recently received attention in natural geomaterials and waste stabilization. Geopolymers are inorganic polymers that have been gradually attracting world attention as potentially revolutionary materials. Similar to natural zeolite minerals, geopolymers form a class of three-dimensionally networked alumino-silicate materials. According to previous research [1-7], a wide range of natural Al–Si minerals, wastes, and slag could serve as potential source materials for the synthesis of geopolymers. Unlike conventional organic polymers, geopolymers are non-combustible, heat-resistant, formed at low temperatures, and fire/acid resistant. Geopolymer products are being developed in different areas such as: Mining, Civil Construction, Building Materials, Environment/Waste Management ... The objective of this paper is to summarize the ongoing research activities related to low-cost geopolymers at University of Jordan and Vrije Universiteit Brussel, and to present the recent results of geopolymers produced from natural geomaterials from Jordan. These new geopolymeric materials present the characteristic of being low cost and using low energy resources. The advancements in this field are very much dependent on the development and characterization of new functional geopolymeric materials exhibiting high micro pollutants adsorption capacity for water storage and transportation [8-10], high mechanical performance for construction purposes, and heat resistancefor high temperature applications.

2. MATERIALS AND METHODS

2.1. Jordanian kaolinite (JK)

Jordanian kaolinite with a purity of 60% [7] was collected from El-Hiswa deposit, which is located south of Jordan about 45km to the east of Al-Quweira town [11]. The elevation of the upper surface deposit ranges between 850 and 950m above sea level. Thirty-two individual samples were collected from different layers of an outcrop, weighing in total 32 kg. These samples represent 8 m of the upper layers of this outcrop. Preparation of the Jordanian kaolin samples involved crushing (using Jaw crusher, RETCH-BB1A) of an oven dried clay (at 105 °C) with a grain size less than 425μm. Then the samples were mixed in 50 L plastic drum for several times to make them more homogenous. Optimizing the filler/clay/water/NaOH ratios was necessary before the fabrication of geopolymers. Kaolinite samples were tested for plasticity limit. Plasticity limits are important to optimize the amount of water needed for the specimens' preparation. According to the ASTM D4318 [12], the plasticity limit for the Jordanian kaolinite from El-Hiswa was 22%. The chemical composition of the Jordanian Kaolinitic soil (JK) was determined using XRF analysis (Bruker System S4 Pioneer) as shown in Table 1.

Table 1. Chemical composition of Jordanian Kaolinitic soil (JK)

Major Oxides	Al_2O_3	CaO	Cl	Fe_2O_3	K_2O	MgO	MnO	NaO	P_2O_5	SiO_2	SO_3	TiO_2
(wt %)	25.1	0.68	0.1	7.5	1.4	0.21	0.01	0.21	0.16	48.8	2.93	0.86

2.2. Jordanian silica sand (SS)

Silica sand is one of the most well-known inert fillers which are used in cement-based materials. It is characterized by its low price, high stability, and high resistance to chemical attack [13–15]. This filler should be well crystalline and should not include impurities such as organic matter or swilling/shrinking minerals to be utilized as inert filler. Organic matter dissolves easily in the presence of chemical solutions, while the swelling/shrinking minerals experience dramatic changes in their dimensions in presence of water. Impurities even in small percentage could cause disintegration of the end geopolymeric products. However, the silica sand employed in the present work was washed with distilled water and sieved at 100 and 400 μm [16]. The chemical composition of the silica sand (SS) was determined using XRF analysis (Bruker System S4 Pioneer) as shown in Table 2.

Table 2. Chemical composition of Jordanian Silica sand (SS)

Major Oxides	Al_2O_3	CaO	Cl	Fe_2O_3	K_2O	MgO	MnO	NaO	P_2O_5	SiO_2	SO_3	TiO_2
(wt %)	0.46	0.06	0.03	0.05	0.00	0.00	0.02	0.09	0.01	98.9	0.05	0.14

2.3. Jordanian zeolitic tuff (ZT)

The zeolitic tuff was from North–East Jordan deposits (Aritayn area) located around 50km to the east from Amman. The mineral composition of this tuff belongs mainly to phillipsite type of zeolites [17–19]. The chemical composition of the Jordanian zeolitic tuff (ZT) was determined using XRF analysis (Bruker System S4 Pioneer) as shown in Table 3.

Table 3. Chemical composition of the Jordanian zeolitic tuff (ZT)

Major Oxides	Al_2O_3	CaO	Cl	Fe_2O_3	K_2O	MgO	MnO	NaO	P_2O_5	SiO_2	SO_3	TiO_2
(wt %)	11.2	7.39	0.17	13.10	1.090	10.30	0.180	1.96	0.413	38.3	0.102	3.12

2.4. Fabrication of geopolymers' specimens

Two series of geopolymers' specimens were prepared from kaolinitic soil (JK), filler (SS) and Jordanian zeolitic tuff (ZT), and alkali solution. The composition of each series is shown in Table 4.

Table 4. Composition of Geopolymeric Specimens

	Materials ratios compared to clay				
ID	JK	Water	SS	NaOH	JZ
M-JK	100	22	100	14	0
M-JZ	100	30	0	14	100

Kaolinitic soil (JK) was used as a source of aluminumsilicate. Silica sand was used as a filler to provide high mechanical properties. NaOH was used as an alkaline activator for dissolution of aluminoslicate. Water was the reaction medium. The Jordanian zeolitic tuff and silica sand are used as functional and inert fillers respectively. The weighed filler and kaolinitic soil were mixed first and then the sodium hydroxide (GCC, 96%) solution was added. Mixing was done at two different speeds; the first speed at 107rpm for 2 min and the second speed at 198 rpm for 10min. Good mixing is important to obtain homogeneous and comparable specimens and to avoid the agglomeration of the mixture. Each mixture (series) was divided into specimens (50 g each). The mixture was molded immediately after weighing to avoid drying and decrease of the workability of the mixture. The paste was molded in a stainless steel cylinder (diameter of 25mm and height of 45mm) at a pressure of about 15MPa. The molded specimens of each series were cured by placing them in a ventilated oven at 80 °C for 24 h. After this curing time, the specimens were removed and let to cool at room temperature. These fabrication process steps are shown in the following diagram.

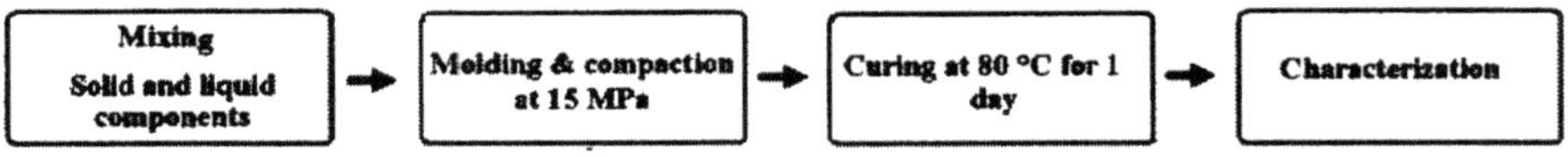

After the specimens' preparation, they were subjected to different testing techniques, experimental procedures, and characterization.

2.5. Mechanical and physical characterization of geopolymers

The fabricated specimens, dry and immersed, for each series were tested for the compression strength. Compression test was performed using CONTROLS testing machine (Model T106 modified to suit with standard testing), where the load was applied and increased by a displacement rate.

2.6. Heat exposure procedure

The effect of heating on the strength of the geopolymers was determined after exposing bulk specimens to various temperatures. The following heat exposure procedure was used in this research:

1. The zero measurements (reference) were recorded immediately after curing at 80°C for 24 hours.
2. The test specimens were placed in the oven.
3. The oven was brought to the required temperature.
4. The required temperature was maintained during the specified period of time (usually 24 hours).
5. After the exposure period, the specimens were removed from the oven, and then brought back to ambient temperature before further testing.
6. The properties of interest were measured (e.g. length, strength ….)
7. After placing the same specimens in the oven, the temperature was increased to the next required temperature and steps 4 to 7 were repeated with the same specimens until the highest required temperature (1000°C) was reached.

An average of at least three specimens was used for each test. The specimens were examined visually before and after each exposure step. Any change in the appearance of the tested specimens before and after exposure was observed, compared and recorded.

2.7. Ageing, and wet-drying cyclic test

In order to understand the long-term stability and durability of geopolymers when used as a material for different applications, M-JS specimens were prepared for evaluation. Two series of specimens were prepared for ageing and wet-drying cyclic tests. The first series was aged under dry conditions, and the second series were aged under immersed conditions in distilled water for 180 days. Measurements of compressive strength were plotted as a function of time.

The last series underwent cycles of drying and wetting. In other words, the specimens were immersed- for approximately one day and then dried in an oven at a temperature of 40 °C for one day. This cyclic test was repeated 50 times. These specimens were tested on compressive strength immediately after a certain number of cycling.

2.8. Adsorption properties of geopolymers and their starting materials

2.8.1. Sample preparation

About 100 g sample of each geopolymer sample M-JS, and M-JZ were crushed using Retsch RM 100-crusher. The crushed samples as well as the filler (zeolitic tuff) and the clay materials (Jordanian kaolinite) were sieved (Fema-SAIZGITTER) at different meshes (<250, 250 –500, and >500μm). The portion with a particle size of 250–500 μm was washed seven times with distilledwater (100mL each) to ensure complete removal of excess NaOH (checked by measuring pH). The washed sample was dried in the oven overnight at 105 ∘C and kept in a desiccator with anhydrous $CaCl_2$.

2.8.2. Adsorption of Cu^{2+} cations

Ten standard solutions were prepared. dissolving (10–100 ppm) of copper acetate (98%, Hopkin and Williams) in 0.1M NaClO4 solution (98%, Eurostar Scientific LTD) at two pH values, i.e. 4 and 6, were prepared (pH adjustment using NaOH/HClO4). In order to draw the calibration curve, the absorbances of several standard solutions (10, 30, 50, 70, and 100 ppm) were measured using atomic absorption spectrometer (Varian, AA-250 plus). For isotherm studies, a 100-mL polyethylene bottle was filled with 50mLofCu2+ solution of varying concentrations (10–100 ppm, ten standards) and 0.05 g of adsorbent (JK, ZT, M-RS, and M-JZ). A 10.0mL portions from each reagent bottle were withdrawn after 24 h of shaking, filtered by microfilters (0.45 μm Nylon), and centrifuged. The metal concentrations were determined using the atomic absorption spectrometer, for further information see [9].

3. RESULTS AND DISCUSSION

3.1. Effect of temperature on geopolymers

The effect of various temperatures on the compressive strength of the M-JS series was studied and evaluated. The heat exposure procedure, section (2.6), was performed on the specimens to detect the effect of various temperatures on a range of properties.

Compressive strength was performed on specimens with surface area = $(12.5\times12.5\times\pi)$ mm^2 and height = 50 mm, while the speed of the machine head during testing was 2 mm/ minute. Figure (1) shows the compressive strength of M-JS at RT after heating at various temperatures up to 1000°C. It is observed that the compressive strength increases with heating up to 400°C. Afterward, the compressive strength decreases with increasing the temperature up to 800°C. The compressive strength of the immersed specimens shows the same trend of dry specimens by heating up to 1000°C. Compared to the reference specimens, the specimens tested exhibit high mechanical performance after heating at elevated temperatures, up to 600 °C.

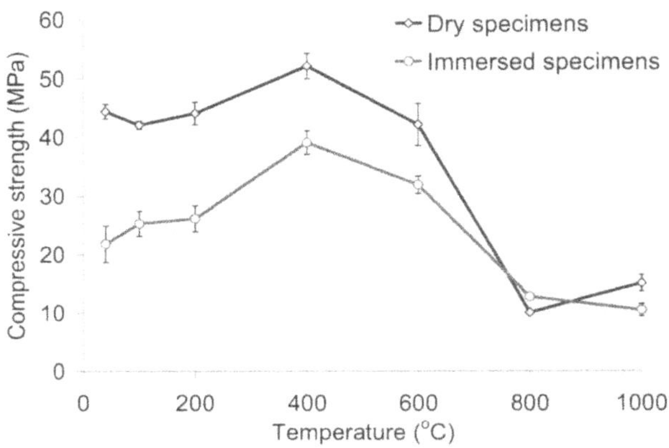

Figure 1. Compressive strength of geopolymers (M-JS) at RT after heating the specimens at different temperatures.

3.2. Ageing, and wet-drying cyclic

Two groups of geopolymers series, M-JS, were prepared and the effects of ageing under dry and immersed conditions on their strength were evaluated. Each series consists of three specimens. The different series were aged under dry and immersed conditions for: 7, 30, 60, 90 and 180 days as shown in Figure 2. The immersed specimens exhibit the same trend of compressive strength for those dry specimens, but with overall reduction varies between 50% and 60%.

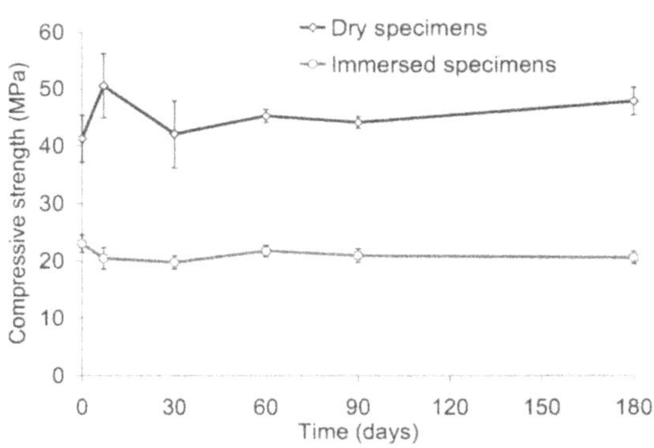

Figure 3. Compressive strength of geopolymers (M-JS) with ageing for 180 days at ambient conditions and after immersion.

It was observed that the geopolymers, M-JS, conserve their compressive strength under dry and immersed conditions for 180 days compared with non-aged specimens as shown in Figure (2). The compressive strength of dry specimens varies between 40MPa and 50MPa, and the compressive strength of immersed specimens is around 20MPa during this period of ageing. The wet-drying cyclic specimens show little changes in compressive strength after 50 cycles, Figure (3). Their strength is slightly higher than the strength of immersed specimens as shown in Figure (2). In general, the strength of cyclic specimens is stable with wet-drying cycles. These results confirm that these geopolymers exhibit long term stability and good durability under different environmental conditions.

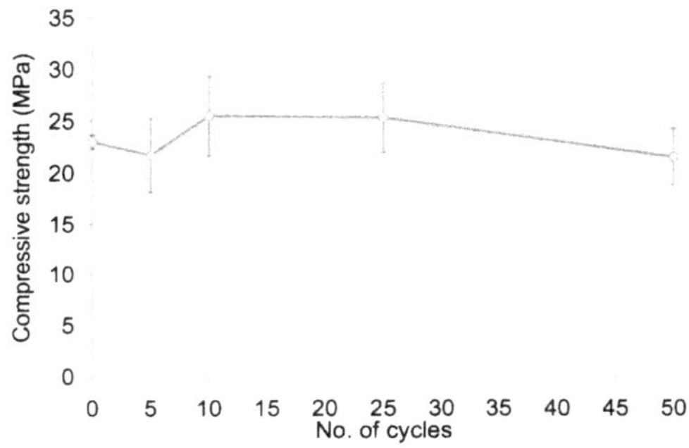

Figure 3: Compressive strength of wet geopolymers (M-JS) after wet-drying cycles.

3.3. Adsorption capacity towards micropolutants

The adsorption capacity of geopolymers (M-JS, and M-JZ) towards micropollutants such as cadmium, copper and lead was investigated in previous studies [8-10]. The summary of the adsorption towards Cu^{2+} as a case study will be will present. In order to compare the adsorption properties of the geopolymers with the minerals from which they are made, the adsorption capacity of zeolitic tuff (ZT), Jordanian kaolinite (JK), and silica sand (SS) towards Cu^{2+} were also investigated at pH=4 and 6. Geopolymers ,M-JS, with sand filler exhibit very low adsorption capacity toward Cu^{2+} as shown in Figure (4) . Geopolymerization, M-JZ geopolymers, using Jordanian zeolitic tuff (ZT) as a functional filler increases sharply the adsorption capacity ,55 mg Cu/g adsorbent at pH=6, of the end geopolymeric materials towards Cu^{2+}. This adsorption capacity is much higher even than the adsorption capacity of zeolitic tuff (ZT), 25 mg Cu/g adsorbent at pH=6. These geopolymers , M-JS, exhibit compressive strength of 20MPa and 9MPa under dry and immersed conditions respectively. Therefore these types of geopolymeric materials, M-JZ, be used for both: construction and treatment of water at the same time.

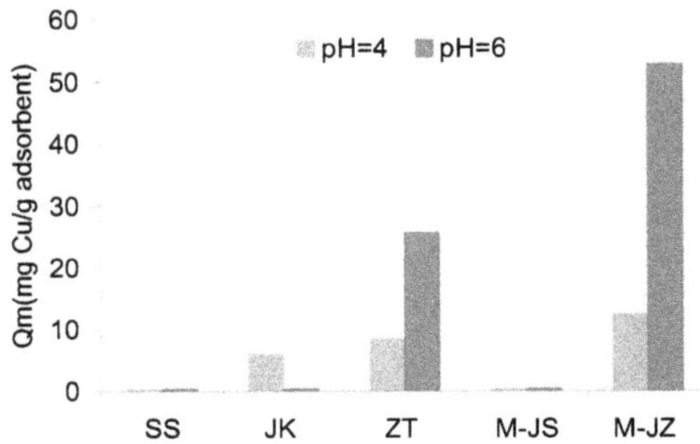

Figure 4. Adsorption capacity of precursors (SS, ZT&JK) and geopolymers (M-JZ & M-JS) towards Cu^{2+}, for further information [9]

4. CONCLUSIONS AND RECOMMENDATIONS

Jordanian kaolinitic deposits from El-Hiswa areas satisfy the criteria to be used as precursors of low-priced, stable, durable, reliable and high quality construction materials. The material has maximum compressive strengths of 41 MPa and 23 MPa, tested under dried and immersed conditions, respectively, by using around 6.2% NaOH. Also these geopolymers exhibit high strength with heating up to 600 °C.

Geopolymers could be used for both construction and water treatment if natural Jordanian zeolitic tuff is used instead of silica sand as filler. The produced geopolymeric materials exhibit high mechanical performance and high adsorption capacity of micro pollutants (such as heavy metals) at the same time.

ACKNOWLEDGMENTS

The financial support of the University of Jordan "Functional Materials for Water Purification Unit" and the project "Chemical stabilization of natural geomaterials for construction and industrial applications" funded under the contract ZEIN2006PR333 by the Flemish (Belgium) Vlaamse Interuniversitaire Raad (VLIR) within the "Own Initiatives" program is gratefully acknowledged.

REFERENCES

[1]O.G. Ingles, Mechanism of clay stabilization with inorganic acids and alkalis, J. Soil Res. Aust. 81–95, (1970)

[2]G.A. Patfoort, J. Wastiels, Use of Local Resources in Construction Materials, Vrije Universiteit Brussel, Brussels, Belgium, pp. 12–14 (1989)

[3]M. Rowles, B. O'Connor, Chemical optimisation of the compressive strength of aluminosilicate geopolymers synthesised by sodium silicate activation of metakaolinite, J. Mater. Chem. **13** 1161–1165, (2003)

[4]G.M. Gemerts, R. Mishre, J. Wastiels, Stabilization of Kaolinitic Soils from Surinam for Construction Purposes, Vrije Universiteit Brussel, Brussels, Belgium, (1989)
[5]J. Davidovits, Geopolymers and geopolymeric new material, J. Therm. Anal. **35,** 429–441, (1998)
[6]M. Alshaaer, Stabilization of kaolinitic soil from Jordan for construction purposes, M.Sc. thesis, Vrije Universiteit Brussel, Brussels, Belgium, (2000)
[7]M. Alshaaer, M. Esaifan, H. Khoury, J. Wastiels, H. Cuypers, Chemical stabilization of local materials in Jordan for construction purposes, in: Proceedings of the 8th International Conference on Concrete Technology in Developing Countries, Hammamat, Tunisia, 8–9 November (2007)
[8]B. El-Eswed, R. I. Yousef, M. Alshaaer, F. Khalili, H. Khoury, Alkali solid-state conversion of kaolin and zeolite to effective adsorbents for removal of lead from aqueous solution, Desalination Publications, European Desalination Society, Via Antica Arischia, 1- 67100 L'Aquila, Italy, **8**, 124–130 (2009)
[9]R. I. Yousef, B. El-Eswed , M. Alshaaer, K. Fawwaz, H. Khoury, The influence of using Jordanian zeolitic tuff on the adsorption, physical, and mechanical properties of geopolymers products, J. Hazardous Materials, **165**, Issues 1-3, 379-387, (2009)
[10]M. Alshaaer, B. El-Eswed, R. I. Yousef, F. Khalili, H. Khoury, Low-cost Solid Geopolymeric Material for Water Purification, Environmental Issues and Waste Management Technologies in the Materials and Nuclear Industries XII: Ceramic Transactions, **207**, ISBN: 978-0-470-40848-3, (2009)
[11]H. Khoury, Clays and Clay Minerals in Jordan, The University of Jordan, p.116, (2002)
[12]ASTM D 4318, Standard Test Methods for Liquid Limit, Plastic Limit, and Plasticity Index of Soils, vol. **04.08**, American Society for Testing and Materials, (2003)
[13]H. Justnes, Thaumasite formed by sulfate attack onmortar with limestone filler, Cem. Concr. Compos. **25**, 955–959 (2003)
[14]V. Bonavetti, H. Donza, G. Menéndez, O. Cabrera, E.F. Irassar, Limestone filler cement in loww/c concrete: a rational use of energy, Cem. Concr. Res. **33**, 865–871 (2003)
[15]H. Justnesa, L. Elfgrenb, V. Ronin, Mechanism for performance of energetically modified cement versus corresponding blended cement, Cem. Concr. Res. **35**, 315–323 (2005)
[16]International Silica Industries Co., Jordan, http://www.jiig.com/.
[17]M.K. Nawasreh, S.M. Yasin, N.A. Zurquiah, Zeolitic Tuff: Geological Survey Administration, Natural Resources Authority, Jordan (2006)
[18]H. Khoury, K. Ibrahim, A. Ghrir, T. Ed-Deen, Zeolites and Zeolitic Tuff in Jordan, The University of Jordan, p. 124 (2003)
[19]R.I. Yousef, M. Tutanji, G. Derwish, S. Musleh, Chemical and structural properties of Jordanian zeolitic tuffs and their admixtures with urea and thiourea: potential scavengers for phenolics in aqueous medium, J. Colloid Interface Sci. **216,** 348–359 (1999)

GREEN TECHNOLOGY FOR EXTRACTION OF IRON FROM ORES AND OTHER MATERIALS

Allen W. Apblett and Kevin Barber,
Department of Chemistry, Oklahoma State University,
Stillwater, OK, USA 74078

ABSTRACT

Acetylacetone (2,4-pentanedione) forms strong complexes with metal ions that are reasonably volatile and easy to sublime or crystallize. Thus, reaction of hematite ore with refluxing acetylacetone results in extraction of iron as the tris-pentadionate complex. The iron complex can be readily separated and converted to iron oxide. Furthermore, using similar processes, iron oxide can be obtained from clay minerals and waste materials.

INTRODUCTION

Worldwide production of four metals, aluminum, iron, copper, and zinc, leads to the generation of over 2,000 megatons of waste per year. An allied environmental concern is the gradual depletion of mineral resources and increased damage to ecosystems by mining activities as replacement ores are aggressively sought. The optimal use of mineral resources requires the application of green science and technology leading to reduction or elimination of generated waste, recovery of useful commodities from stockpiled waste and non-conventional ores, and generation of valuable co-products where possible. Hydrometallurgy provides a particularly promising approach for the greening of ore extraction processes [1].

The principle ores for iron are hematite (Fe_2O_3) and magnetite (Fe_3O_4). The iron is generally extracted from these ores by carbothermal reduction in a blast furnace at high temperatures. While the iron thus produced is a convenient starting material for production of iron-containing chemicals and ceramics, the synthesis of these materials by this route is unnecessarily energetically costly. Therefore, solvent extraction of iron ores to give useful iron-containing chemicals can be quite attractive, particularly if they are also amenable to obtaining iron from inferior ores or waste materials. Extraction of ores with acetylacetone (2,4-pentanedione) has previously been demonstrated to be very useful for a variety of metals including uranium [2, 3], manganese [3], and aluminum [4] from their ores. Acetylacetone also proved useful for recovery of aluminum [5] and gallium [3] from waste materials, and iron and aluminum from soils [6]. In the case of extraction of iron from ore, it was previously demonstrated that aluminum and iron could be extracted from leucite (a potassium aluminosilicate) using acetylacetone [7]. However, this process involved dissolution of the ore in strong acids followed by extraction of the target metals into trichloroethylene/acetylacetone.

The extraction of iron from ores and other materials by reaction with acetylacetone can be very useful since the resulting iron(III) acetylacetonate, $Fe(acac)_3$, is extremely useful for production or iron containing materials. This includes chemical vapor deposition of iron oxide thin films [8] and the sol-gel synthesis of photocatalytic iron-doped titania powders and electrodes [9]. Thermolysis of $Fe(acac)_3$ provides an facile route to magnetite nanoparticles [10-13], amorphous nanoscopic iron(III) oxide nanoparticles, [14], iron-containing ternary nanoparticles [15], and Fe_3O_4/ZnO core-shell nanoparticles [16]. Iron acetylacetonate is also a highly efficient catalyst and reagents for a wide variety of organic transformations [17, 18].

EXPERIMENTAL

All reagents were commercial products (ACS Reagent grade or higher) and were used without further purification. Water was purified by reverse osmosis and was deionized before use. M. X-ray powder diffraction (XRD) patterns were recorded on a Bruker AXS D-8 Advance

X-ray powder diffractometer using copper K_a radiation. The diffraction patterns were recorded for a 2 range of 17-70° with a step size of 0.02° and a counting time of 18 seconds per step. Crystalline phases were identified using a search/match program and the PDF-2 database of the International Centre for Diffraction Data [19]. Dynamic light scattering (DLS) measurements were performed using aqueous suspensions and a Malvern HPPS 3001 instrument. X-ray fluorescence spectroscopy (XRF) was performed on a Thermo Electron QuanX XRD spectrometer.

RESULTS AND DISCUSSION

Great Western Iron Ore Properties Inc. has mineral claims on approximately 2,000 acres in the Tonto National Forest, in Gila County, about 8 miles north of Young, Arizona. The iron ore deposits are estimated to be in 284,200,000 tons distributed in six separate iron ore deposits. This is arguably the best deposit of iron ore in the Southwest United States. Hematite is the primary iron mineral along with lesser amounts of a variety called specular hematite. A finely ground sample of hematite ore received from Great Western Iron, was analyzed by X-ray fluorescence spectroscopy (Figure 1). It was found to be rich in iron oxide (88.1% Fe_2O_3) with minor amounts of silica (9.8%), potassium (1.0 % as K_2O), and calcium (0.7% as CaO). X-ray powder diffraction identified the main minerals present to be hematite and quartz (Figure 2). These analyses indicate that the ore is of exceptional quality and an excellent source of iron.

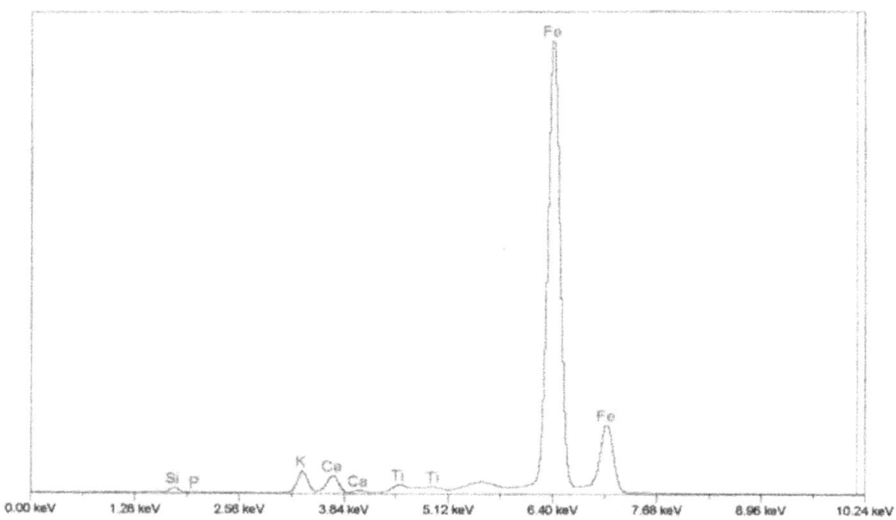

Figure 1. X-ray Fluorescence Spectrum of Great Western Iron Hematite

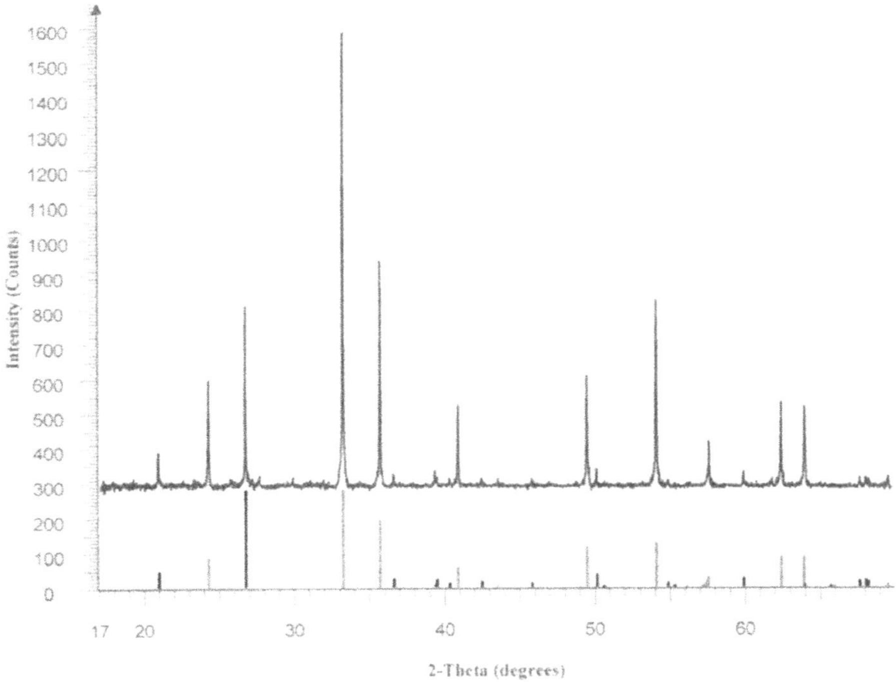

Figure 2. X-ray Diffraction Pattern for Great Western Iron Hematite. Black lines are the JCPDS database XRD pattern for quartz and grey lines are those of hematite.

In this investigation, the conversion of hematite, Fe_2O_3, into Iron(III) acetylacetonate was used as a means for extraction of iron from the ore. Typical processes for preparing iron(III) acetylacetonate have involved the electrochemical oxidation of iron in acetylacetone [20] or metathesis reactions between sodium acetylacetone and $FeCl_3$ [21]. However, Chaudhari *et al.* previously reported that metal acetylacetonates, including the $Fe(acac)_3$, could be produced via a reaction of the corresponding metal hydroxide, hydrated metal oxide, or metal oxide with a stoichiometric amount of acetylacetone. Therefore, the acid/base reaction between iron oxide (Figure 3) and acetylacetone was investigated as the means for extraction of iron.

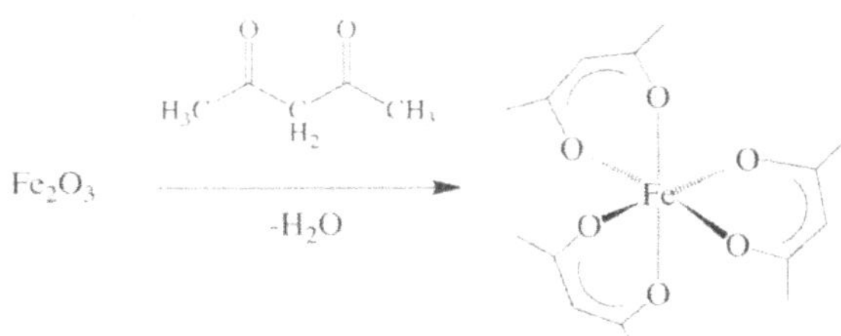

Figure 3. Schematic for Conversion of Hematite into Iron (III) Acetylacetonate.

Hematite ore (10.00 g) was heated at 140°C with 400 ml of a 1:9 V:V water:acetylacetone mixture in a sealed media jar for 48 hours. In previous work, it was found that the presence of a small amount of water catalyzed the extraction of metals by acetylacetone, possibly by forming hydronium ions through an equilibrium reaction. Water has sufficient miscibility with acetylacetone such that a single-phase extractant liquid was obtained. Furthermore, $Fe(acac)_3$ is extremely soluble in this mixture, making it appropriate for extraction of the ore. At the end of two days, the dark red solution was séparated from the insoluble residue by filtration. The solid was washed and dried and its weight was determined to be 1.15 g. X-ray fluorescence spectroscopy (Figure 4) demonstrated that the latter material was mainly silica (69.6%), titanium (0.7% as TiO_2), potassium (17.7% as K_2O), and calcium (10.5% as CaO). Only 1.5 % of Fe_2O_3 remained in the extract giving it a light pink color as opposed to the dark reddish purple of the hematite ore. X-ray powder diffraction (Figure 5) confirmed that the extraction residue was mainly silica with traces of other minerals. An exact match to a known mineral could not be found, but a fairly close match was found for orthoclase, a potassium aluminum silicate mineral with the formula $KAlS_3iO_8$. Orthoclase is known to form admixtures with iron, sodium, rubidium, and calcium ions [22] so it is possible that the iron that failed to be extracted is trapped within this mineral or some other recalcitrant aluminosilicate phase.

Evaporation of the dark red liquid extract in a rotary evaporator under vacuum yielded 39.0 g of iron acetylacetonate, $Fe(acac)_3$ as dark red large hexagonal plate-like crystals. The product's identification as $Fe(acac)_3$ was made by X-ray powder diffraction and was confirmed by infrared spectroscopy. The latter identification was predicated on the spectrum of the crystalline product being identical to that of an authentic sample purchased from a commercial source. The extraction solvents were saved from the rotary evaporation step and were reused in subsequent reactions after replenishment of the acetylacetone that had reacted away. No loss of reactivity was observed in three cycles of use and no degradation products from the acetylacetone could be detected by analysis of the recovered solvent by gas chromatography/ mass spectroscopy.

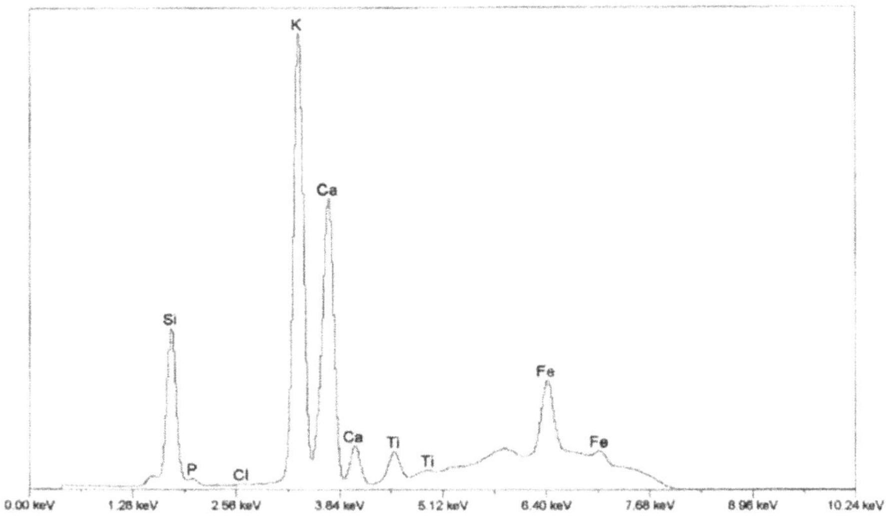

Figure 4. X-ray Fluorescence Spectrum of Extraction Residue

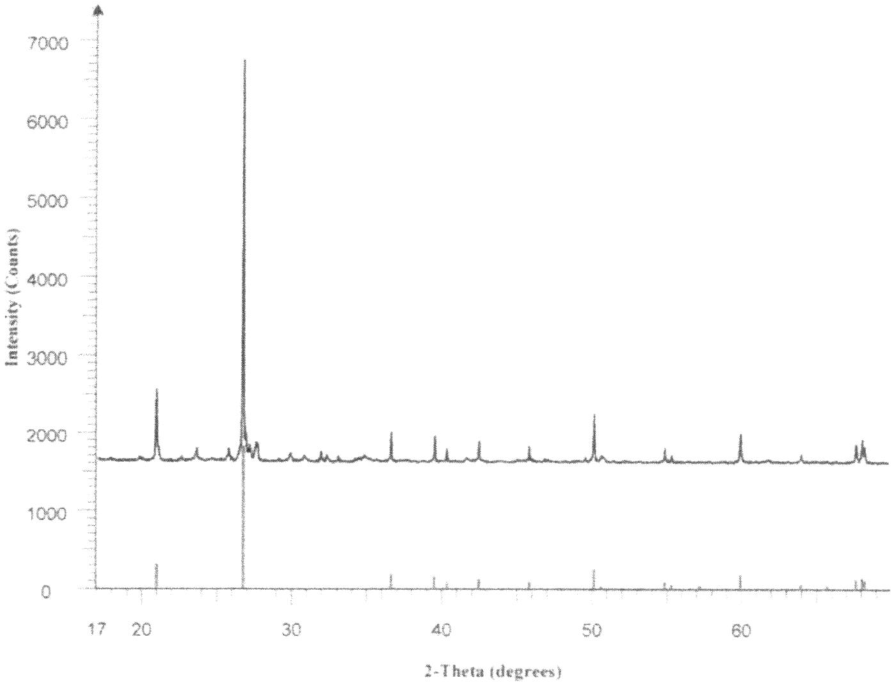

Figure 5. X-ray Diffraction Pattern for Extraction Residue. Grey lines are the JCPDS database XRD pattern for quartz

During the Permian period a vast prehistoric sea that covered what's now northwestern Oklahoma colored the soil a red as it deposited iron-rich clays. This has led to the distinctive red soil colors with hues ranging from reddish brown to cinnamon. These Kirkland soils (Permian Red Beds clayey deposits) have a clay mineral content ranging from 35 to 50% montmorillonite and 20 to 25% illite [23]. Analysis by X-ray fluorescence showed that a sample of dark red soil collected in Payne County, OK contained 22.7% by weight of iron(III) oxide, making it a reasonable source for iron provided that the metal could be easily extracted. Therefore, the ore was treated with a 1:9 mixture of water:acetylacetone (V:V)in the same manner as described for hematite above. After 48 hours at 140°C, the extractant had attained the dark red coloration of iron(III) acetylacetonate. After filtration, it was discovered that the soil was now white but XRF spectroscopy showed that there was still a fair amount of iron oxide (6.9%) that remained unextracted. Nevertheless, a good yield of iron was obtained, indicating that iron-rich clay materials can be a useful source of metals. The white color of the extracted clay could make it suitable for use in manufacturing ceramics.

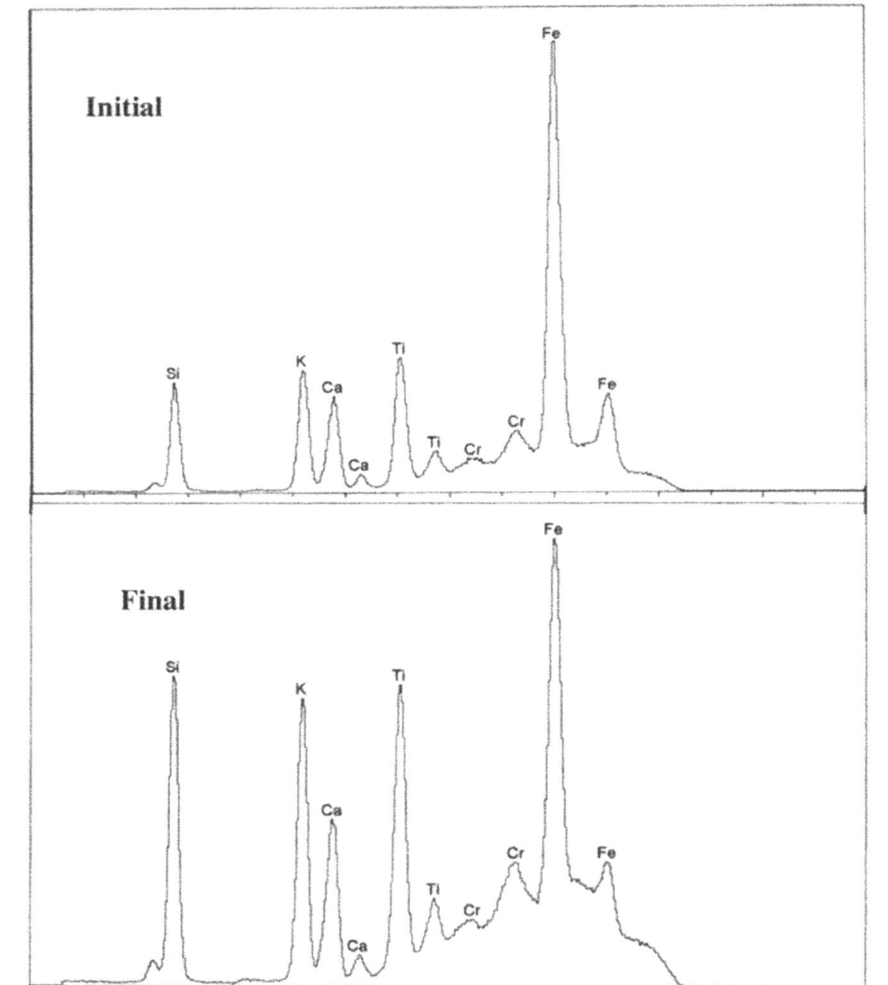

Figure 6. XRF Spectra of Oklahoma Soil Before and After Extraction with Acetylacetone.

Treatment of $Fe(acac)_3$ with ammonium hydroxide produced $Fe(OH)_3$ that, upon heating, was converted to pure hematite. The X-ray fluorescence spectrum is shown below in Figure 7. Alternatively, the $Fe(acac)_3$ can be reduced to iron metal using a variety of reagents such as sodium borohydride, hydrogen, scrap aluminum, etc. For example, treating the dried hematite ore extract sodium borohydride produced a fine iron metal powder.

Since $Fe(acac)_3$ is soluble in organic solvents, it could also be electrolyzed to produce elemental iron. It may be worth investigating an electrolysis procedure for iron production in the future.

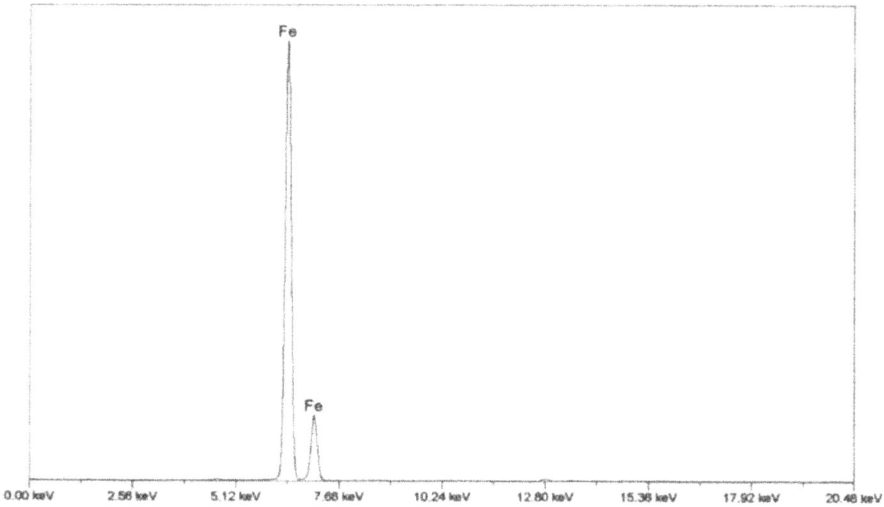

Figure 7. X-ray Fluorescence Spectrum of Purified Hematite

CONCLUSIONS

Extraction of hematite ore and clay minerals with acetylacetone provides an excellent green method for isolation of iron. The extractant is easily recaptured and reused while the $Fe(acac)_3$ that is produced can be used for numerous applications including manufacturing of iron, iron-containing metals and ceramics.

REFERENCES

1. A. R. Burkin, *Chemical Hydrometallurgy Theory and Principles*. Imperial College Press: London, 2001.
2. A. Catino, Uranium extraction from Italian ores. *Ann. Chim. (Rome, Italy)* **1962,** 52, 160-83.
3. H. Lamprey, Properties and applications of metal acetylacetonates. *Ann. N. Y. Acad. Sci.* **1960,** 88, 519-25.
4. H. Ivekovic; A. Janekovic, Aluminum sulfate hydrate and alumina production through bauxite extraction by chelating organic solvents. *Trav. Com. Int. Etude Bauxites, Alumine Alum.* **1981,** 16, 141-3.
5. T. C. Droste Chelation process for recovery of metal values after acid leading of aluminosilicate ores or wastes. US Patent 90-9682 9009682, 19901203, 1991.
6. A. P. Hamblin; A. M. Posner, The Use of Acetyl Acetone as a Selective Extractant of Organically Bonded Metals in Soils. *European Journal of Soil Science* **2006,** 30, (1), 175 - 181.
7. A. E. Paolini; E. Sebastiani; G. Variali, Process for the exploitation of leucite ores. Separation of constituents by solvent extraction. *Chim. Ind. (Milan)* **1980,** 62, (3), 200-7.

8. H. G. Cha; C. W. Kim; Y. H. Kim; M. H. Jung; E. S. Ji; B. K. Das; J. C. Kim; Y. S. Kang, Preparation and characterization of alpha -Fe_2O_3 nanorod-thin film by metal-organic chemical vapor deposition. *Thin Solid Films* **2009,** 517, (5), 1853-1856.
9. J. Marugan; P. Christensen; T. Egerton; H. Purnama, Synthesis, characterization and activity of photocatalytic sol-gel TiO_2 powders and electrodes. *Appl. Catal., B* **2009,** 89, (1-2), 273-283.
10. D. Maity; S.-G. Choo; J. Yi; J. Ding; J. M. Xue, Synthesis of magnetite nanoparticles via a solvent-free thermal decomposition route. *J. Magn. Magn. Mater.* **2009,** 321, (9), 1256-1259.
11. D. Maity; J. Ding; J.-M. Xue, Synthesis of magnetite nanoparticles by thermal decomposition: time, temperature, surfactant and solvent effects. *Funct. Mater. Lett.* **2008,** 1, (3), 189-193.
12. X. Wu; J. Tang; Y. Zhang; H. Wang, Low temperature synthesis of Fe_3O_4 nanocrystals by hydrothermal decomposition of a metalloorganic molecular precursor. *Mater. Sci. Eng., B* **2009,** 157, (1-3), 81-86.
13. R. Shi; G. Gao; R. Yi; K. Zhou; G. Qiu; X. Liu, Controlled synthesis and characterization of monodisperse Fe_3O_4 nanoparticles. *Chin. J. Chem.* **2009,** 27, (4), 739-744.
14. J. Pinkas; V. Reichlova; R. Zboril; Z. Moravec; P. Bezdicka; J. Matejkova, Sonochemical synthesis of amorphous nanoscopic iron(III) oxide from Fe(acac)3. *Ultrason. Sonochem.* **2008,** 15, (3), 257-264.
15.. I. Bilecka; I. Djerdj; M. Niederberger, One-minute synthesis of crystalline binary and ternary metal oxide nanoparticles. *Chem. Commun. (Cambridge, U. K.)* **2008**, (7), 886-888.
16. J. Wan; H. Li; K. Chen, Synthesis and characterization of Fe_3O_4@ZnO core-shell structured nanoparticles. *Mater. Chem. Phys.* **2009,** 114, (1), 30-32.
17. M. W. Zettler, Tris(acetylacetonato)iron(III). *e-EROS Encycl. Reagents Org. Synth.* **2001**.
18. M. Soria-Sanchez; A. Maroto-Valiente; J. Alvarez-Rodriguez; I. Rodriguez-Ramos; A. Guerrero-Ruiz, Efficient catalytic wet oxidation of phenol using iron acetylacetonate complexes anchored on carbon nanofibres. *Carbon* **2009,** 47, (8), 2095-2102.
19. *Powder Diffraction File (PDF-2)* International Centre for Diffraction Data: Newtown Square, PA.
20. S. R. Long; J. J. Lagowski, The Direct Electrochemical Synthesis of Transition Metal Acetylacetonates. *Synth. React. Inorg., Met.-Org., Nano-Met. Chem.* **2007,** 37, (10), 813-815.
21. L. A. Petrova; A. P. Borisov; V. V. Aleshin; V. D. Makhaev, Solid-phase synthesis of iron(III) acetylacetonate upon mechanical activation. *Zh. Neorg. Khim.* **2000,** 45, (3), 390-394.
22. R. Duda; L. Raj, "Orthoclase". In *Minerals of The World*, Arch Cape Press New York, 1990; p 384.
23. G. E. Wilkinson; F. Gray, A Clay Mineralogical Study of Certain Reddish Prairie Soils of Oklahoma, with an Estimation of the Montmorillonite and Illite Content. *Soil Science Society of America Journal* **1954,** 18, 264-268

NANOTECHNOLOGY FOR URANIUM SEPARATIONS AND IMMOBILIZATION

Kevin Barber, Allen W. Apblett, Sulaiman Al-Fadul, Alan Piquette, and Mohamed Chehbouni,
Department of Chemistry, Oklahoma State University,
Stillwater, OK, USA 74078

ABSTRACT

Two different approaches to the removal of uranium using nanotechnology were developed. First, it was demonstrated that nanoparticles produced by exfoliation of molybdenum hydrogen bronze, HMo_2O_6, react rapidly with aqueous uranyl ions to form the mineral iriginite, $UMo_2O_9 \cdot 3H_2O$. The capacity for uranium adsorption is 121% by weight. Secondly, magnetic extractants for uranium were produced by grafting a silsequioxane polymer with ethylenediamine chelating groups on magnetite nanoparticles. These were very effective for removal of uranium from solution via magnetic filtration and also displayed a remarkable adsorption capacity for uranium of 101% by weight.

INTRODUCTION

Uranium is a common contaminant of ground water and can arise from natural and anthropogenic sources. Uranium occurs naturally in the earth's crust and in surface and ground water. When bedrock consisting mainly of uranium-rich granitoids and granites comes in contact with soft, slightly alkaline bicarbonate waters under oxidizing conditions, uranium will solubilize over a wide pH range. These conditions occur widely throughout the world. For example, in Finland exceptionally high uranium concentrations up to 12,000 ppb are found in wells drilled in bedrock [1]. Concentrations of uranium up to 700 ppb have been found in private wells in Canada [2] while a survey in the United States of drinking water from 978 sites found a mean concentration of 2.55 ppb [3]. However, some sites in the United States have serious contamination with uranium. For example, in the Simpsonville-Greenville area of South Carolina, high amounts of uranium (30 to 9900 ppb) were found in 31 drinking water wells [4]. The contamination with uranium is believed to be the result of veins of pegmatite that occur in the area. Besides entering drinking water from naturally occurring deposits, uranium can also contaminate the water supply as the result of human activity, such as uranium mining, mill tailings, and even agriculture [5, 6]. Phosphate fertilizers often contain uranium at an average concentration of 150 ppm and therefore are an important contributor of uranium to groundwater [7]. The Fry Canyon site in Utah is a good example of the dangers of uranium mine tailings. The uranium concentrations measured in groundwater at this site were as high as 16,300 ppb with a median concentration of 840 ppb before remedial actions were taken [8]. Depleted uranium ammunition used in several military conflicts has also been demonstrated as a source of drinking water contamination [9].

Animal testing and studies of occupationally exposed people have shown that the major health effect of uranium is chemical kidney toxicity, rather than radiation hazards [10]. Both functional and histological damage to the proximal tubulus of the kidney have been demonstrated [11]. Little is known about the effects of long-term environmental uranium exposure in humans but there is an association of uranium exposure with increased urinary glucose, alkaline phosphatase, and ß-microglobulin excretion [12], as well as increased urinary albumin levels [13]. As a result of such studies, the World Health Organization has proposed a guideline value of 2 ppb for uranium in drinking water while the US EPA has specified a limit of 30 ppb.

Current municipal treatment practices are not effective in removing uranium. However, experimentation indicates, that uranium removal can be accomplished by a variety of processes such as modification of pH or chemical treatment (often with alum) or a combination of the two [14]. Several sorbants have been shown to be useful for removal of uranium from water. Activated carbon, iron powder, magnetite, anion exchange resin and cation exchange resin were shown to be capable of adsorbing more than 90% of the uranium and radium from drinking water. However, two common household treatment devices were found not to be totally effective for uranium removal [4].

Besides treatment of well water, there is also a strong need for prevention of the spread of uranium contamination from concentrated sources such as uranium mine tailings. Commonly used aboveground water treatment processes are not cost-effective and do not provide an adequate solution to this problem. However, permeable reactive barriers (Figure 1) have been demonstrated to be financially viable and elegant alternatives to active pump and treat remediation systems. Such barriers composed of metallic iron, ferric oxyhydroxide, and bone char phosphate have been designed and proven effective for uranium [8]. Iron metal performed the best and consistently lowered the input uranium concentration by more than 99.9% after the contaminated groundwater had traveled 1.5 ft into the permeable reactive barrier.

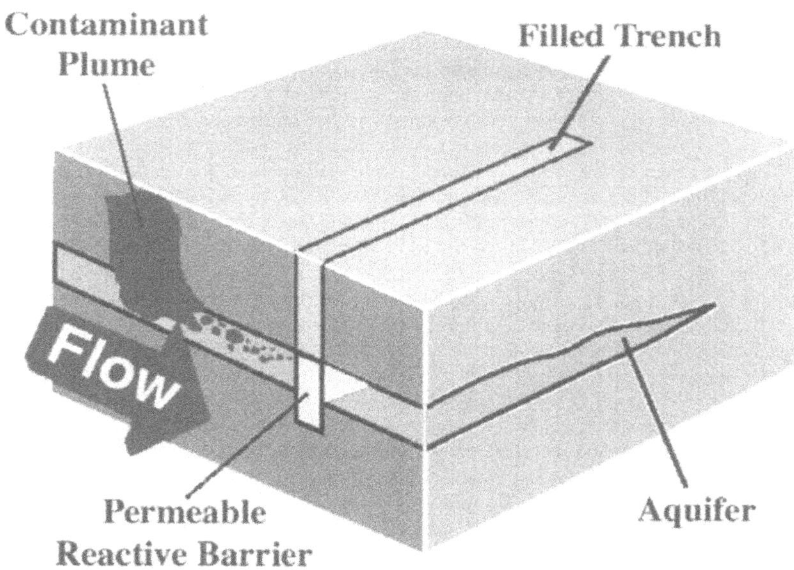

Figure 1. Operation of a Permeable Reactive Barrier

Molybdenum hydrogen bronze (also called molybdenum blue), HMo_2O_6, has been investigated for applications in removal of uranium from aqueous solutions and possible use in a cyclic process for uranium recovery [15]. In the latter respect, it was thought that the protons would provide ion exchange sites on the material in its reduced form while oxidation would remove these sites and release the uranium in concentrated form. However, it was found that oxidation of the blue reagent occurred during the adsorption process so that the reagent turned from blue to yellow. At the same time, it was found that the uptake of uranium was 122 percent by weight and far exceeded the expected capacity based on the number of protons present (the proton concentration is 3.46 mEq/g while the uranium adsorption was 5.14 mEq/g). Analysis of the product from uranium uptake revealed that the bronze had reacted with uranium to produce the mineral iriginite, $UMo_2O_9{\bullet}3H_2O$. This reaction was found to be very selective for uranium, other actinides, and heavy metals. For example, no reaction was observed with aqueous calcium and the uptake of uranium was only slightly hindered by the presence of calcium ions at five times the concentration of uranyl ions. The oxidation of the Mo(V) centers in the bronze was found to be due to reaction with molecular oxygen as the layered structure was disassembled by reaction with uranyl ions.

Nanoparticles show exceptional promise for both *in situ* and *ex situ* environmental remediation, water purification, and pollution-prevention applications. Their small size and ease of suspension in liquids allows them to be readily deployed in slurry reactors for the treatment of contaminated soils, sediments, and solid wastes. Alternatively, they can be anchored onto a solid support such as activated carbon, polymers, or zeolites for enhanced treatment of water, wastewater, or gaseous process streams. Nanotechnology also offers effective alternatives to permeable reactive barriers to prevent spread of pollution in underground aquifers. For such *in situ* remediation, a nanoparticle/water slurry can be injected under pressure or by gravity feed to the contaminant plume where treatment is needed. The small size and propensity to remain suspended for prolonged periods of time allows the nanoparticles to be effectively transported by the flow of groundwater allowing the establishment of an *in situ* treatment zone. Injection of nanoparticles of iron has been demonstrated to effectively remediate chlorinated organic compounds [16] and the technology also holds great promise for immobilizing heavy metals and radionuclides.

The promising results for uranium remediation exhibited by bulk molybdenum hydrogen bronze suggested that nanoparticles of this material would also be very useful for groundwater purification either in the ground or at the tap. It was previously shown that nanoparticles of HMo_2O_6 in butanol could be used to detect and deactivate peroxide-based explosives [17].
Nanoparticles of magnetic materials offer a different approach for removal of uranium from water that is more applicable to above ground treatment of drinking water or wastewater, or for actinide separations. The application of efficient magnetic filtration to actinide decontamination and waste treatment operations is attractive because it can provide very rapid separation of pollutants from aqueous waste streams [18]. This coupled with the ability to switch the filter on and off electronically (avoiding any need for mechanical contact) allows the safety of remote operation. However, since aqueous actinides are not magnetic, magnetic filtering aids must be developed that bind the materials and allow their magnetic separation. These are magnetically-active particles that carry some sort of binding agent for the actinides. In order to maximize the capacity for actinides and prevent permanent magnetization of the particles (which would interfere with electromagnetic manipulation), it is necessary that the particles have length scales

in the nanoscale regime. This results in particles that are superparamagnetic that have high surface areas.

EXPERIMENTAL

All reagents were commercial products (ACS Reagent grade or higher) and were used without further purification. Water was purified by reverse osmosis and was deionized before use. Magnetite particles were prepared via a literature method [19]. X-ray powder diffraction (XRD) patterns were recorded on a Bruker AXS D-8 Advance X-ray powder diffractometer using copper K_α radiation. The diffraction patterns were recorded for a 2θ range of 17-70° with a step size of 0.02° and a counting time of 18 seconds per step. Crystalline phases were identified using a search/match program and the PDF-2 database of the International Centre for Diffraction Data [20]. Colorimetry was performed on a Spectronic 200 digital spectrophotometer using 1 cm cylindrical cuvettes. The uranium concentrations in the treated solutions were analyzed at λ= 415 nm after 5 mL solutions were treated with concentrated nitric acid (1.0 mL) to ensure no speciation of metal ions would interfere with the measurement. For infrared spectroscopic measurements, roughly 10 mg of the sample was mixed with approximately 100 mg FTIR-grade potassium bromide and the blend was finely ground. Spectra in the range of 4000 to 400 cm^{-1} were collected by diffuse reflectance of the ground powder with a Nicolet Magna-IR 750 spectrometer. Typically, 128 scans were recorded and averaged for each sample (4.0 cm^{-1} resolution) and the background was automatically subtracted. Dynamic light scattering (DLS) measurements were performed using aqueous suspensions and a Malvern HPPS 3001 instrument

Preparation of Molybdenum Hydrogen Bronze

Molybdenum trioxide (30.0 g) and 56.8 g of granular zinc metal, 5 mL of water, and a large stirring bar were placed in a 1000 mL filtration flask. A paraffin oil-filled bubbler was attached to the flask's hose connector with a short length of Tygon tubing. Next a dropping funnel containing 200 mL of concentrated hydrochloric acid was placed in the mouth of the flask using a one-hole rubber stopper to make an airtight connection. The flask was then placed in a water bath and the hydrochloric acid was added drop-wise to the magnetically-stirred reaction mixture (caution must be exercised with this reaction since it vigorously evolves hydrogen gas, an explosion hazard). The white solid rapidly turned blue. After the HCl addition was complete and hydrogen evolution had stopped, the reaction mixture was filtered through a 45 µm nylon membrane filter and was thoroughly washed with water. After drying in a vacuum oven at room temperature the yield of dark royal blue solid was 29.50 g (98%). The XRD pattern of the product corresponded to that of HMo_2O_6 in the ICDD database (#14-0041). The infrared spectrum had peaks at: 3450 cm^{-1} (w); 998 cm^{-1} (m); 857 cm^{-1} (m); and 572 cm^{-1} (s).

Reaction of Molybdenum Bronze Nanoparticles with Aqueous Uranyl Nitrate

HMo_2O_6 (0.579 g, 2,00 mmol) was added to 500 mL of a pH 10 solution of sodium hydroxide. The mixture was stirred until the solid had completely suspended (typically 12-24 hr). The resulting, clear dark blue suspension had a final pH of 3.0. A uranium stock solution was prepared by dissolving 1.26 g uranyl nitrate in 50 g of water. A portion of the resulting solution (41.88 g corresponding to 2.00 mmol of uranium) was added to the molybdenum hydrogen bronze suspension and the mixture stirred for 24 hr. The color of the mixture gradually changed during this period of time from dark blue to green and then to yellow. At the completion of the reaction the pH was 2.4. The bright yellow product was isolated by centrifugation and was dried at 60°C to give a yield of 0.987 g. Infrared spectrum (DRIFTS, solid diluted in KBr): 1631 cm^{-1} (vs); 980 cm^{-1} (vs); 899 cm^{-1} (vs); 849 cm^{-1} (s); 732 cm^{-1} (m); and 539 cm^{-1} (s).

Preparation of N-(2-aminoethyl)-3-aminopropyl silsesquioxane coated magnetite nanoparticles

Magnetite nanoparticles (3.0 g) were treated with a solution of N-(2-aminoethyl)-3-aminopropyl]trimethoxysilane (2.0 g) in toluene (30 mL). The reaction mixture was stirred with a mechanical stirrer at 80°C for 24 hours. Afterwards, the material was isolated by filtration, washed extensively with toluene. Drying in an oven at 65 °C yielded 3.63 g of product. IR (KBr, cm^{-1}): 3474(w); 3297(w); 2223(w); 2083(w); 1790(m, br); 1669(m); 1587(m); 1449(m); 1357(w); 1201(m); 1117(s); 1027(s); 776(m); 579(m); and 445(s).

Testing of Magnetic Extractant for Uranium Removal from Water

Different concentrations of stock solutions of UO_2^{2+} (0.040, 0.050 and 0.060 M) were prepared and treated with the magnetic extractant (0.10 g). The mixtures were stirred for two hours and then filtered via magnetic filtration employing a glass pipette filled with medium steel wool attached to a 170 lb pull extended reach electromagnet. UV visible spectroscopy was used to measure the equilibrium concentration of UO_2^{2+} in the permeate solution after acidifying with nitric acid. The Langmuir isotherm equation was applied to calculate the adsorption capacity of the magnetic extractants.

RESULTS AND DISCUSSION

Aqueous suspensions of nanoparticles of HMo_2O_6 are readily prepared by agitating a mixture of the bulk molybdenum hydrogen bronze in water. The formation of a suspension is accompanied by a decrease in pH suggesting that the suspensions result from deprotonation of the molybdenum hydroxyls followed by exfoliation of the resulting negatively charged sheets. Since the bronze has a layered structure (Figure 2) it is expected to behave in the same manner as clays and thus exfoliate and disperse in water. Notably, atomic force microscopy shows that when a suspension of bronze particles dispersed in water is evaporated on mica, the resulting particles have discrete thicknesses of 1.8, 2.7, and 3.6 nm that likely correspond to 2, 3, and 4 layers (Figure 3). The amount of hydrogen bronze that can be suspended in water can be enhanced by reaction with sodium hydroxide to increase the negative charge on the particles. Therefore, in this investigation, the molybdenum hydrogen bronze was produced using a pH=10 solution and HMo_2O_6 in a ratio that corresponded to deprotonation of 2.5% of the bronze. The pH of the resulting suspension is still acidic (3.0). Particle sizing of the suspension showed an average particle size of 180 nm.

Reaction of the suspension of molybdenum hydrogen bronze with a solution of uranyl nitrate leads to formation of a yellow product over a period of 24 hours. As previously reported [15], the fading of the blue color of the bronze is secondary to the absorption of uranium and appears to be due to oxidation of Mo(V) centers as the lattice of the bronze is reconstructed to form a uranium molybdenum oxide. The X-ray powder diffraction pattern (XRD, Figure 4), of the solid reaction product showed that it was the mineral iriginite $UMo_2O_9{\cdot}3H_2O$. The formation of iriginite from HMo_2O_6 indicates a theoretical absorption capacity for uranium of 121% by weight. This is sufficient to remove all of the uranium in the experiment reported herein. Therefore, the uranium content of the treated solution was below the detection limit (6.7×10^{-3} M) of the colorimetric method used for analysis.

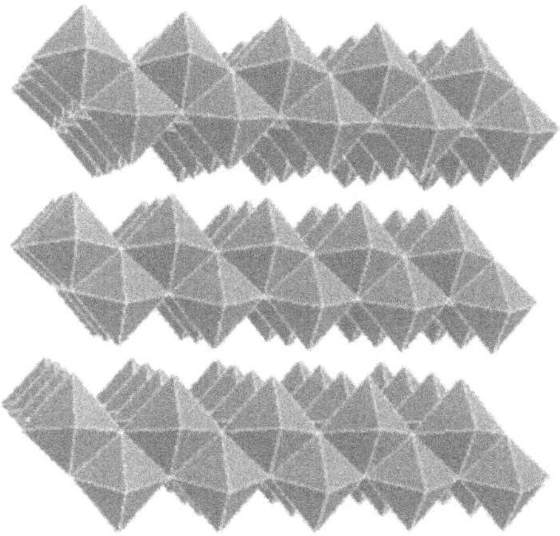

Figure 2. Layered structure of HMo_2O_6.

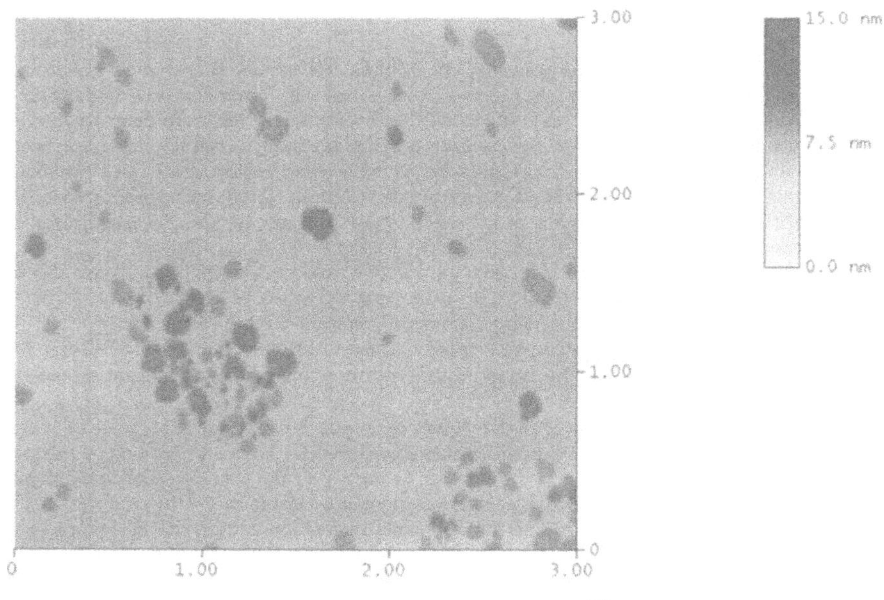

Figure 3. AFM Image of Molybdenum Hydrogen Bronze Nanoparticles

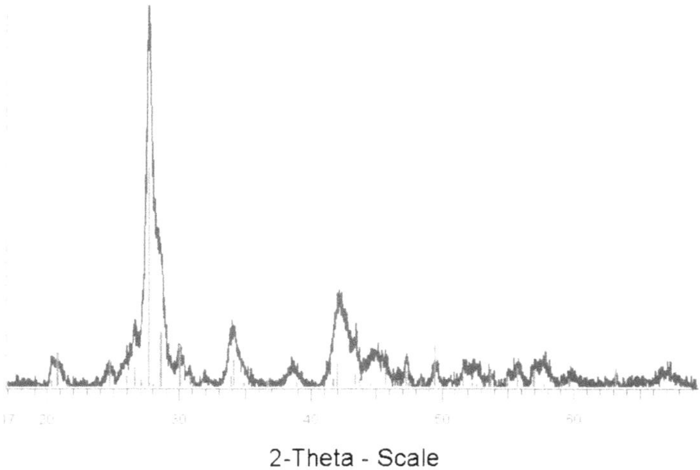

Figure 4. XRD Pattern of Uranium/Molybdenum Hydrogen Bronze Product. Vertical bars are the ICDD database pattern for iriginite , $UMo_2O_9{\cdot}3H_2O$, (ICDD # 29-1372)

It is notable that the infrared spectrum of the product (Figure 5) is very similar to that of bulk HMo_2O_6 with minimal shifting of the peaks. This suggests only minor rearrangement of the molybdenum-oxygen network during adsorption of uranium. There is one new peak at 899 cm^{-1} that can be attributed to the asymmetric stretching vibration of uranyl ions [21].

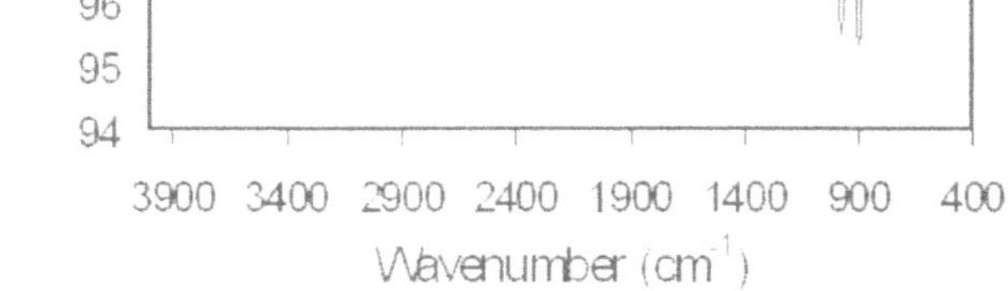

Figure 5. Infrared spectrum of Uranium/Molybdenum Hydrogen Bronze Product

Nanoparticles of magnetic materials offer a different approach for removal of uranium from water that is more applicable to above ground treatment of drinking water or wastewater, or for actinide separations. The application of efficient magnetic filtration to actinide decontamination and waste treatment requires magnetically-active nanoparticles that carry a binding agent for the actinides. Therefore, magnetite nanoparticles with a size of 8-10 nm were prepared and were the coated with a silsesquioxane polymer that contained ethylenediamine chelating groups that can bind uranium and other metals quite strongly (Figure 6).

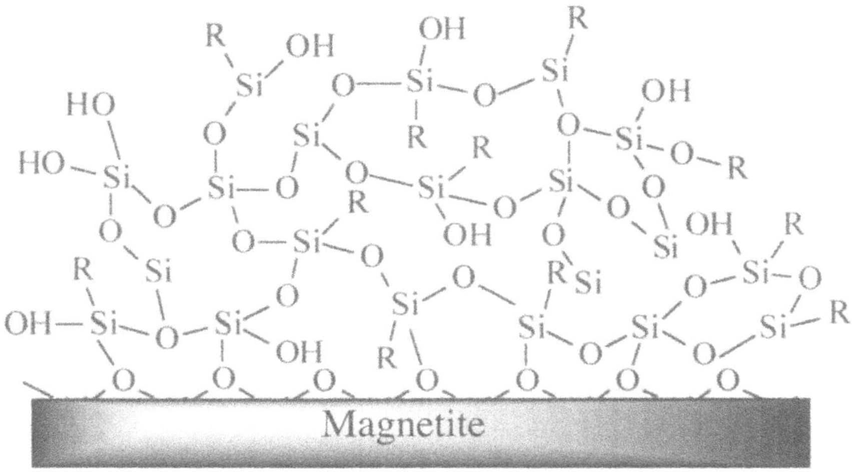

Figure 6. Surface of Magnetic Extractant R= $-CH_2CH_2CH_2NH\ CH_2CH_2NH_2$)

The magnetic extractants were synthesized by reaction of the magnetite nanoparticles with [N-(2-aminoethyl)-3-aminopropyl]trimethoxysilane in toluene at 80°C. In the initial stages of this reaction, hydroxyl groups on the surface of magnetite particles condense with a methoxy group of a silane molecule producing $R(MeO)_2Si\text{-}O\text{-}Fe$ linkages and methanol. Ambient moisture further hydrolyzes the alkoxy groups to produce silanols that subsequently condense to form siloxane (Si-O-Si) bonds. The latter linkages form between both surface-bound silanols (leading to cross-linking) and between the surface and dissolved species in solution leading to the growth of a thick layer of silsesquioxane. The gain in weight of the magnetite particles corresponds to a 22% by weight loading of a silsequioxane polymer with the general formula [(N-(2-aminoethyl)-3-aminopropyl)$SiO_{1.5-x}(OH)_x]_n$. The XRD pattern of the product showed that it still contained nanoparticles of magnetite with a crystallite size of 6.7 nm according to Scherer's equation. The IR spectrum shows strong absorption bands that appear at roughly 1623 cm^{-1} and 1070 cm^{-1} that are due to the NH_2 group vibration and the Si-O stretching vibrations, respectively. The coating of the magnetite nanoparticles leads to some aggregation such that the magnetic extractants have an average size of 72 nm in aqueous suspension, as demonstrated by dynamic light scattering (Figure 7).

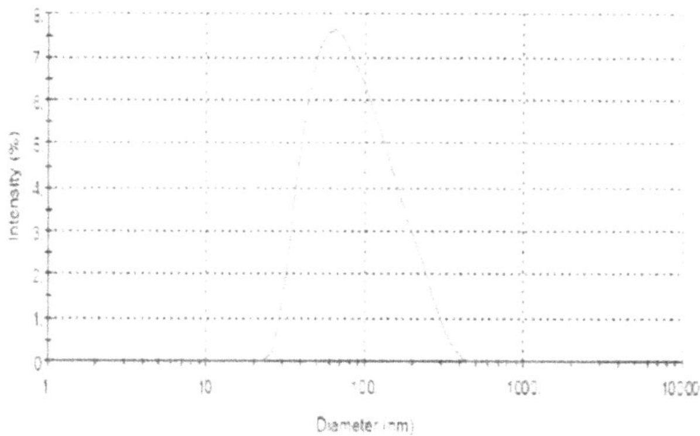

Figure 7. Particle Size Distribution of Magnetic Extractant Suspended in Water

Treatment of 10 mL of a 0.040 M solution of uranyl nitrate with 0.1 g of the magnetic extractant for 24 hr followed by magnetic filtration reduced the concentration of uranium to 0.014 M indicating a rather high capacity for uranium adsorption. Under the same conditions, the uncoated magnetite nanoparticles had no detectable influence on the concentration of uranium. In order to quantify the uranium capacity of the magnetic extractant, two additional uptake experiments were performed with 0.050 and 0.060 M uranyl nitrate starting solutions and the data was analyzed using a Langmuir isotherm (Figure 8). The uptake capacity was found to be 4.3 mmol of uranium per gram of extractant. On a weight basis, the magnetic particles can adsorb a little over 100% of their weight of uranium (1.01 g of uranium/g of magnetic extractant. This high capacity bodes well for the use of this material to rapidly purify contaminated drinking water or wastewater, or for its use in the separation of actinides.

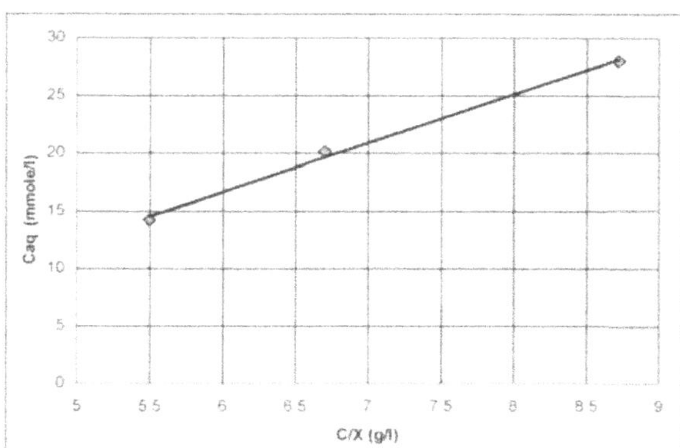

Figure 8. Langmuir Isotherm for Uranium Adsorption

CONCLUSIONS

Nanotechnology research will have a major impact in the area of nuclear and hazardous waste management due to the development of novel materials and processes for remediation of pollution and for separations processes. In this investigation, HMo_2O_6 nanoparticles were shown to have the ability to adsorb 121% of their weight of uranium. Furthermore, they lock the actinide away in a mineral that is a natural repository for uranium. HMo_2O_6 nanoparticles have tremendous potential for *in situ* remediation of uranium-contaminated ground water. They could also be useful for uranium refinement or separation because the uranium is readily recovered (as ammonium uranate) from the molybdate salt by reaction with aqueous ammonia [22]. Even more promising materials for uranium separation can be produced by grafting a silsequioxane polymer with ethylenediamine chelating groups on magnetite nanoparticles. These were very effective for removal of uranium from solution via magnetic filtration with an adsorption capacity for uranium of 101% by weight,

ACKNOWLEDGEMENT

This paper is based in part upon work supported by the National Science Foundation under Grant Number ECCS-0731208 Any opinions, findings, and conclusions or recommendations expressed in this paper are those of the author(s) and do not necessarily reflect the views of the National Science Foundation.

REFERENCES

[1]L. Salonen "238U Series Radionuclides as a Source of Increased Radioactivity in Groundwater Originating from Finnish Bedrock" In: *Future Groundwater Resources at Risk* (International Association of Hydrological Sciences, Wallingford, UK., 1994), 71-84.

[2.]M.A. Moss, R.F McCurdy, K.C. Dooley, M.L. Givner, L.C. Dymond, J.M. Slayter, and , M.M. Courneya "Uranium in drinking water - report on clinical studies in Nova Scotia". In: *Chemical Toxicology and Clinical Chemistry of Metals*. S.S. Brown and J. Savory (eds.): Academic Press, London. pp. 149-152 (1983).

[3]United States Environmental Protection Agency "Occurrence and Exposure Assessment for Uranium in Public Drinking Water Supplies" **EPA 68-03-3514** (1990).

[4]J.D. Navratil, "Advances in Treatment Methods for Uranium Contaminated Soil and Water" *Archive of Oncology*, **9**, 257-60 (2001).

[5]C.R. Cothern and W.L. Lappenbusch, "Occurrence of Uranium in Drinking Water in the U.S." *Health Phys*., **45**, 89 (1983).

[6]D.R. Dreesen, J.M. Williams, M.L. Marple, E.S.Gladney, and D.R. Perrin, "Mobility and Bioavailability of Uranium Mill Tailings Constituents. *Environ. Sci. Technol*., **16**, 702 (1982).

[7]R.F. Spalding and W.M Sackett "Uranium in Runoff from the Gulf of Mexico Distributive Province: Anomalous Concentrations" *Science*, **175**, 629 (1972)

[8]United States Environmental Protection Agency "Field Demonstration Of Permeable Reactive Barriers To Remove Dissolved Uranium From Groundwater, Fry Canyon, Utah" **EPA 402-C-00-001** (2000)

[9]United Nations Environment Programme "Depleted Uranium in Bosnia and Herzegovina Post-Conflict Environmental Assessment Revised Edition" Switzerland, 2003.

[10]D.M. Taylor and S.K. Taylor "Environmental Uranium and Human Health. *Rev Environ Health* **12**, 147-157 (1997).
[11]D.P. Haley. "Morphologic Changes in Uranyl Nitrate-Induced Acute Renal Failure in Saline- and Water-Drinking Rats. *Lab Investig* **46**,196-208 (1982).
[12] M.L. Zamora, B.L. Tracy, J.M. Zielinski, D.P. Meyerhof and M.A. Moss "Chronic ingestion of Uranium in Drinking Water: a Study of Kidney Bioeffects in Humans. *Toxicological Science*, **43**, 68-77 (1998).
[13]Y. Mao, M. Desmeules, D. Schaubel, D. Bérubé, R. Dyck, D. Brûlé, and B.Thomas, Inorganic Components of Drinking Water and Microalbuminuria. *Environ. Res.*, **71**, 135-140 (1995).
[14]S.K. White, and E.A.Bondietti, Removing Uranium by Current Municipal Water Treatment Processes. *J. Am. Water Works Assoc.*, **75**, 374 (1983)
[15]B.P. Kiran, A.W. Apblett, and M. Chehbouni,. "Selective Absorption of Heavy Metals and Radionuclides from Water in a Direct-to-Ceramic Process' Ceramic Transactions, **143**, 385-394 (2003)
[16] D. Elliott and W. Zhang' "Field Assessment of Nanoparticles for Groundwater Treatment". *Environ. Sci. Technol.* 35, (2001) 4922–4926
[17]Allen W. Apblett, B.P. Kiran, Srilatha Malka, and Alan Piquette "Nanotechnology for Neutralization of Terrorist Explosives" *Ceramic Transactions*, 172, (2006).
[18] Y. A. Liu, Ed, Industrial Applications of Magnetic Separation. (Institute of Electrical and Electronics Engineers, New York, 1979).

[19]X. Liu, Z. Ma, J. Xing, and H, Liu, "Preparation and Characterization of Amino-Silane Modified Superparamagnetic Silica Nanospheres" *J. Magn. & Magn.Mater.* 270 (2004) 1-6.
[20]"Powder Diffraction File (PDF-2)" (International Centre for Diffraction Data, Newtown Square, PA).
[21] S. P. McGlynn, J. K. Smith, and W. C. Neely, "Electronic structure, spectra, and magnetic properties of oxycations. III. Ligation effects on the infrared spectrum of the uranyl ion." *J. Chem. Plays.* 35, (1961) 105-116.
[22]Allen W. Apblett, B.P. Kiran and Mohamed Chehbouni "Molybdenum-Oxide Based Sorbants for Toxic Metals" Ceramic Transactions. **176**,(2006) 15-23.

HOW THE CLASSIC MATERIALS SCIENCE STOOL IS BEING CHANGED BY THE SUSTAINABILITY STOOL

Marsha S. Bischel, Ph.D.
Amy A. Costello, LEED-AP
Armstrong World Industries, Inc.
Lancaster, PA, USA

ABSTRACT

A classic way to view materials science is to envision a stool: the three legs are composition, process and structure, while the seat represents the final properties. Sustainability can also be represented by a stool whose legs are economic, social and environmental issues. This presentation will examine materials that have changed because of the requirements that have been imposed by the sustainable building community. While some of these materials perform functions that lower the environmental impacts of buildings, many are not inherently sustainable themselves because the holistic impacts of the raw materials and processes used in their manufacture were not addressed. It is necessary to simultaneously address the three legs of the sustainability stool and the three legs of the materials stool in order to create materials that are truly sustainable.

INTRODUCTION

The broad influence of the construction industry on the world of materials is frequently overlooked and underestimated. However, it has a great impact on many materials-related industries, from metals to photovoltaics. As the building industry embraces "sustainable" practices, the influence on the materials community may become even greater.

The construction sector is vital to the US economy and contributes approximately 13% of the nation's GDP, with buildings alone accounting for nearly half that total.[1] In 2008 the construction and engineering sector of the Fortune 500 was ranked second in revenue growth; in comparison, metals were ranked eighth.[2] Excluding food and fuel, construction materials make up 60% of the total materials flow within the US economy.[3] And despite the recent economic downtown, the US market for "sustainable" building materials is expected to exceed $80 billion by 2013, growing at an annual rate of more than 7%.[4]

World-wide, buildings account for 30-40% of the total energy used,[5] 30% of the carbon dioxide emissions,[5] and use approximately 40% of the total raw materials consumed.[1] In the United States, it is estimated that construction-related waste is approximately 40% of the total non-residential waste generated.[6,7] Consequently, any initiative that reduces the energy footprint of buildings, reduces raw material usage in buildings, or reduces waste during construction is likely to have a large impact on broad sustainability efforts. For example, the Intergovernmental Panel on Climate Change has determined that the building sector has the greatest potential for reducing greenhouse gas emissions.[5]

Due in large part to the large impact that buildings have on the environment and the use of resources, sustainable building practices have grown rapidly, and are now driving much of the broader sustainability movement. Globally, more than 200 sustainable building standards and guidelines have been developed. Although these differ in detail, the focus of these systems is on the development of buildings practices that reduce the negative impacts of the built environment. These standards influence all aspects of the building process, from site location, resource usage in construction and operations, maintenance, energy efficiency and materials content. As a result of these standards, the requirements for various materials used in buildings are changing. Because the individuals involved in creating and promoting these green building standards have traditionally been architects, public officials, and "futurists,"[8] many of the changes associated with materials have been made without the input of materials scientists and engineers. Thus, the materials science world is being rapidly changed by forces outside the materials community.

In response to these changes, the materials science community must embrace new models for designing sustainable materials, technologies and processes. It must also more fully understand the vast array of issues associated with the sustainability movement. Finally, the materials science community must become actively involved in the development of sustainability standards and conduct more research into the holistic impacts of materials and processes.

MATERIALS ISSUES AS THEY RELATE TO SUSTAINABILITY

The sustainable building movement has the potential to profoundly impact the realm of materials science. One way to address this impact is to consider the way in which the classic Materials Science Stool is being changed by sustainability issues. In the classic materials model, a material can be modeled as a three-legged stool: the seat of the stool represents the properties of the material; the three legs represent raw materials, process and structure. Coincidentally, sustainability is also easily represented by a three-legged stool. In this case, the seat of the stool is "sustainability," while the three legs represent social, economic and environmental factors.

In both cases, all three legs of the stool must be balanced if the stool is to be functional, producing an optimized product or process. The interaction and intersection of the two stools are of interest so that the Materials Stool can be further balanced to account for sustainability issues, and materials science can begin to drive the issues of sustainability.

Sustainability Issues in the Materials Stool

Each aspect of the Materials Stool is potentially impacted by sustainability issues. A table of issues is provided in Table I. For example, sustainability issues related to raw materials include the incorporation of recycled or rapidly renewable bio-based materials, and the production or use of toxins. The former is generally considered to be a positive outcome, while the later is not desired. A factor not commonly addressed at the academic level is the source of the raw materials: are the materials coming from an environmentally sensitive area, or from a country with no child labor laws?

Processing considerations related to sustainability include the amounts of energy and potable water needed to produce the material, the amount of waste generated, and the production of any toxins during processing. Social issues such as worker safety and fair labor are also factors that should be addressed.

Material structures may be greatly impacted by the sustainable building movement. While optimization of the material's microstructure is a common practice, for example performing annealing processes to remove defects, other factors listed in Table I may be less familiar. For example, optimization of the macrostructure (which in this case may include any macroscopic scale up to and including the final object) may mean that an assembly is fully manufactured in a factory and then taken to a building site for installation, thus reducing waste; pre-engineered trusses are an excellent example of such a system. It may also mean that a different "shape" may produce the same strength properties while reducing the amount of material used in the system; producers of hollow-core wooden doors and metal based systems have had great success using this technique.[9,10]

Table I: Sustainability Issues in Materials Science

Raw Materials	Process	Structure	Properties
Recycled content Recyclable materials Toxins (used or created) Emissions created during processing of raw materials Source of materials Bio-based	How much energy needed to produce? How much water needed to produce? How much waste generated? Toxic by-products Worker safety Emissions generated	Optimized at micro level? Optimized at macro level? Designed for disassembly? How many components? Macrostructures that use compatible materials?	Durable Strength Generates or stores energy Reduces energy use in buildings Enhances indoor environment Life safety

The concept of "design for disassembly" is gaining traction within the sustainable building movement. Essentially, this encourages products to be easily deconstructed at the end of the initial use phase so that the components can be reused or recycled easily.[7,8] This requires taking a holistic view of the final product or assembly as it is intended to be used, and impacts both the physical design and the materials selection. The best practices for materials selection include:[11]

- Marking parts so as to quickly and clearly identify the material type, similar to the numbering system currently used for many consumer plastic products. This simplifies any recycling.
- Reducing the number of different materials used in the product to simplify material separation during the reclamation phase. This enhances the recycling process.
- Using "compatible" materials that do not need to be separated during recycling. Mixtures of materials tend not be easily recyclable. For example, if non-ferrous metals are used within an otherwise all-steel assembly, the non-ferrous components must be removed prior to recycling to avoid contamination of the steel. Plastic laminated with metal foil is another material that presents challenges for recycling.

Clearly, it may not be possible to achieve all of the above design criteria simultaneously and still have the material meet the required performance; however, it is important to optimize as many factors as possible.

Finally, sustainability influences the final material properties in a number of ways. In general, building materials must be strong and durable, and meet the performance criteria required by the life safety codes. A product made from a rapidly renewable resource is only sustainable if it meets or exceeds the performance characteristics of the material it replacing: a floor that lasts only a few months before needing to be replaced would not be sustainable, for example.

The energy efficiency and self-sufficiency of a building are also directly linked to the materials used in its construction; typically, the energy efficiency of the building is the most heavily weighted area of concern in the sustainable building standards. Many new buildings depend on natural light to provide much of the light needed by the occupants; therefore, proper glazing materials are critical. Reflective interiors are also needed to effectively disperse the natural light within the space.[12] Proper insulation, thermal mass and roofing materials are needed to ensure that energy losses are minimized. Photovoltaics, fuel cells, and other energy harvesting technologies are required to meet the requirements for renewable energy that are found in the sustainable building standards. However, these materials should be as sustainable as possible in terms of their raw materials, structure and process.

Materials Issues in the Sustainability Stool

If one examines the sustainability stool, there are many materials-related issues. Some of these are shown in Table II. The factors which are currently addressed by most traditional design and manufacturing processes are indicated in *italics*.

Table II: Materials-related issues in Sustainability

Economic	**Social**	**Environmental**	**Sustainability**
Cost of process *Cost of raw materials* Cost of end-of-life *Viability of product in market place* Use of regional materials Increases productivity Reduces operational costs	*Worker safety* No toxins used or produced Use of regional materials Local manufacture Increases productivity Positive health benefits Responsible labor	*Reduces use of natural resources* Non-polluting Impact of obtaining of raw materials Impact of transportation End-of-life issues Reduces green house gasses No toxins	Life cycle assessment *Durable* *Lowers carbon footprint* Increases quality of indoor environment *Decreases energy usage*

Among the economic factors which are not commonly addressed is the cost to handle or process materials at the "end-of-life": what happens when a product has reached the end of its intended useful life? Will it be easily recycled into the same material, as is the case with many metal products? Can it be turned into a product with a higher value, as with plastic bottles or bags being used to make carpeting and floor products?[13,14] Or must it be "downcycled,"[8] turned into a product of lesser value, as when concrete is turned into fill or dry wall is used as landfill cover? What are the costs associated with this next phase of the material's life?

Another factor which is frequently not addressed by the materials community is the use of "regional" raw materials. This has broad implications for sustainability, as the use of local materials should reduce costs to the manufacturer (economic), supports the local economy (social), and reduces the environmental impact of the material that is related to transportation of the raw materials (environmental).

An issue that relates materials to both social and environmental factors is whether or not toxins are used in the manufacturing process, or are produced by the material or the manufacturing process. The elimination of these materials was ranked second in importance as an attribute of a sustainable building in a 2007 survey of architects, behind energy management needs.[15] Several agencies, including the US Environmental Protection Agency (EPA), the International Agency for Research on Cancer (IARC), and others list various chemicals of concern that can adversely affect humans. Among these chemicals are benzene, arsenic, cadmium, and lead. If one examines the list of components used in photovoltaics, photoconductors, semiconductors, and batteries, many contain one or more of these chemicals of concern. Some sustainable building standards require third party certification that certain volatile organic chemicals are not emitted in quantities exceeding certain threshold levels as mandated by such authorities as the State of California.[16]

The ability of a material, or the process by which it is made, to reduce greenhouse gas emissions is also sustainability issue. The use of materials and systems that help capture renewable energy is clearly a critical component in reducing the emissions currently associated with heating and cooling buildings; however, the processes used in the production of these materials should also reduce greenhouse gas emissions.

Perhaps the most difficult aspect of sustainable development is fully understanding all the impacts that a material has on the environment. Life cycle assessments, or LCA's, can be used to understand and compare the full range of environmental impacts associated with a material or product. These assessments look at a broad array of factors include raw material production, manufacture, distribution, use, and end-of-life including all transportation steps necessary or caused by the material's existence.[17] The sum of all those steps, or phases, is the life cycle of the material. In addition to understanding the materials environmental impacts, the LCA can be used to optimize the environmental performance of a material or product. When an LCA is conducted, it may be apparent that a material that has optimized a single attribute, such as embodied energy, but may be deficient with regards to other environmental impacts.

A New Model for Sustainable Materials

It is clear from the preceding discussion that the existing "stool" model for optimizing materials can be related to sustainability models. However, if sustainability attributes are to be more fully recognized and adopted by the materials science community, a new model may be needed that fully addresses both the materials aspects and the sustainability aspects. Therefore, a new model is proposed that simultaneously optimizes both areas of concern.

As shown in Figure 1, this model resembles a wheel, with "Sustainable Materials Properties" at its hub. The three legs from each of the two stools form the wheel: social issues, raw materials, economic issues, process, environmental issues and structure must all be addressed and be balanced if the wheel is to function properly, creating a sustainable material with six well-defined spokes. Such a model directly links the sustainability issues to the materials issues, and allows them to be addressed proactively during the initial phases of design, rather than as a secondary issue. Proactive, holistic sustainable design of materials and their manufacturing processes is necessary for the materials community to move from being reactive to the needs of the sustainability movement to a state where they are a driver of the movement.

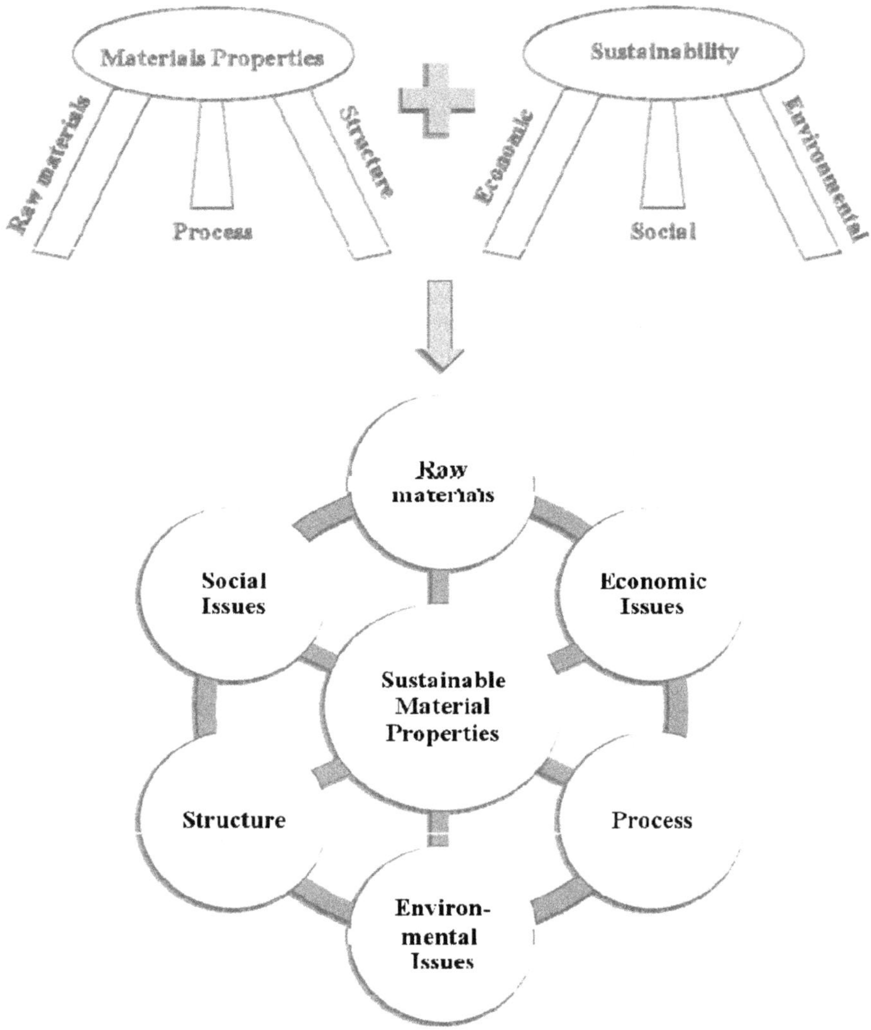

Figure 1: Sustainable Materials Model as represented by a wheel with six spokes

EXAMPLES OF SUSTAINABLE BUILDING MATERIALS

Many examples of sustainable building materials exist in today's building market. Some are traditional materials that have a long history of using recycled or reclaimed materials, such as steel and aluminum. Other materials have been re-engineered to be more environmentally friendly, such as urea-formaldehyde free plywood and low-VOC coatings. Some material categories have changed their raw material streams to include more recycled content; examples include carpeting made from recycled plastic bottles or old carpeting, and concrete with fly ash. Still other materials are needed to meet today's energy efficiency goals: glazing materials, reflective roofing, vegetative roofing systems, photovoltaics, fuel cells, and high efficiency light sources. Materials that help optimize thermal mass, such as concrete, are also being used.

As new materials are developed, the model for sustainable materials should be followed, to ensure that all the critical aspects related to both materials issues and sustainability issues are addressed.

EXAMPLES OF MATERIALS THAT HAVE BEEN CHANGED BY THE SUSTAINABLE BUILDING MOVEMENT

There are numerous examples of materials that have changed to meet the demands of the sustainable building movement. A brief overview of two of these will serve to demonstrate the impact that sustainability issues have had on the materials community, and how outside forces have redefined specific materials.

Elimination of Toxic Chemicals from Plywood

Because plywood and other engineered wood panels frequently use scrap wood material as a primary ingredient, giving them a high recycle content, these would likely be considered to be environmentally-friendly materials. However, until recently, urea formaldehyde was the binder of choice for these materials when used for interior applications. Urea formaldehyde contains relatively high levels of free formaldehyde, which can later escape from the material into the surrounding environment. Formaldehyde is a known carcinogen.

Concerns over formaldehyde emissions from building materials began in the US in the 1970's when formaldehyde-emitting insulation came into wide-spread use. By 1985, the US Department of Housing and Urban Development had begun to limit formaldehyde emissions from products used in prefabricated and mobile homes.[18] Eventually, states such as California imposed more sweeping limitations. Most sustainable building standards place severe restrictions on formaldehyde emissions; many will not allow materials containing urea formaldehyde to be used in buildings.[16,19]

As a result of these mandates, manufacturers of plywood and similar products were forced to redesign their products. Initially, many switched to phenol formaldehyde resin systems, as these emit much lower levels of free formaldehyde, and were already being used for exterior applications. Others use methyl diisocyanate, a polyurethane binder, and a new class of soy-based binder systems.[20,21]

Plywood, therefore, is an example of how an entire product category was redesigned to meet environmental mandates placed on the industry.

Fly Ash as a Means to Reduce the Environmental Footprint of Concrete

One of the key components of concrete has traditionally been Portland cement. However, the production of cement has been identified as a major contributor to greenhouse gas emissions: some estimates are that global production of Portland cement generates 6-7% of the yearly carbon dioxide released.[22] The carbon dioxide is produced during the combustion of the fuels needed to produce the cement, and when calcium carbonate is converted to calcium oxide and carbon dioxide during calcining.[23] The US EPA estimates that in 2007, cement production resulted in the release of 44,500 tons of CO_2 Equivalent (25% of all industrial emissions).[23] The processing of the cement component has the greatest environmental impact associated with the production and use of concrete.[22] Thus, the reduction of cement in concrete has been targeted as a way to decrease greenhouse emissions.

Significant amounts of the Portland cement used in concrete can be replaced by fly ash, a waste product from coal-fired power plants. The US EPA has identified the increased use of fly ash and other coal combustion products as an environmentally beneficial practice, and has identified the following benefits associated with the use of fly ash in concrete: reduction in the amount of fossil fuels used; reduction in carbon dioxide emissions; reduction in the amount of virgin materials used; and reduction in the amount of fly ash that is land filled.[24] The EPA has set goals of increasing the use of fly ash in cementitious materials by 50% in the 10 year period from 2001 to 2011.[24]

The American Coal Ash Association has released figures indicating that for every ton of fly ash used in concrete to replace a ton of Portland cement, 376 liters of water are saved, 718 kg of CO_2 equivalent is avoided, 310 liters of gasoline are not consumed, and 202.5 liters of oil are not consumed.[25] At the same time, the use of fly ash results in a stronger concrete that is more impervious to water, has a lower heat of hydration, shows less thermal cracking, and can be more resistant to sulfate attack.[22] In many areas, fly ash would also be a locally available material.

As a result of figures such as these, concrete with fly ash would appear to be a highly preferred product from a sustainability perspective. In fact, its use is encouraged in some sustainable building standards.[16,26] However, there are concerns from some about the presence of potentially toxic chemicals in the fly ash, and whether these would have an adverse affect on building occupants. Of particular concern are heavy metals and Radium-226.[22] However, the EPA believes that any heavy metals, including mercury, are encapsulated in the concrete and are unlikely to leach out. New EPA requirements to reduce the amount of mercury in emissions from coal fired plants could increase the amount of mercury that remains with the fly ash.[24]

The EPA is monitoring this effect and some sustainable building standards have placed restrictions on the type of fly ash that can be used in concrete. For example, the Green Guide for Healthcare, a guide for building sustainable hospitals and other healthcare facilities, explicitly requires documentation that any fly ash used cannot have been co-fired with hazardous waste, medical waste, or tire-derived fuels.[19]

In addition, the push by the EPA and others to use increasing amounts of fly ash in concrete may have the unintended consequence of directing resources away from investigating other potential fillers and modifiers. Again, outside forces have played a large role in defining this material.

OTHER SUSTAINABILITY ISSUES FACING MANUFACTURERS

There are other issues related to sustainability that face manufacturers of materials that must be addressed. Several important ones include social issues that may not always have been addressed during the early phases of product development or technology evolution; these issues must be addressed if all sustainability issues are to be balanced. In addition, there are issues related to recycled content that must be understood, due to their potential impact on raw materials selection.

Social Issues Related to the Sourcing of Raw Materials

One of the key issues facing any manufacturer is the source of its raw materials. Traditionally performance, price, quality and availability are the primary drivers in selecting raw materials. However, in a sustainable process, social issues must also be addressed.

Rapidly renewable bio-based resources are highly valued by the sustainable building community, as they can be easily replanted and regenerate in 10 years or less; examples include bamboo, cork, wool, and straw. However, the rapid expansion of the use of wood and bamboo has led to concerns over the impact of the harvesting of these materials on local eco-systems. For example, most bamboo is grown and processed in China, and there are concerns about forestry practices, the toxicity of binders, worker safety and impact on endangered species. Non-sustainable wood harvesting practices, such as clear cutting, are also a concern. As a result, various certification programs for wood harvesting have evolved, such as the Forest Stewardship Council (FSC), which accredits forests managed to meet the social, economic, ecological, cultural, and spiritual needs of present and future generations. These certifications are required by some of the sustainable building systems. Yet, some certified products, such as bamboo, are still not widely available in the US, creating a potential imbalance between demand and supply.

Another issue that does not seem to be commonly addressed as new technologies are developed is the reliance on raw materials that come from potentially unreliable sources. The National Research Council of the Academy of Sciences (NRC) has released two reports in recent years that have highlighted the lack of stable sources for certain critical raw materials.[27,28]

Rare earths are a class of material that has grown enormously in recent years, fueled in part by their use in the newest generation of magnets used in motors, generators, lighting devices, switches, and batteries.[28-30] Global demand for these materials is expected to grow by at least 10% annually.[29] Although the US was the leading producer of rare earths in the 1970's and 1980's,[27-29] it now has no active rare earth mines; the market is dominated by China, which provides 97% of the world supply. As internal demand for rare earths increases within China, the amount of these materials exported to other countries is expected to drop by nearly 50% in the seven years from 2002 to 2009.[29] In 2008, the NRC placed rare earths on its list of 11 minerals that are "critical" to the US economy.[27]

As another example, consider cobalt. Fifty percent of the world's supply of cobalt is in the Congo,[31] an area that has been recently ravaged by civil war. Indeed, control of cobalt and other mineral supplies was a key issue in the conflict.[32,33] Despite the uncertainty that comes with having a raw material supply that is linked to a recent war zone, the use of cobalt in batteries and fuel cells[27] has grown greatly in the last decade,[34] and will continue to grow.[32,34,35]

These examples demonstrate the need to more fully understand the entire supply chain as new products and materials are developed: the supply must be sustainable economically, environmentally, and socially.

Issues related to Recycled Content

The use of materials with recycled content is a strong driver in the sustainable building movement. The most commonly used building assessment schemes, such as those from the US Green Building Council and Green Globes, award significant points to the use of materials that have high recycled content. A recent survey found that the use of recycled materials as part of a building project was cited as the third most common sustainability strategy used by architects.[15] However, the use of recycled materials is not as straight forward as it might seem.

First, not all recycled content has positive environmental impacts. This is particularly true if there are concerns around residual toxins that might be added to the end product, or if the recycled material takes more energy to process than would virgin material. As noted previously, some sustainable building standards place restrictions on the types of fly ash waste that can be used in concrete. There are similar restrictions for slag from blast furnaces. There are also concerns about the use of heavy metals in plastics.[36] However, if recycled plastics are used in making new materials,[36] such as carpet, fabric or plastic lumber, it is likely that some amount of these chemicals of concern will be introduced into the final product.[8] It is therefore important to understand the exact requirements of the end-user, the specifier, and the body writing the standards or regulations before adding recycled material to the raw material stream.

The second concern related to recycle content is that not all recycled content is considered equally. For example, the US Green Building Council's Leadership in Energy and Environmental Design (LEED) Green Building Rating System™ is the most widely used sustainable building rating system used in the US, and is used as a model for many other similar systems around the world. The calculation that is used to determine recycle content for LEED certification is not a simple sum of the components. Instead, the LEED system assesses post-consumer and post-industrial waste differently,[26] even if the properties associated with the two sources are equivalent. Post-consumer waste would include materials such as plastic bottles, aluminum cans, construction waste, and glass bottles. Post-industrial sources include waste generated from industrial processes such as fly ash from coal generating power plant, synthetic gypsum, slag from a steel plant, and wood chips from a saw mill; internally generated scrap that can be re-used in an existing process does not qualify. The equation that is required by the LEED system is as follows, and is performed on a mass basis:

$$\textit{LEED Recycle Content} = \textit{100\% Post consumer} + \textit{50\% Post industrial} \tag{1}$$

As a result of this formula, concrete with a fly ash content of 15% is considered to have a recycled component due to the ash of 7.5%, while a wallboard product made with 100% synthetic gypsum would be considered to have a recycled content of only 50%. Thus, some sources of recycled materials may be favored over others by manufacturers seeking to enhance the recycled content calculation for products destined for LEED-rated buildings.

By appearing to favor post-consumer content, such schemes may have the unintended consequence of driving manufacturers towards more problematic raw material streams: post-consumer materials are less likely to be well characterized and more likely to be contaminated than equivalent post-industrial streams. For example, if a carpet manufacturer is using reclaimed carpeting as a raw material, the chemical composition of the fibers is less likely to be known than if fiber cut-offs from another manufacturer are used. The uncertainties surrounding these raw material streams add complexity and costs to the manufacturing process: safeguards against contamination must be implemented, and materials must undergo more extensive testing to ensure quality and composition. In comparison, a post-industrial waste stream is likely to be well characterized chemically, and free from contaminants, and is therefore more easily and cost-effectively used.

WHAT ELSE IS NEEDED FROM THE MATERIALS SCIENCE COMMUNITY

The challenges facing society overall as it addresses the broad and complex issues of sustainability will continue to impact the materials science community. Just in the realm of sustainable buildings, demand for more environmentally sensitive materials is expected to dramatically: it is estimated that the demand for "green" building materials will grow at an annual rate of 7.2% through 2013.[4] What then is needed from the materials community to successfully and proactively address this demand?

First, materials scientists and engineers must become more educated about the broad needs of the sustainability movement. They must embrace more holistic models, such as the one presented, in order to incorporate sustainability into all phases of material development. Attention must be paid to details that traditionally have not been of concern, such as the supply chain for raw materials, and end-of-product-life issues. The need for processes and materials that minimize resources, waste and environmental impacts is one of six goals identified in the 2008 "Federal Research and Development Agenda for Net-Zero Energy, High Performance Green Buildings" (National Science and Technology Council).[3] The areas of focus for achieving this goal, all of which directly impact the materials science community, are: developing processes that minimize the waste generated during the phases of building construction and demolition; developing more life cycle assessments to fully understand the environmental and health impacts of materials; and developing new building materials that have a minimal impact on the environment and on public health through the entire life cycle.[3]

In some cases, there may be pressure to re-invent traditional materials manufacturing processes using sustainable, non-toxic reactions, relying on the twelve principles of so-called "green chemistry."[37] Such processes are designed to reduce the hazards associated with chemical processes, thus reducing the potential risks and impacts on the environment, workers and end-users of products.

A more complete understanding of the total impact a product has on the environment is also critical as life cycle assessments become more prevalent. These assessments address the impact of a material from its raw material extraction, through the manufacturing process, the use phase, and its re-use or disposal at end-of-life. The USGBC's Research Committee released "A National Green Building Research Agenda" in 2007; this blueprint for sustainability research cites a need for "robust and reliable" data for use in LCA's.[6] As the technical experts who best understand a particular material, its processing needs, and its performance characteristics, materials scientists should play an important role in the generation of this data and the development of the LCA's.

A particular need for LCA's that has been identified by the National Institute of Standards and Technology (NIST) is for reliable data on the service life of materials.[38] The service life of a material is a critical factor in developing a meaningful life cycle assessment: a material that has a service life of 10 years has a very different impact than one that has a service life of 50 years. It is NIST's belief that the science needed to generate quantitative, accurate predictions of service life is in its infancy.[38] Since tests of material lifespan are commonplace in the realm of materials science (fatigue, creep, etc.), it seems logical that this is a field that should be ripe for materials science researchers.

Materials scientists must also become engaged in the processes that are being used to develop the codes, standards and guidelines that address sustainable materials, buildings and development around the world. Without the input of the proper experts, these standards may contain requirements that are not technically sound or achievable. In addition, if the materials community does not participate, there will be continued opportunities for outside forces to change material requirements, and it will be forced to be reactive rather than proactive.

CONCLUSION

The realm of materials science is being profoundly impacted by the sustainability movement; this movement is in turn being largely driven by the desire for sustainable buildings. The result is that the traditional models used to develop and optimize materials are no longer sufficient, as they frequently do not adequately address the three pillars of sustainability: economics, social concerns, and environmental impacts. A new model that simultaneously addresses the sustainability concerns along with the traditional materials concerns of raw material, structure and process is needed; one such model has been presented here.

As the sustainability movement continues to grow, the materials community must more fully understand all the requirements that are being placed on materials and their manufacturing processes. The community must become actively engaged in the processes that are being used to define sustainable materials, and they must help conduct the research that is needed to provide sound technical data on environmental impacts.

Finally, the materials community needs to change the current dynamic, and move from being reactive to the demands of the sustainability movement, and become instead a driver of the movement. Only in this way will society as a whole have sustainable materials that are technologically sound and environmentally benign.

REFERENCES

[1]US Green Building Council, Green Building Facts (April 2009).
[2]Fortune 500 Annual Ranking, *Fortune Magazine online* (July 2009), http://money.cnn.com/magazines/fortune/fortune500/2009/performers/industries/fastgrowers/index.html
[3]National Science and Technology Council Committee on Technology, *Federal Research and Development Agenda for Net-Zero Energy, High-Performance Green Buildings*, October 2008.
[4]US Demand for Green Materials to Exceed $80 Billion in 2013, *AWCI's Construction Dimensions* 10 (April 2009).
[5]Cheng, C., Pouffary, S., Svenningsen, N., Callaway, M., *The Kyoto Protocol, The Clean Development Mechanism and the Building and Construction Sector – A Report for the UNEP Sustainable Buildings and Construction Initiative*, United Nations Environment Programme, Paris, France (2008).
[6]US Green Building Council Research Committee, *A National Green Building Research Agenda*, (November 2007).
[7]US Environmental Protection Agency, Lifecycle Building Challenge (June 2009).
[8]W. McDonough and M. Braungart, *Cradle to Cradle*, New York: North Point Press (2002).
[9]Lynden Door, *Architectural and Commercial Interior Doors* (2008).
[10]Dietrich Industries, http://www.dietrichindustries.com/dietrich_leed.asp

[11]Environmental Defense Fund, End-of-Life Vehicle Management: Design-for-Recyclability, 1999, http://www.edf.org/documents/905_GC_recyclability.htm
[12]Yoon, Y-J and M. Moeck, "Impact of Ceiling and Wall Reflectance on LEED Credits," *Journal of Architectural Engineering*, **11**, 81-90 (2005).
[13]Interface, Closing the Loop: ReEntry® 2.0, http://www.interfaceglobal.com/Sustainability/Sustainability-in-Action/Closing-the-Loop.aspx
[14]Armstrong World Industries (Australia) Commercial Flooring, http://www.armstrong-aust.com.au/commflrpac/content/files/49495.pdf
[15]Green Buildings Research White Paper, *Building Design and Construction* (October 2007).
[16]California Collaborative for High Performance Schools, *CHPS Best Practices Manual 2009*, 2009.
[17]ISO 14044 Environmental Management, Life Cycle Assessment - Requirements and Guidelines
[18]US Environmental Protection Agency, An Introduction to Indoor Air Quality – Formaldehyde (April 2008), http://www.epa.gov/iaq/formalde.html
[19]GGHC, *The Green Guide for Health Care Version 2.2* (2003).
[20]Environmental News Network, Going Beyond Formaldehyde Binders in Manufactured Wood Products (October 2, 2007), http://www.enn.com/top_stories/article/23563
[21]US Environmental Protection Agency Green Chemistry, Greener Synthetic Pathways Award 2007: Development and Commercial Application of Environmentally Friendly Adhesives for Wood Composites, http://www.epa.gov/oppt/greenchemistry/pubs/pgcc/technology.html
[22]Build It Green, *Fly Ash Concrete* (August 17, 2005), http://www.builditgreen.org/files/uploads/Resources/Build%20It%20Green%20fact%20sheets/Fly-Ash-Concrete.pdf
[23]US Environmental Protection Agency, Inventory of US Greenhouse Gas Emissions and Sinks: 1990 – 2007 (April 15, 2009).
[24]US Environmental Protection Agency, Waste and Materials-Flow Benchmark Sector Report: Beneficial Use of Secondary Materials – Coal Combustion Products (February 12, 2008).
[25]American Coal Ash Association Educational Foundation, *Coal Combustion Products* (April 17, 2009).
[26]US Green Building Council, *Leadership in Energy and Environmental Design (LEED) Green Building Rating System™ Version 3* (2009).
[27]National Research Council of the National Academies, *Managing Materials for a Twenty-first Century Military*, Washington, D. C.: The National Academies Press (2008).
[28]National Research Council of the National Academies, *Minerals, Critical Minerals and the US Economy*, Washington, D. C.: The National Academies Press (2008).
[29]K. Garber, America's New Energy Dependency: China's Metals, *US News and World Report online* (July 1, 2009), http://www.usnews.com/articles/news/national/2009/07/01/americas-new-energy-dependency-chinas-metals.html?PageNr=4
[30]A. Grant, The Secret Sauce of Hi-Tech: Obscure Metals, *Discover Magazine online* (August 17, 2009), http://discovermagazine.com/2009/jul-aug/17-secret-sauce-of-hi-tech-obscure-metals
[31]US Geological Survey, 2008 Mineral Commodity Annual Survey (January 2008), http://minerals.usgs.gov/minerals/pubs/commodity/cobalt/mcs-2008-cobal.pdf
[32]P. Salopek, The River Congo: Torrents of Civil War Pound Ravaged Congo, *Chicago Tribune* (December 10, 2000).
[33]A. Zajtman, DR Congo's Cobalt Mountain, *BBC News Online* (September 20, 2000).

[34]J. Louis, Cobalt Outlook Looks Blue, *Resource Investor online* (February 4, 2008), http://www.resourceinvestor.com/News/2008/2/Pages/Cobalt-Outlook-Looks-Blue.aspx
[35]S. Sutton, Raw & Manufactured Materials: 2009 Overview, *Ceramic Industry online* (January 1, 2009), http://www.ceramicindustry.com/CDA/Articles/Feature_Article/BNP_GUID_9-5-2006_A_10000000000000491077
[36]Nordisk Miljömärkning, *Nordic (Swan) Ecolabelling of Furniture and Fitments, v 3.5* (2003).
[37]US Environmental Protection Agency, Green Chemistry (2009), http://www.epa.gov/greenchemistry/
[38]National Institute of Standards and Technology Building and Fire Research Laboratory, *Measurement Science for Sustainable Infrastructure Materials*, http://www.bfrl.nist.gov/goals_programs/SIM_goal.htm

IMPACT OF MATERIALS SELECTION ON THE SUSTAINABILITY OF WIND ENERGY

Davion M. Hill, Narasi Sridhar
Det Norske Veritas, Research and Innovation USA
5777 Frantz Rd. Dublin, OH 43017 USA

ABSTRACT

In 2008, the U.S. Department of Energy released a national feasibility study on using wind power to produce 20% of the U.S. electricity supply in 2030. To reach this goal, a projection of over 290GW of new wind turbines will be needed, requiring an average linear manufacturing rate of over eight thousand 1.6MW turbines per year until 2030. Wind turbine materials include composite blade structures, steel support towers, other metallic components in the nacelle, and copper and semiconductor components. All of these components require an energy and carbon investment (i.e. CO_2 production during manufacturing). The economic and environmental benefits of wind power are well researched, yet the effect of materials performance on sustainability of wind power requires more study. This paper will address the environmental impact of materials in wind turbines, including the ROI from a carbon perspective, or the "ROCI" (Return on Carbon Investment) arising from the energy and raw materials inputs. The effect of increasing the lifetime of materials on ROCI will be examined.

INTRODUCTION

Wind energy is presently the most cost competitive large scale renewable energy source, but the manufacture of wind turbines requires energy and materials input. In addition, their reliability is a concern for the industry. The reliability of the wind turbine rests on the performance of materials in its constituent components. To evaluate the materials factor in wind turbine reliability, probabilistic risk models based on existing wind turbine reliability data can be constructed. The models can then be applied to the projected wind turbine production in the United States over the next two decades and the sustainability of wind energy can be estimated.

Sustainability of new technology is the optimized balance of *social*, *environmental,* and *economic* benefits whose relationship is often represented by a Venn diagram of these three overlapping circles. In the business context, it is best summarized by the triple bottom line, also known as "people, planet, profit" as stated by Elkington.[1, 2] While most wind turbine feasibility models are based on economics, a Return on Carbon Investment (ROCI) approach evaluates sustainable environmental energy production, and this parameter is used to illustrate the importance of materials reliability in the growing renewable energy sector.

The ability of a low carbon energy source to produce energy depends on the materials used to build it, and CO_2 is emitted in manufacturing these materials. The CO_2 emitted is related not only to the energy consumed in producing the materials, but also the direct production of CO_2 in extracting metals from ores or building polymers from hydrocarbon feedstock. Therefore, an amount of CO_2 is invested in the use of these materials that must be recovered through the prolonged generation of CO_2-reduced energy. If those materials are not fulfilling their design life, then they must be replaced, resulting in additional CO_2 production. If a system requires maintenance, repair, and replacement, this is also an energy - and therefore carbon - investment. Every material failure is lost money and spent carbon.

This paper examines two dimensions of the sustainability of wind energy from a materials performance perspective through a simple case study. Materials issues for wind turbines are introduced. Then, the first dimension of sustainability is analyzed, i.e. the economics of wind energy based on materials performance. The second dimension of sustainability is then addressed; the environmental impact of wind energy through a carbon life cycle assessment (LCA) of wind turbine blades using life cycle inventory (LCI) databases.

MATERIALS ISSUES IN WIND ENERGY

As indicated by the U.S. Department of Energy (DOE), demands on manufacturing materials and operation of wind turbines is expected to increase, and turbine materials consumption will continue to be dominated by steel.[7] Generators and gearboxes may evolve to use large permanent magnet motors which will place demand on naturally magnetic materials. Future gearbox designs will include variable speed or continuously variable generators which will require high strength metal alloys. Opportunities exist for introducing aluminum and new composite materials, provided that cost, strength, and fatigue requirements are met. Glass-reinforced fiber is the dominant material for blades, though there are some developments in stronger carbon fiber blades, provided that cost requirements can be met.

Martinez, et al. have determined that the energy payback for a 2 MW wind turbine is less than one year, and that the energy invested into the turbine materials is paid back many times over; as much as 34 times in their assessment.[19] If wind energy is expected to be a substantial portion of U.S. electricity generation in the coming decades, the DOE feasibility study proposes that over 300 GW of capacity will need to be installed between the present date and 2030. The estimated annual US electricity demand in 2030 by the Energy Information Agency (EIA) is between 5.2 and 5.8 billion MWh. The wind turbine installation growth rate is projected to start at rates of 4,000 MW/y (2600 turbines of 1.5 MW size), and that rate will gradually increase between 2012 and 2020 to a capacity of 16,000 MW per year (over 10,600 1.5 MW turbines), and then reduce slightly to eventually meet 20% electricity demand by 2030. Scaled-up wind turbines of higher power are expected to reduce the number of turbines required to meet these targets, but the materials invested in those turbines will obey the Square-Cube Law, i.e., a proportional increase in size requires a new materials volume that is proportional to the cube of the scale factor, and the new materials surface area is proportional to the square of the scale factor. A project of this magnitude, if executed, will place significant demand on manufacturing materials.

Wind turbines are large, with blade lengths exceeding 35 meters. Present day megawatt scale wind turbines rest on steel towers that are 80 meters tall, and the overall turbine height is twice that of existing high voltage electric transmission towers. The 3 MW turbines of the near future will approach two and a half times that height, with a swept diameter near 90 meters. Offshore turbines are expected to be even larger, approaching 5 MW and perhaps even 10 MW in one to two decades.[7] Common turbine designs of today have nameplate ratings of 1.5 MW, but 2.3 MW and 3.0 MW turbines also exist.

RELIABILITY FACTORS AND THE COST OF ENERGY

Since the wind turbine is intended to be a carbon-free (or carbon-reduced) source of energy, failures and the consequent downtime represent lost potential carbon savings, i.e., for every kWh that is not produced due to failure, there is a quantity of CO_2 savings that is lost. In addition, if materials are involved in the failure (replacement and loss), then there is a CO_2 investment in those replacement materials as well.

Reliability of wind turbines may be described by a "bathtub curve" which is the sum of early failures or infant mortality during the break-in period, random failures continuously throughout its operational life, and late failures due to wear out. Warranties are designed to cover the infant mortality period during the first 3-5 years of operation, and the expected end of useful life is determined by some threshold of wear out failures. Reliability of wind turbines is a concern to the industry, and while public access to long term reliability data is limited for the US, some European wind farms have collected and published incomplete sets of reliability data.[3, 4, 11] It is interesting to note the following conclusions from these reliability studies:

1. Gearbox, generator, and drivetrain failure rates are relatively low but the consequence of failure in the form of downtime and cost is high. On a normalized scale, of all recorded failures the drivetrain failure frequency is less than 0.1 (<10% of occurrences), but more than 5 days of downtime are associated with drivetrain failure, on average.
2. Electrical system failures are common and probable with minor downtime consequence per event. On a normalized scale, of all recorded failures the failure frequency of electrical systems is near 0.6 (60% of occurrences), but the down time is less than 2 days per event.
3. Failure related to blades occurs with medium frequency with median downtime and replacement costs. On a normalized scale, blades and blade subsystems fail with a frequency near 0.2 (about 20% of recorded failures in the study) with an associated downtime of 4 days.

From these studies, gearbox failures are high risk because of their cost and consequence, even though the failures do not necessarily occur often. Electronics failures occur often but with low consequence, so they are of low to moderate risk. Operators appear to view electronics failures as more of a nuisance than a major issue. Blade failures, however, occur more often than gearboxes and represent downtime that may be something more than just a nuisance. The qualitative picture of these risks is presented in a risk matrix as shown in Figure 1.

The cost of energy (COE) calculation for a wind turbine is adapted from conventional utilities, except that the operations and maintenance costs for wind turbines do not include a fuel cost. A generally accepted interpretation of the COE calculation is given below. The COE is usually in units of cents per kilowatt-hour or dollars per kilowatt-hour. For the purposes of this paper, it can be assumed that the COE is in $/kWh. The COE is then given by **Equation 1**.[6, 5]

In the COE calculation in **Equation 1**, AEP_{max} and CF are sometimes combined into a single Annual Energy Production (AEP) term. The capacity factor is indicative of the wind turbine availability. As stated by Walford[5], capacity is affected by equipment reliability as much as it is affected by reliability of the wind source. The O&M contribution consists of both scheduled and unscheduled maintenance costs, and LRC is spread over the life of the turbine to cover major overhauls or component replacements. It is assumed that wind turbine components will last the life of the turbine itself, but it is indicated by some reliability studies that this is not always the case and the LRC factor is an acknowledgement of this fact. Therefore, Equation 1 illustrates how reliability directly affects the cost of operation of a wind turbine.

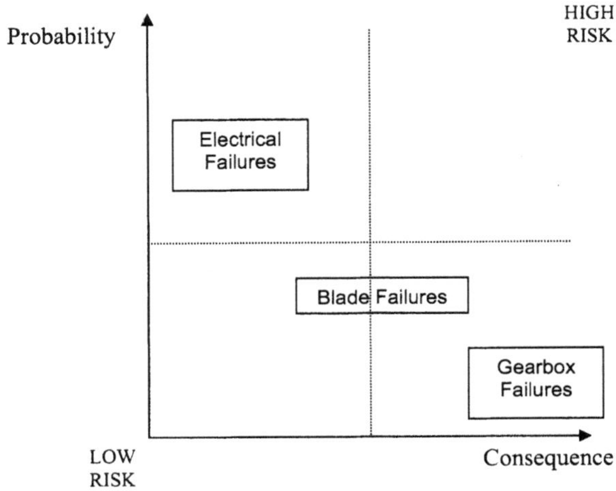

Figure 1 A qualitative risk assessment of wind turbine failures illustrates the effect of consequence on an event, where the red area is high risk, the green area is low risk, and the yellow areas are moderate risk..

$$COE = \frac{ICC * FCR + LRC}{AEP_{max} * CF} + O\&M \qquad \textbf{Equation 1}$$

COE	Cost of Energy (\$/kWh)
ICC	Initial Capital Cost (\$)
FCR	Fixed Charge Rate (%/y)
LRC	Levelized Replacement Cost (\$/y)
O&M	Operations and Maintenance Costs (\$/kWh)
AEP_{max}	Maximum Annual Energy production (kWh/year)
CF	Capacity Factor (%)

BLADE FAILURES

Reliability studies of wind turbine blades have indicated that blades are experiencing cracking before meeting their 20 year design life. Though not always catastrophic, blade failure outside of the warranty period is an unexpected cost that increases O&M costs for the wind farm. Wind energy is expected to be cost-competitive because it should have lower O&M costs due to the absence of fuel purchase, however, industry data shows that the O&M costs steadily increase with time for wind farms, due in part to unexpected component malfunction or repair and failures.[7] The expected O&M costs of a wind farm are targeted near \$0.005/kWh when new, but it can double within the first decade of operation, and quadruple by the end of the service life. An additional \$0.02/kWh of O&M costs can represent as much as a 20-30% increase in the total cost of electricity (COE) from the wind farm over its life for an assumed competitive COE of 6-7¢ per kWh.

Loads on turbine blades are simultaneously varying in the tensile, compressive, torsional, and flexural modes as the mass of the blade experiences rotational, gravitational, and lateral wind forces. Other complications such as weather, environment, and even the wake generated by other turbines can add to the complexity of the wind blade loading.[8] Impulse loading from turbulent or sudden wind gusts are difficult to predict and model, and it is expected that the blade should endure these unpredictable loads while also accomplishing its design function. When turbine blades have failed in the past, cracking has been observed at the root of the blade. The cracking event is signaled by the breaking and delaminating of fibers.[9, 9] According to long term data from Tavner, et al, an average 5.8% rate of blade-related failures in operating wind farms across Europe has been observed, and as much as 7% of the failures in Germany are blade-related. These failures are direct blade cracking failures that are different from the total blade subsystem failures cited earlier.[4,11] If a blade failure is to occur in the US, it can be assumed that replacement material cost of the blade alone can range from $50,000 to over $100,000.[12] Based on average electricity costs, downtime incurs lost revenue ranging on average from $0.06/kWh to as much as $0.15/kWh in some states, depending on the time of day or year. The average numbers obscure the opportunity cost of lost on-peak revenue that depends entirely on real-time wholesale electricity rates, where short-lived price peaks of as much as $1.00/kWh are possible. In the event of a blade failure, experienced field personnel may be needed to guide the blade replacement, and high hourly rates for crane rentals incur heavy expenses that may be borne by the operator. Wind turbines are expected to have a capacity factor between 25 and 40% which is adjusted to account for weather patterns, expected failures, and maintenance, but the capacity factor does not always account for all unexpected failures such as blades.[12]

Based on present day megawatt turbine design features, each turbine will be equipped with 3 blades constructed of composite fiber-reinforced materials. Most blade manufacturers use an E- or S-type fiberglass reinforced with a polyester resin selected for its relative strength and durability. The production of ethylene or propylene monomers are produced from hydrocarbon feedstock by high temperature cracking reactions and are subsequently treated with a catalyst to create synthetic plastics and resins such as polyester and epoxy. Natural gas is one possible feedstock and petroleum is another. The ethylene and propylene monomers are cracked in a high temperature process with a catalyst to create synthetic plastics and resins.[16]

The methodology for assessing the impact of blade failures and replacements over the course of the projected 20 year growth of the U.S. wind industry on the total life cycle is shown conceptually in Figure 2. The life cycle CO_2 inventory - the amount of CO_2 quantified and assigned - is calculated for a turbine blade. This CO_2 value is then projected onto many turbine blades. As blades fail and are replaced, the downtime and CO_2 impact from the materials replacement are subtracted from the CO_2 savings of the renewable and CO_2-free wind power.

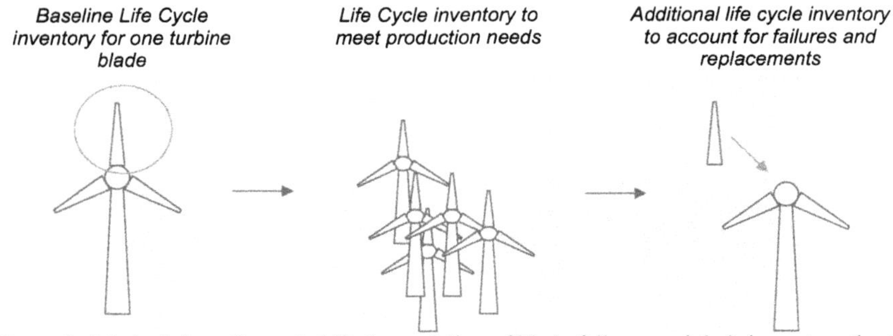

Figure 2 Methodology for probabilistic modeling of blade failures and their impact on the total life cycle of the growing U.S. wind industry.

It is estimated that over 1000 kg of resin are used to manufacture a typical blade for a MW-scale turbine.[13] However, this number may be larger. In 2007, it has been stated that over 43,000 blades were produced.[14] Assuming each turbine uses 3 blades, these numbers indicate that perhaps 270-540 million kg of resin will be required over the next 23 years to fulfill the manufacturing demands for turbines. However, these numbers can be quantified somewhat more succinctly by using life cycle databases such as those from NREL or the Swiss Ecoinvent database. There are many components to the total life cycle of resin, but one NREL citation indicates that 1 kg of methylene diphenyl diisocyanate resin has associated life cycle costs of 0.41 kg of CO_2 emitted to air from its manufacture, in addition to an input of 5 L of diesel, 1.23 kWh of electricity, 0.91 m^3 of combusted natural gas and 0.51 m^3 of processed natural gas, and a required input of 362 L of water.[15]

PROBABILISTIC MODEL OF WIND FLEET PERFORMANCE WITH BLADE FAILURES

It is difficult to place appropriate bounds on the life cycle carbon cost of these resins because there are associated transport costs, VOC emissions, and other costs that occur over the entire life cycle. To keep it simple, an estimate can be made and the uncertainty in that estimate can be addressed in a probabilistic model. The total carbon footprint of this single kg of resin-including the additional life cycle contributions – is likely greater than what is estimated here. Instead of quantifying these costs with incrementally greater deterministic precision with likely diminishing returns on the effort, the probabilistic model is designed to acknowledge uncertainty in all factors simultaneously. After the model is constructed and executed, it is only a small extra step to perform a sensitivity analysis as a way to "double check" these estimates and quantify possible sensitivity of the calculation to various assumed parameter ranges.

The model uses some input data from various sources to determine the energy production scenarios. First, the known installed capacity of turbines from 2007-2009 is written into the model. In 2007, 11,600 MW of turbines were installed. By 2008, 5,200 MW had been added to increase that capacity to 16,800 MW. By 2009, total capacity has increased to 25,300 MW. For the remaining model calculations, the projection was linearized to meet 20% of electricity demand in 2030 per the DOE report. The EIA estimates the present annual US electricity demand to be 3.99 billion MWh. Projections indicate electricity demand to be 5.28 to 5.8 billion MWh annually in 2030.[7] Deterministically, this represents an average needed linear growth of

15,305 MW per year. Assuming that all turbines are rated at 1.6 MW (nameplate capacity) and accounting for a capacity factor correction – as the nameplate rating is always higher than the actual rating. The capacity factor is the ratio of time that the turbine is actually producing power out of all possible operating times. From the projections to 2030, this represents approximately 13,696 turbines installed per year until 2030. It is of course possible that the turbines will increase in capacity from 1.6 MW to perhaps 2.0 or 3.0 MW, but again the model can account for this uncertainty if needed. For the purpose of this demonstrative calculation, 1.6 MW is assumed.

With the turbine growth rates established, known failure data was superimposed on the growth scenario. According to long term data from Guo, et al, of all failures observed during the observation period, 5.8% were blade-related failures in operating wind farms across Europe.[4, 11] The data was based on three different observation periods lasting between 720 and 744 hours, monitoring between 2,036 and 2,164 turbines at any given time. Total observed failures ranges from 130 to 175 during the observation periods, and of those failures, as few as 6 and as many as 15 were blade related. Using this data, extrapolating it to a period of one year, and dividing by the number of turbines, a blade failure rate was determined to be 0.1 per turbine per year. Similarly, downtime due to blade failure per turbine per year averaged 3.7 hours per turbine per year.

Probability distributions were added to the following inputs. The capacity factor (CF) was given a normal distribution around 32% with a standard deviation of 0.035, accounting for small probabilities of turbines having a CF as low as 22% and as high as 42%. Coal CO_2 emissions were determined from the NREL LCI database, and a uniform distribution from 0.994-1.010 ton/MWh was adopted.[17] The downtime per blade failure was estimated to be a normal distribution around 4 days with a standard deviation of 1, assuming a +/- 1 day error on the Hahn data.[3]

The size of the turbines was deterministically assumed to be 1.6 MW, the approximately average turbine size as of 2008. This average was chosen to simplify the calculation. If instead the turbine size were varied, then there would be variable outputs on the fleet nameplate capacity as well as the total number of turbines in the fleet. For the purpose of this calculation, the desire is to determine whether blade failures have a significant impact on the performance of the fleet as a whole.

With the expected energy demands of the US annually in 2030 of 5.28 billion MWh, 20% of that demand is approximately 1.06 billion MWh. Blade failures are pseudo deterministic in this model, but the downtime is probabilistic. The reason why the number of blade failures is not assigned a probability distribution is because there is not enough information from the literature to accurately reflect the distribution, but the framework of this model is adaptive enough to account for this once more data is available. Based on the Hahn data, three separate observation periods produced an average 0.1 blade-related failures per turbine per year. Using a normal probability distribution around 4 days per failure with a standard deviation of 1, the hours of downtime per year were calculated. The blade failures per turbine per year factor was applied to the total US fleet in this scenario. The total fleet population was counted by dividing the total fleet nameplate rating by the nameplate rating per turbine. The structure of the model is shown in Figure 3.

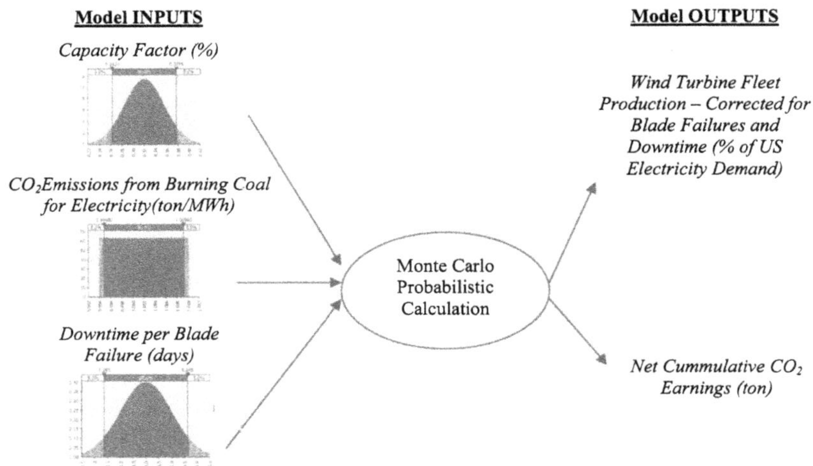

Figure 3 Using @Risk software, the probabilistic inputs for the model were run through a Monte-Carlo type calculation to estimate the range of operational values that could be assumed for the US wind turbine fleet, accounting for blade failures over the next two decades.

RESULTS

Probabilistic models were constructed using @Risk software from Palisade Corporation. The corrected output (accounting for blade failures) is shown in Figure 4. In the figure, it is shown that blade failures allow for some variability in the reliability of wind turbines and their ability to meet the 20% goal, though the variation is not too significant. By the probabilistic model, the 20% goal may be reached as early as 2027 and as late as 2029. The absolute numbers are not as important as the variability – the model essentially shows that there may be as much as a 2 year variation in when the goal could be reached depending on the frequency of blade failures. The DOE 20% report does not address what would be done to maintain and sustain the wind fleet after 2030, so the model illustrates what would happen to the wind fleet if production of all new turbines stopped. As portions of the fleet meet the end of their design life, production would gradually decrease until 2051. This projection is carried through to the cumulative carbon savings over time, assuming that the fleet is not rejuvenated. While it is likely that some action will be taken to rejuvenate the fleet, this model does not yet endeavor to predict how that might happen.

The data shown in Figure 4 is sensitive to two variables: capacity factor and downtime. The sensitivity of the model to these variables can be estimated with a regression chart, which shows the variable and the length of the bar in the chart is a measure of the relative "strength" of that variable on a scale from -1 to +1. By the end of 2030, the first turbines that were installed in 2010 will be reaching the end of their design life, so 2030 is a transition where the number of turbines in service switches from a growth trend to a decreasing trend. The effect on the net percentage of electricity production in the US is then dominated by the decreasing fleet size, and any blade failures or increased downtime due to the capacity factor is negligible compared to the rate at which turbines are reaching the end of their design life. Before that point, however, the capacity factor is a more dominant factor in determining the effectiveness of the wind turbine project rather than downtime due to blade failures, as shown in Figure 5.

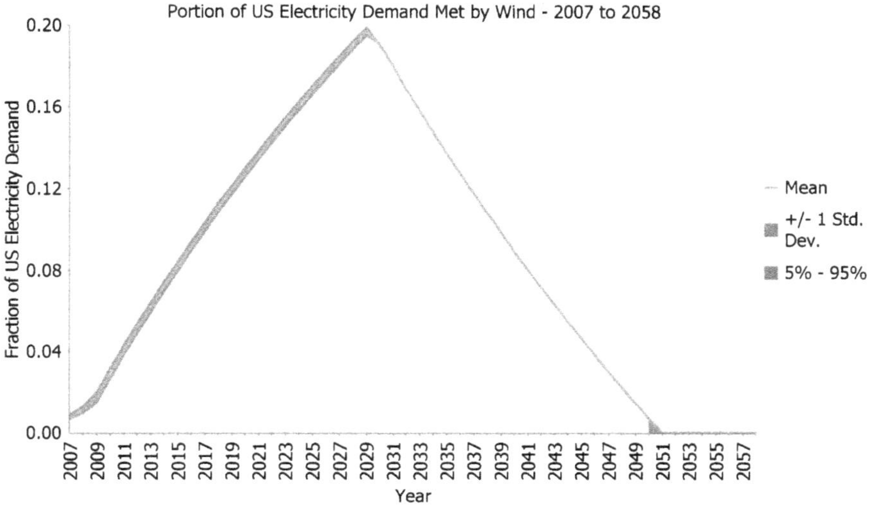

Figure 4 The projection for the US wind fleet electricity production peaks near 2030 and is within 1-2% of reaching the 20% goal when accounting for uncertainty in CF and downtime.

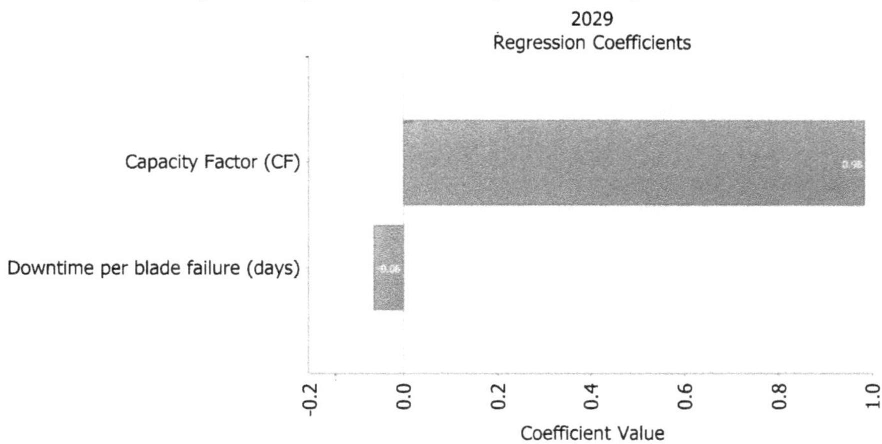

Figure 5 Capacity factor is a more dominant variable than downtime per blade failure in determining the overall performance of the future US wind fleet.

If downtime due to blade failures were more significant, then the strength of that variable in the regression analysis would be greater.

In Figure 6, it is illustrated that as the total wind fleet ages, it will produce more energy and therefore "earn" more CO_2 as it continues to produce carbon-free electricity. However, as downtime becomes more of an issue and units are retired, the life cycle actual CO_2 earnings will vary anywhere between approximately 23 and 24 billion tons. Uncertainty in the output with the age of the fleet increases as blade failures occur and the fleet is reduced because of end of life. Near 2030, the cumulative CO_2 earnings curve undergoes an inflection from an increasing slope to a decreasing slope as the production of wind turbines peaks and the earliest turbines reach the end of their useful life. As the fleet continues to operate and more turbines are decommissioned, and blade failures continue (at an assumed constant rate), then more uncertainty in the carbon payback causes the curve to become more broad near 2050 and beyond.

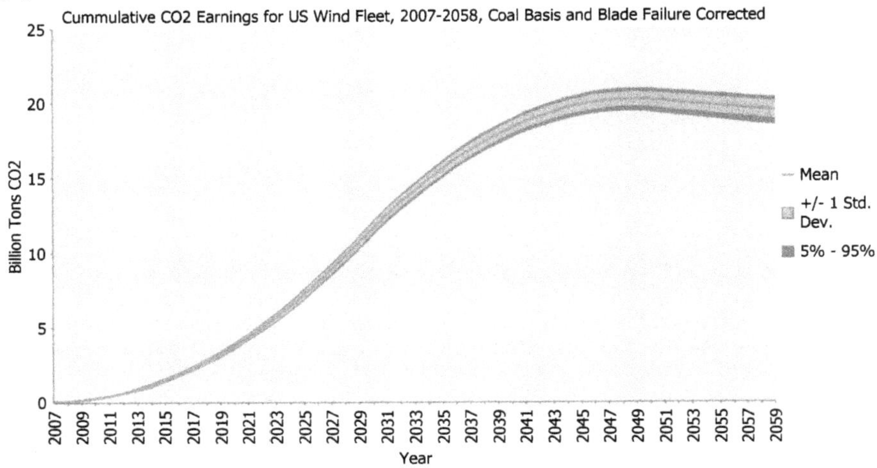

Figure 6 The total CO_2 earnings over the life cycle of the US wind project, which shows greater uncertainty farther in the future in the value of CO_2 "earned" for each MWh of wind energy produced when compared to coal.

What is shown in Figure 6 is that wind turbines earn back their value in CO_2 almost instantaneously; the ROCI is less than 1 year. The return on carbon investment is sensitive to capacity factor and downtime, but is also sensitive to the CO_2 value of a MWh of coal-powered electricity. Capacity factor and CO_2 value dominate the determination of wind energy carbon savings in this model, with blade reliability playing a minor role, especially early in the fleet construction (see Figure 7). As the fleet ages, blade reliability increases in importance slightly (see Figure 8), but the simplest rule wins out: wind turbines are most effective if they are effective at catching the wind.

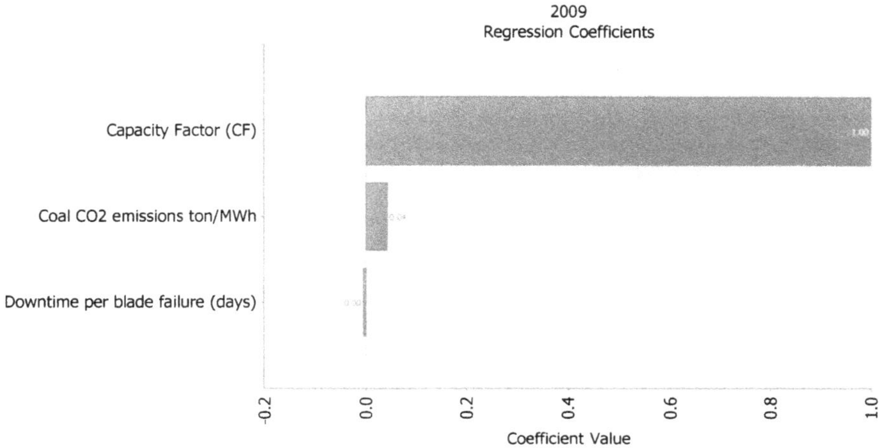

Figure 7 Early in the wind initiative (2009), capacity factor strongly affects wind turbine ROCI, with smaller contributions from the value of CO_2 on a coal basis and downtime.

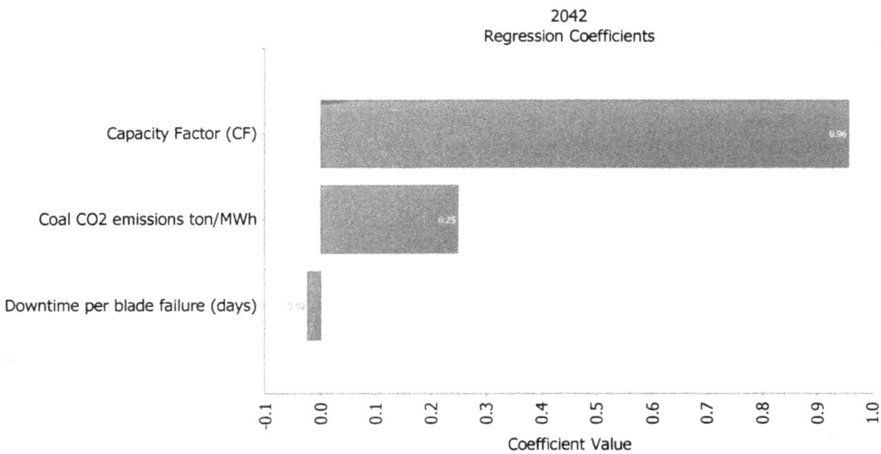

Figure 8 Later in the wind initiative (2042), capacity factor continues to dominate ROCI, but with a diminishing fleet, the value of CO_2 on a coal basis becomes more important, and the reliability of the turbines has slightly increasing importance.

REMARKS AND CONCLUSIONS

This model is limited in a few aspects. First, it only addresses one dimension of the sustainability argument: environment (from a CO_2 perspective). In addition, only blade failures have been considered but there are other failures that can increase downtime, such as gear box failures. Sub-system failures like sensors and supporting electronics do not force considerable downtime per failure incident, but they occur often. Failures that happen more rarely but with greater consequence such as gearbox failures are often more emphasized because of their evident impact on downtime. The statisitics used here are borrowed from one publication for one rare and consolidated reliability study; conditions for other wind farms may vary. Therefore the study presented here is incomplete in that it only addresses one subsystem and assumes global conditions and behavior for the US wind fleet. If all subsystem failures were incorporated into the model, perhaps a more interesting picture of the wind fleet effectiveness could be revealed.

There are some findings from this study, however, that are of note. If the total downtime due to *all* subsystem failures is significant, then model perhaps indicates that the overall downtime for wind turbines should be considered more carefully when assessing the overall performance of a wind project, and that the assumption of a 20 year design life is not always warranted. In addition, the ROCI for wind turbines is evidently immediate and substantial. And finally, this study appears to confirm that capacity factor is the most dominant factor in determining wind turbine performance, i.e., wind turbines must be placed in areas where they can maximize their ability to harvest energy from the wind. In retrospect and from the perspective of someone familiar with the wind industry, this conclusion is not necessarily profound. However, this model is perhaps the first that attempts to construct a framework for assessing total system reliability in a probabilistic fashion and can therefore provide direction for future work.

While this paper studies Return on Carbon Investment, the next step would be a study of how these materials inputs influence costs, i.e., this framework provides a natural pathway to evaluate the next dimension of Elkington's sustainability: profit. The financial picture likely looks different, as the installation and operation of wind farms only appears attractive to investors when federal tax credits are in effect, as historical production trends indicate.[7] The financial return on investment is likely not as immediate as ROCI as other studies have indicated. In addition, if hydrocarbon feedstock prices are volatile, then the replacement cost of materials for blades is also volatile. If a source for permanent magnet materials is monopolistic, then the wind turbine market will be held hostage to price control mechanisms in that market. If the average sale price of electricity is only $0.06/kWh and coal power remains a cheap, then the profit margins for a wind farm will always be tight. Adding the costs of failure and the lost revenue from downtime is yet another step to be considered in this model.

As a final remark, evaluation of offshore wind farms (as opposed to onshore) may add a different perspective as the environment is more harsh and therefore failures are perhaps more common, yet the winds are stronger and the capacity factor is generally higher. Extending this study to other renewable energy systems may be beneficial. This probabilistic framework can be applied to reliability studies of any renewable energy system, provided that operational data is available and there is a way to evaluate the life cycle carbon and energy inventory of the materials used for construction.

REFERENCES

[1] *The Future of Sustainability: Re-thinking Environment and Development in the Twenty-First Century.* The World Conservation Union, January, 2006.

[2] Elkington, John. *Governance for Sustainability.* Corporate Governance Volume 14, No. 6, November 2006.

[3] Hahn, et al. *Reliability of Wind Turbines: Experiences of 15 Years with 1,500 Wind Turbines.* Institut fur Solare Energieversorgungstechnik (ISET). Kassel, Germany.
[4] Tavner, Peter. *Reliability and Availability of Wind Turbine Electrical and Electronic Equipment.* Durham University. Presented at "2009 Sandia Wind Turbine Reliability Workshop". Hosted by Sandia National Laboratory (SNL) and National Renewable Energy Laboratory (NREL). June 17-18, 2009.
[5] Walford, Christopher. *Wind Turbine Reliability: Understanding and Minimizing Wind Turbine Operation and Maintenance Costs.* Sandia National Laboratories SAND2006-1100. 2006.
[6] Dismukes, et al. *The Industrial Life Cycle of Wind Energy Electrical Power Generation: ARI Methodology Modeling of Life Cycle Dynamics.* Technological Forecasting and Social Change 76 (2009) 178-191.
[7] *20% Wind Energy by 2030: Increasing Wind Energy's Contribution to U.S. Electricity Supply.* U.S. Department of Energy, Energy Efficiency and Renewable Energy (EERE). July 2008.
[8] Sorenson, J.D. *Effective Turbulence Models and Fatigue Reliability in Wind Farms.* Probabalistic Engineering Mechanics 23 (2008) 531-538.
[9] Marin, J.C., et al. *Study of Fatigue Damage in Wind Turbine Blades.* Engineering Failure Analysis 16 (2009) 656-668.
[10] Marin, J.C., et al *Study of Damage and Repair of Blades on a 300 kW Wind Turbine..* Energy 33 (2008) 1068-1083
[11] Guo, Watson Tavner, Xiang. *Reliability Analysis for Wind Turbines with Incomplete Failure Data Collected from After the Date of Initial Installation.* Reliability Engineering and System Safety. February 2009.
[12] Fingersh, L., et al. *Wind Turbine Design Cost and Scaling Model.* L. Fingersh, et al. NREL/TP-500-40566. December 2006
[13] Kanaby, Gary. *Manufacturing Blades for Turbine Reliability.* Knight and Carver Wind Group. Presented at "2009 Sandia Wind Turbine Reliability Workshop". Hosted by Sandia National Laboratory (SNL) and National Renewable Energy Laboratory (NREL). June 17-18, 2009.
[14] Yarla, Rahul. *Large Blade Testing Facility in the U.S.*" Massachusetts Wind Technology Testing Center., MA-DOE/NREL Large Blade Testing for 2011. Presented at "2009 Sandia Wind Turbine Reliability Workshop". Hosted by Sandia National Laboratory (SNL) and National Renewable Energy Laboratory (NREL). June 17-18, 2009.
[15] "Methylene diphenyl diisocyanate resin, at plant, US PNW". National Renewable Energy Laboratory, U.S. Life Cycle Inventory Database.
[16] *EPA Office of Compliance Sector Notebook Project: Profile of the Plastic Resin and Manmade Fiber Industries.* EPA/310/R-97/006. Plastic Resin and Manmade Fiber Sector Notebook Project. Sept. 1997.
[17] *U.S. Life Cycle Inventory Database.* National Renewable Energy Laboratory (NREL). www.nrel.gove/lci/database/default.asp
[18] *Advanced Materials for Turbine Blade Manufacture.* Reinforced Plastics. April 2007.
[19] Martinez, et al. *Life Cycle Assessment of a 2MW Rated Power Wind Turbine: CML Method.* Int J Life Cycle Assess (2009) 14:52-63.

PRECIPITATION BEHAVIOR OF CHROMIUM IN CHROMIUM(III)-BEARING SLAG

Xingrong WU*, Liaosha LI, Ping Wang, Zhaojin Wu and Yuanchi Dong

Anhui Provincial Key laboratory for Metallurgical Engineering & Resources Recycling, Anhui University of Technology, Maanshan, Anhui, 243002, China. * *E-mail: music@ahut.edu.cn*

ABSTRACT

The two chromium(III)-bearing slag glasses (chrome slag and stainless steel slag) were synthetified by chemical agents. The precipitation behaviour of chromium were investigated in the solidified slags with additional 5% Al_2O_3 cooled from melting temperature of 1550°C. The effect of slag compositions on the precipitation of chromium was clarified by SEM and EDX and XRD. Results show that 5 wt% Al_2O_3 addition can enrich most of chromium from the two slags by formation of spinel phase. The spinel phase contains 30 wt% Cr_2O_3 in the presence of solid solution in stainless steel slag, whereas 7 wt% Cr_2O_3 in spinel solid solution in chrome slag, the differences are also discussed.

1. INTRODUCTION

Chromium is the major alloying element in stainless steel and the important composition in chemical engineering. Chromium deposit in nature is chromite. Using chromite to product stainless steel, chromium-bearing slag are generated as waste material or byproduct, containing EAF stainless steel slag and Argon Oxygen Decarburization (AOD) stainless steel slag. Another significant quantities of chromium slag are from the production of chromium salt, which is formed by calcining the mixture of calcined soda, dolomite, limestone and chromite at 1100 ~ 1200°C and by leaching the sodium chromate. These slags usually contain a quantity of valuable metals of Cr, which can be recovered and reused in a judicious manner.

Chromium is usually encountered in the environment at oxidation states of (III) and (VI). Concerns regarding trivalent chromium are less stringent, both because it has a much lower human toxicity rating and because it is less environmentally mobile—due to its low solubility and its tendency to become absorbed or complexed by organic molecules.[1-3] Therefore, a volume of papers have been published describing to transform Cr(VI) in chromium-containing slag to Cr(III) in order to detoxify.[4-6] However, studies by authors[7,8] have presented evidences that for certain slag compositions atmospheric oxidation of trivalent to hexavalent chromium can nevertheless occur at ambient conditions. So, utilisation and disposal practices for wastes containing trivalent chromium (which by implication would be as solids) are considered to be of acceptable environmental risk, only detoxification of chromium-containing slag is not a method for long.

A sustainable method for waste processing is to convert the wastes into useful products. In order to make these products from heavy metal contaminated waste streams, the metals have to be removed or immobilized. Considerable attentions have been paid to the management and recycling of the slags. Therein, one important means is to reduce metal-from-slag recovery technique.[9,10]

However, the chromium oxide in the slag are much lower, the reducing costs based on overall slag has been the problem. This studies, inspired by selective enrichment of vanadium from steelmaking slag with low concentratation of V_2O_5,[11,12] ensue to investigate the distribution chromium and further to precipitate the most of chromium in one phase by modification of chromium(III)-bearing slag. The experimental results on the enrichment behavior of Cr_2O_3 in slag were analyzed using XRD, SEM and EDX, which will be helpful to separate chromium-concentrating phase from slag and further recover the chromium from the phase.

2. EXPERIMENTAL PROCEDURE

Two kinds of chromium-bearing slag samples used in this study for comparion. One was chrome slag discharged during the production of chromium salt. The other was stainless steel slag generally discharged from a converter. The main components of the two slags are MgO, SiO_2, Al_2O_3, Cr_2O_3 and FeO shown in Table 1. Laboratory reagent-grade oxide powders, CaO, SiO_2, Fe_2O_3, Al_2O_3, MgO and Cr_2O_3 were carefully weighed and blended to produce the compositions represented in Table 2. Assuming that the content of Cr(VI) in the Stainless steel slag are ignored, Cr(VI) in chrome slag is not introduced for comparion.

During the preparation of the two mother slags, additional 5% Al_2O_3 were used as modifiers. About 150 g of each powder mixture was placed into a Al_2O_3 crucible, which was loaded in a furnace with heating elenments of $MoSi_2$ and heated to 1550°C. The temperature was monitored by a B type thermocouple (Pt–6Rh/Pt–30Rh) placed in direct contact with the crucible bottom. The furnace had a constant temperature zone (±2 °C) over a length of 80 mm. After the slags were melted completely for 30min, the crucibles with slags were removed out quenched in water. The two mother slags were dried and then ground to powder.

10g of the two crushed slags were placed in two alumina crucibles and heated to the liquid state at 1550°C in air. After an isothermal time of 30min, cooling experiments were performed from the isotropic melt at cooling rate of 3 °C /min to the crystallization temperature of 1300°C, held for 30min before quenched in air.

Quenched samples were cut using a low speed diamond saw. The samples were then ground with 600-1000 grit SiC paper, and finally polished to 1μm using diamond paste. Carbon was evaporated on the sample surface to provide a conducting layer for micro-analysis. The microstructures were observed using a scanning eletron microscope (Philips XL30) operated at 20kV. The SEM was equipped with an Energy Dispersion X-ray Spectroscopy (EDX or EDS) unit (EDAX DX4i) that allowed compositional measurements to be done. The identification of the precipitated crystalline phases was performed by powder X-ray diffractometer (Rigaku D/max-r) using Cu Kα radiation in 2θ range from 10° to 80°.

3. RESULTS AND DISCUSSION

3.1. X-ray diffraction analysis

Fig. 1 shows X-ray diffraction parttems of the milled slags solidified at 1300°C. It could be seen from these diffraction patterns that the two samples are highly crystalline. The diffraction patterns also contain a slight halo due some glassy Ca–Mg–Al silicate phase present. Having indexed these patterns in stainless steel slag, spinel is identified as the main constituent, The spinel phase represents a solid solution of $MgAl_2O_4$ and $MgCr_2O_4$. Whereas, in the spectrum of chrome slag, two compounds were identified: spinel and Gehlenite ($2CaO{\cdot}Al_2O_3{\cdot}SiO_2$), and calcium-aluminium silicates also exist in the form of glassy. In chrome slag sample, the gehlenite ($2CaO{\cdot}Al_2O_3{\cdot}SiO_2$) was presented as main phase and spinel is the minor phases.

The difference of main crystalline phase in the two slag samples can be explained by chemical compositions. MgO and Al_2O_3 are much higher in chrome slag than in the stainless steel slag, so spinel ($Mg(Al,Cr)_2O_4$) can be preicipitated as main phase. But in stainless steel slag, CaO and SiO_2 is much higher than that in chromium slag, gehlenite can be produced as main phase , thereby, the amounts of spinel phase is relatively lower.

3.2. SEM and Microprobe analysis by EDS of samples

The spinel group (AB_2O_4) comprises a wide variety of oxide minerals where A and B represent the R^{2+} and R^{3+} ions, respectively.[14] Spinels exhibited a wide-ranging solid solution. In this series, substitution of Al and Cr (both R^{3+} ions) has extensively occurred in different locations to a certain extent.

In order to observe the different phases precipitated in chromium slag and in stainless steel slag, a SEM study was carried out, as shown in Fig.2 and Fig.3. From the images, we can see that three phases were precipitated in both slags.

In chrome slag, the phase labeled by 2 are calcium alumina-silicates (the atomic ratio of Ca: Si :Al =2 :2:1), recognized as $2CaO \cdot Al_2O_3 \cdot SiO_2$. The phase labeled by 1 is composed by O, Mg, Al and Cr(the atomic ratio of Mg: (Al+Cr) =2:1), recognized as spinel solid solution ($Mg(Al,Cr)_2O_4$). The third phase labeled by 3 is matrix, mainly consisted of O, Al, Ca and Fe. These results can be refined by XRD data. By calculating from EDS results, only about 7 wt% Cr_2O_3 exists in the presence of solid solution in spinel phase, Cr is not found in exists in matrix phase, that is mean that almost all of Cr_2O_3 were precipitated into $MgAl_2O_4$ phase from the slag modified by Al_2O_3. The content of Cr_2O_3 in $Mg(Al,Cr)_2O_4$ phase is not high, we suppose that spinel is main phase with a large amount due to MgO is high in chrome slag, accordingly Cr_2O_3 content in spinel phase is low.

Whereas, in stainless steel slag (see Fig. 2), large grey grains are $2CaO \cdot Al_2O_3 \cdot SiO_2$ (labeled by S1), minor amounts of white grains are spinel (labeled by S2). But EDS studies indicate that Cr_2O_3 content in the spinel phase is more than 30 wt%. There is a small amount of Cr found in matrix phase (Cr_2O_3 ~ 2wt%).

As described above, Cr can be precipititated in spinel solid solution in both slags modifying by Al_2O_3. By comparion, although the differences are that spinel phase is not the main phase in stainless steel slag, the content of Cr_2O_3 (30 wt%) is higher in the spinel phase in stainless steel slag than that (7 wt%) in chrome slag. The reason is that stainless steel slag has lower content of MgO (10wt% in Table 2), and solidifies on cooling to form $2CaO \cdot Al_2O_3 \cdot SiO_2$ from main constituents of silicates and aluminium oxides. Therefore, the spinel phase from MgO, Al_2O_3 is minor phase and has less volume, and Cr_2O_3 in stainless slag is precipitated in spinel solid solution, the content of Cr_2O_3 in the spinel phase is higher. Otherwise, in chrome slag, MgO content is higher(24wt% in Table 2), spinel phase forms from main constituents of MgO and Al_2O_3, and has larger volume, so, the content of Cr_2O_3 in the spinel phase is lower.

According to Lopez's results, [15] charge chrome slag usually contains 8-12% Cr, which is higher than in stainless steel slag. Cr appears mainly in chromite and metallic phases. A small amount of Cr is also found in spinel (Cr_2O_3 2-5%) and olive (Cr_2O_3 <1%). Other author[8] found that the chromium is normally bound in the slag matrix in various silicate or spinel phases in stainless steel slag. Therefore, Cr is distributed in more than one phase, that is to say that Cr cannot be precipitated in one phase, and cannot be enriched.

Unlike to steel slags, stainless steel slags do not consist of dicalcium and tricalcium silicates, dicalcium silicates, and periclase except that they contain chrome-nickel-magnesium oxide solution in the form of chromites in which Cr and Ni exist.[15] The steel slag composition is found to vary over a wide range of basicity (%CaO/%SiO_2) and iron oxide content. This paper investigates that the main phase is $2CaO \cdot Al_2O_3 \cdot SiO_2$ in stainless steel slag modified by Al_2O_3 and Cr is mainly precipitated in spinel solution. The mechanism of modifying by Al_2O_3 will be studied further.

4. CONCLUSIONS

In order to clarify the distribution behavior and the preicipitation mechanism of Cr_2O_3 in chromium-containing slag, the effect of compositions of the two slags on the precipitation of Cr_2O_3 were investigated and compared in the slags solidified at 1300°C.

We can draw the following conclusions from the experiments:

(1) In the steelmaking steel slag and chrome slag both modified by 5 wt% Al_2O_3, Cr is mainly precipitated in spinel phase $Mg(Al,Cr)_2O_4$.

(2) Cr_2O_3 content in spinel phase precipitated from steel slag is higher than that in chrome slag.

ACKNOWLEDGEMENTS

The authors would like to thank the financial supports for the research by Anhui Provincial Natural Science Research Project in Colleges and Universities (Project No. KJ2009A69) and Anhui Provincial Natural Science Foundation (Project No. 090414150).

REFERENCES

1) D.C. Schroeder and G.F. Lee, Potential Transformations of Chromium in Natural Waters, *Water. Air, & Soil Pollution*, 4[3-4] 355–365 (1975).

2) D. Rai, B.M. Sass and D.A. Moore, Cr(III) Hydrolysis Constants and Solubility of Cr(III) Hydroxide, *Inorg. Chem.*, 26[3] 345-349 (1987).

3) R.J. Bartlett and B.R. James, pp. 267-304 in *Chromium in the Human and Natural Environment*, Edited by J. Nriagu, John Wiley and Sons, New York, 1988.

4) H. W. Kilau and I. D. Shah, Chromium Bearing Waste Slag: Evaluation of Leachability When Exposed to Simulated Acid Rain; pp. 61–81 in Third Symposium, *Hazardous Industrial Waste Management and Testing*, Edited by L. P. Jackson and A. R. Rohlik, Philadephia, ASTM STP, 851, 1984.

5) K. Pillay, H. Blottnitz and J. Petersen, Ageing of Chromium(III)-bearing Slag and Its Relation to the Atmospheric Oxidation of Solid Chromium(III)-oxide in the Presence of Calcium Oxide, *Chemosphere*, 52[10] 1771-1779 (2003).

6) S. Etsuro, E. Shin and N. Takashi, Reduction Behavior of Chromium Oxide in Molten Slag Using Aluminum, Ferrosilicon and Graphite, *ISIJ Int.*, 42[6] 609-613 (2002).

7) N. Takamitsu, N. Kunihiko and M. Katsumi, Recovery Rate of Chromium from Stainless Slag by Iron Melts," *ISIJ Int.*, 44[4] 665-672 (2004).

8) C. Y. Rha, S. K. Kang and C. Kim, Investigation of the Stability of Hardened Slag Paste for the Stabilization/Solidification of Wastes Containing Heavy Metal Ions, *J. Hazard. Mater.*, B73 255-267 (2000),.

9) S. A. Khan, G. Ray, M. K. Sinha and P. V. T. Rao, Recovery of Charge Chrome from Mixed Metal and Granulated Slag, *Tata Search (India)* , 4 115-118 (2001).

10) H. P. Mashanyare, R. N. Guest, The Recovery of Ferrochrome Slag at Zimasco, *Miner. Eng.*, 10[11] 1253-1259 (1997).
11) X. R. Wu, L. S. Li and Y. C. Dong, Experimental Crystallization of Synthetic V-bearing Steelmaking Slag with Al_2O_3 doped, *J. Wuhan Univ. Technol.*, 20 [9] 63-67 (2005).
12) X. R. Wu, L. S. Li and Y. C. Dong, Influence of P2O5 on Crystallization of V-concentrating Phase in V-bearing Steelmaking Slag, *ISIJ Int.*, 47 [3] 402-407 (2007).
13) J. H. Li, "Recycling and recovery technology of chromium slag," in *Proceedings of the Second Asia-Pacific Regional Workshop on Hazardous Waste Management and Practice*. Asia-Pacific Regional Center for Hazardous Waste Management Training & Technology Transfer, Beijing, 1999.
14) J. R. Smyth, S.D.Jacobson, and R.M. Hazen, "Comparative crystal chemistry of dense oxide minerals," pp. 157–186 in *High-Temperature and High-Pressure Crystal Chemistry*, edited by R. M. Hazen and R. T. Downs, Reviews in Mineralogy and Geochemistry vol. 41.,Mineralogical Society of America, 2000.
15) F. A. Lopez, A. Lopez-Delgado and N. Balcazar, Physico-chemical and mineralogical properties of EAF and AOD slags, *Associazione Italiana di Metallurgia (Italy)*, 417-426 (1997).

Table I. Chemical Composition of Chromium-containing Slags in References

slag	Components (%)						Reference
	SiO_2	CaO	Al_2O_3	Fe_2O_3	MgO	Cr_2O_3	
chrome slag	8-11	23-36	5-8	7-11	20-33	3-7	13)
Stainless steel slag	26-32	45-48	2-9	3-7	3-7	3-4	14)

Table II. Chemical Composition of Slags Incorporated in the Experiments

slag	Components (%)					
	SiO_2	CaO	Al_2O_3	Fe_2O_3	MgO	Cr_2O_3
chrome slag	12	36	12	11	24	5
Stainless steel slag	30	45	4.5	6.5	10	4

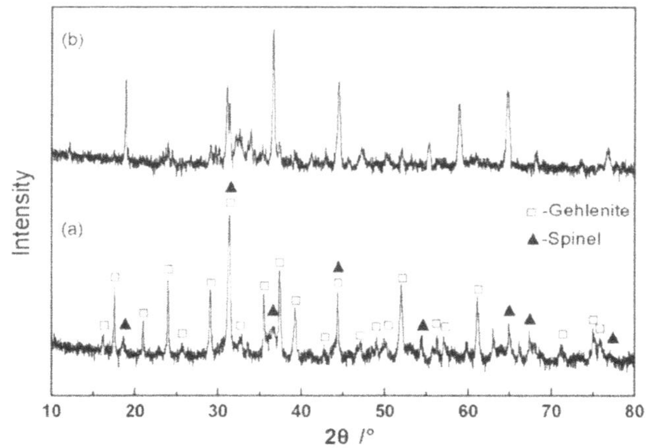

Figure 1. The X-ray diffraction patterns of the two slags. (a) stainless steel slag, (b) chromium slag. Peaks for gehlenite are marked by square, and spinel by triangle.

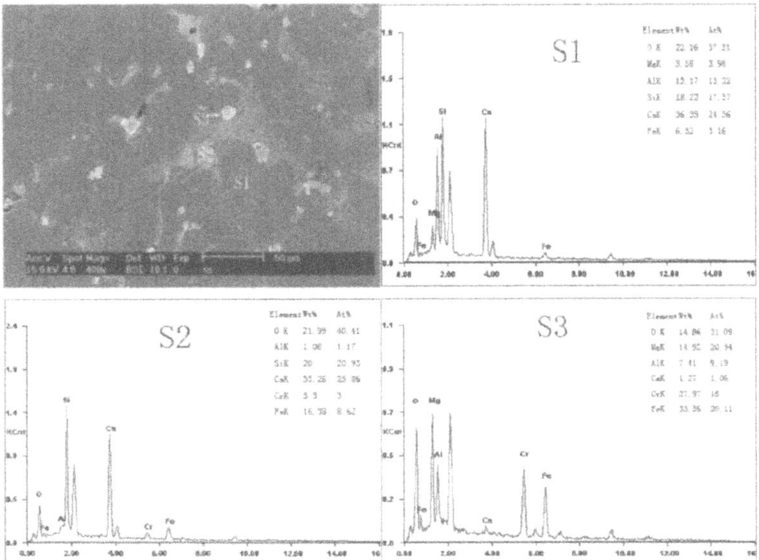

Figure 2. SEM micrograph of stainless steel slag solidified at 1300°C and EDX spectrum patterns of the phases refered by phase S1, S2 and S3.

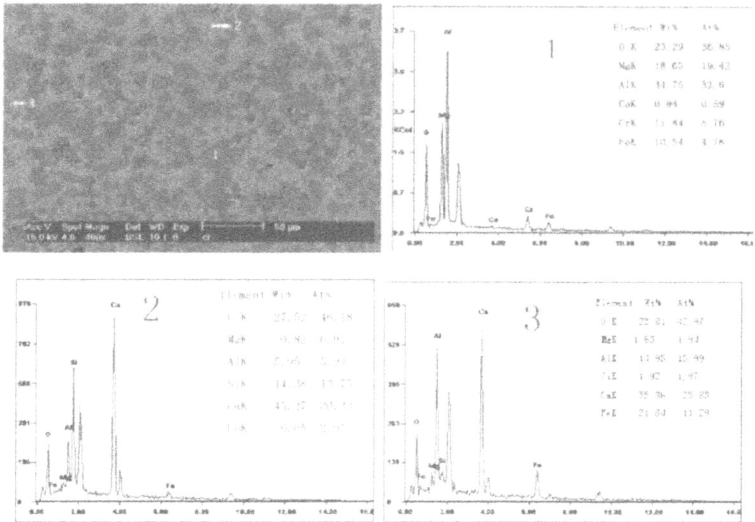

Figure 3. SEM micrograph of chrome slag solidified at 1300°C and EDX spectrum patterns of the phases refered by phase1, 2 and 3.

IMPROVED ENERGY EFFICIENCY AND ENVIRONMENTAL BENEFITS FOR CALCIUM TREATMENT IN STEEL

James Yanker, David Dudek
Minteq International Inc.
Bethlehem, PA USA

ABSTRACT

Increasing demands on steelmakers, especially in regard to energy efficiency and environmental considerations, are pushing steel producers to use more accurate and effective calcium treatment. Solid calcium core wire, particularly when used in conjunction with deep lance injection, is technically superior to alternative powder fill wires, assuring enhanced process control, highest calcium yield and desired steel cleanliness. It results in improved energy efficiencies and environmental advantages such as highest calcium content per package, lowest transport, material handling and storage costs, lowest calcium consumption/waste and fewest overall deliveries. Safety benefits of this technology include less material handling, fewer package changeovers and no breakage of stiff, welded wire. In addition, calcium produced in the United States uses an aluminothermic process that produces significantly lower emissions compared with alternative electrolytic processes used in imported calcium. This paper reviews a means to save energy, improve safety and reduce emissions in calcium treatment of steel.

INTRODUCTION

With all the news about CO_2 emissions standards, greenhouse gasses, carbon footprint, etc., there is increasing demand on the steel industry to adopt new technologies that reduce carbon emissions. At the same time, there is increasing focus on the bottom line, going up-market with higher quality steels and adopting new technologies like thin slab casting whenever possible, all requiring continuous improvement of steel cleanliness. In order to meet increased quality standards, it is important that the size, morphology and distribution of inclusions, particularly oxides be controlled and minimized. In addition, controlled and consistent calcium treatment products, processes and parameters are necessary to maintain good caster quality and throughput.

As expected, a lot of energy and environmental impact can be attributed to the calcium treatment. Higher calcium yields with solid calcium core wire, particularly when used in conjunction with deep lance injection have a direct impact on lowering calcium consumed and overall emissions compared to alternative powder fill wires. In addition, calcium produced in the United States uses an aluminothermic process that produces significantly lower emissions compared with alternative electrolytic processes used in imported calcium.

This paper considers greener ways of calcium treatment, not only in the wire product and injection system used but also the production method of calcium itself. A comparison of CO_2 emissions and reducing overall carbon footprint is presented.

GREENER WAYS OF CALCIUM TREATMENT

The steel industry has long been aware of the beneficial aspects of calcium treatment of molten steel. In addition to desulfurization via injection of calcium silicon powder, desired inclusion modification, steel cleanliness and castability are achieved via treatment with calcium cored wire treatment.[1, 2] The addition of calcium modifies high melting, solid alumina inclusions to a form that is liquid at steelmaking temperatures.

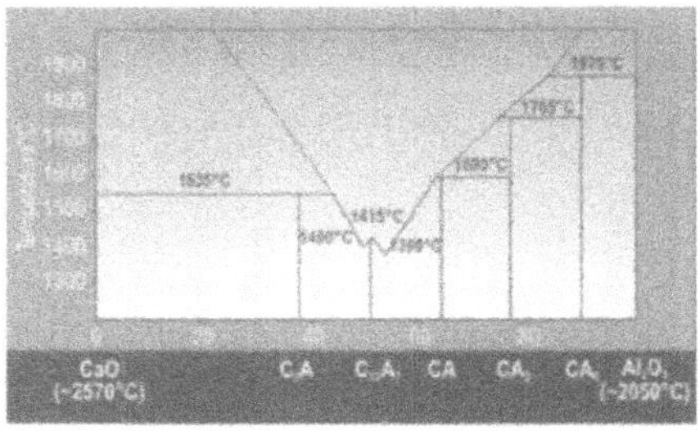

Figure 1. Phase diagram of the CaO – Al_2O_3 System

The most effective wire treatment is achieved when wires made of solid metallic calcium rod encased in steel (as opposed to powdered calcium blends) are used.[3, 4]

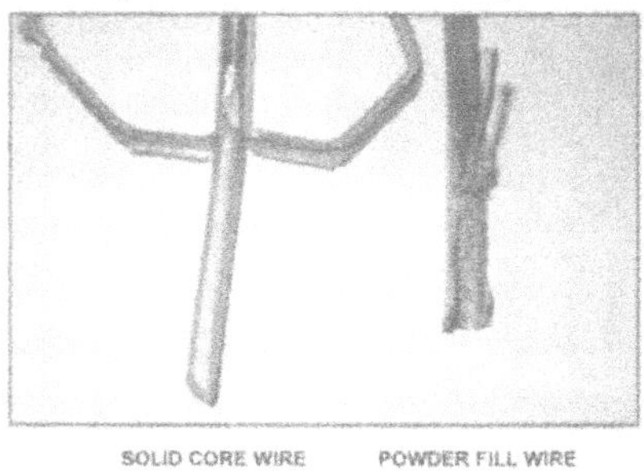

Figure 2. Improved Technology Solid Core Wire vs. Powder Fill Wire

Further benefits are achieved when solid calcium core wire is used in conjunction with deep lance injection which releases calcium deep into the bath at a corresponding ferrostatic pressure to suppress vaporization.[4,5,6]

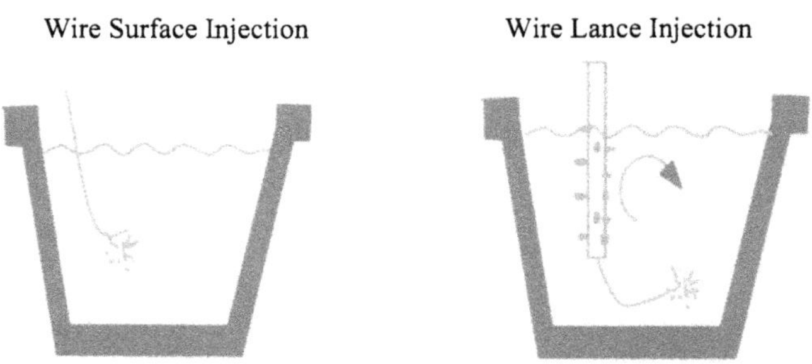

Figure 3. Schematic Diagrams of Wire Surface Feed vs. Wire Lance Injection

TYPICAL RESULTS AND LOWER TRANSPORTATION EMISSIONS USING SURFACE FEED SOLID CORE Ca WIRE VS. POWDER FILL WIRE

Solid core pure calcium products have higher packing efficiency compared to powder fill wires, therefore more pounds of calcium per cage (package), more pounds of calcium per delivery, less transport needed and lower emissions.

The 13mm FeCa powder fill wire is typically 30% calcium / 70% iron powder with each cage weighing approximately 4,200 lbs. wire, 2,700 lbs. FeCa filling or 800 lbs. calcium (approx. 19% of the total wire weight is calcium powder). This compares with the 9mm solid core calcium wire with each cage weighing approximately 4,450 lbs. wire or 1,300 lbs. pure calcium, (29% of the total wire weight is solid calcium core).

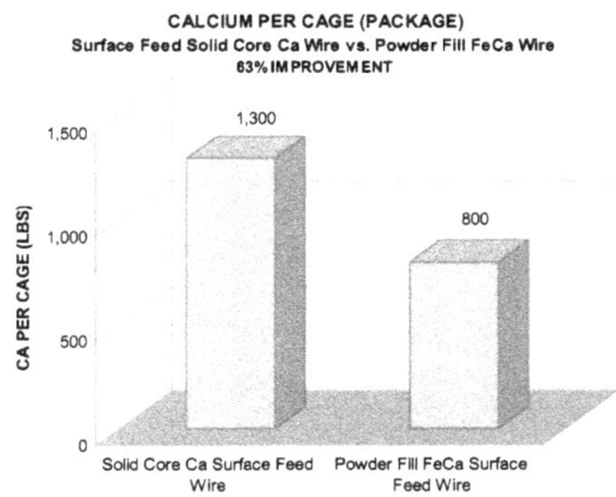

Figure 4a. Example showing 63% more Ca per cage (package), so fewer cages and deliveries needed and lower CO_2 emissions with Surface Feed solid core Ca wire vs. Powder Fill Wire

A typical truck holds 10 cages FeCa powder fill wire (42,000 lbs. wire, 27,000 lbs. FeCa fill, 8,000 lbs calcium) compared to 8 cages solid core wire (35,000 lbs, wire, 10,000 solid core calcium). Although there are less cages of solid core wire per truck, there is over 60% more calcium per cage with the solid core wire which means over 25% more calcium per truck and less than ¾ of the needed transportation (Figure 4b). Less transportation means lower carbon emissions.

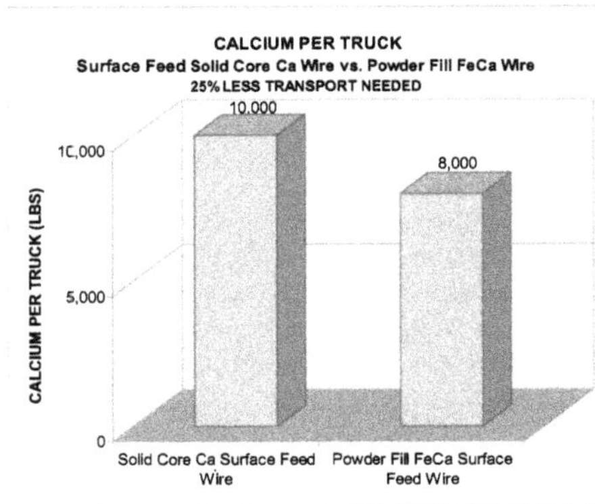

Figure 4b. Example showing 25% more pounds of Ca per truck, so fewer deliveries are needed and lower CO_2 emissions with surface feed solid core Ca wire vs. powder fill wire

Improved calcium yield and less calcium needed is another factor in reducing freight and overall emissions with solid calcium core wire. The following table shows typical solid calcium core wire versus powder fill wire results with over double the calcium yield and over 45% lower calcium consumption.

Table I. Typical Solid Calcium Core vs. Powder Wire

Capacity - 1.8M Ton annual Si-free production		
Heat Size - 220 Ton		
	Solid core Ca Surface Feed Wire	Powder fill FeCa Surface Feed Wire
Ca Yield in Ladle (%)	31%	15%
Ca Add (lbs./ton)	0.16	0.30
Ca Add (lbs./heat)	36	66
Ca Consumed (lbs./year)	290,000	540,000
Annual Deliveries (trucks/year)	28	85
Diesel Fuel Consumed (gals./year)	2,800	8,500
CO_2 Emissions (tons/year)*	30	95

The combination of improved packing efficiency and less calcium needed results in significantly less deliveries and overall emissions. In above example, there is over 70% reduction in annual deliveries, 28 per year with solid core calcium wire versus 85 deliveries per year with the powder fill wire. The 70% reduction in annual diesel fuel consumed (est. 2,800 gals versus 8,500 gals.) has a direct impact on reduced emissions. In this example, overall CO_2 emissions were reduced 70% to 30 tons per year with the solid core calcium wire versus 95 tons per year using the powder fill wire (Figure 4c).*

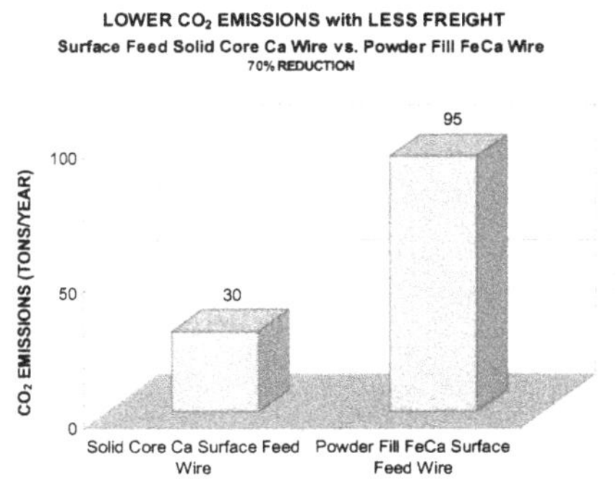

Figure 4c. Example showing 70% lower overall CO_2 emissions due to reduced freight and lower consumption with solid core calcium wire.

One potential improvement involved the use of pure calcium powder in a stiff, welded jacket. In theory, the welded jacket, with 0.045-0.050" thickness permits deep penetration before melting. Unfortunately, the steel cladding accounts for over 80% of the total wire weight and as in the case of iron calcium wires, is a huge disadvantage to both transport (below 7,000 lbs. calcium per delivery) and emissions.

TYPICAL RESULTS AND LOWER TRANSPORTATION EMISSIONS USING SOLID CORE Ca DEEP LANCE INJECTION TECHNOLOGY VS. SURFACE FEED POWDER FILL

The 8mm solid core calcium wire used for deep lance injection is special-packaged on wooden spools with each spool weighing 1,500 lbs. wire, 735 lbs. pure calcium. Because the refractory lance already introduces calcium deep in the ladle, a heavy steel jacket is not required to facilitate penetration. For this reason, a much thinner 0.010" steel cladding is used and 49% of the total wire weight is solid calcium core. A typical truck holds 24 spools, approx. 35,000 lbs. wire, 17,000 lbs. of solid calcium core. In this way, there is over 2 times the calcium per truck and less than half the transportation needed compared with powder fill FeCa fill wire (Figure 5a). Again, less transportation means lower emissions.

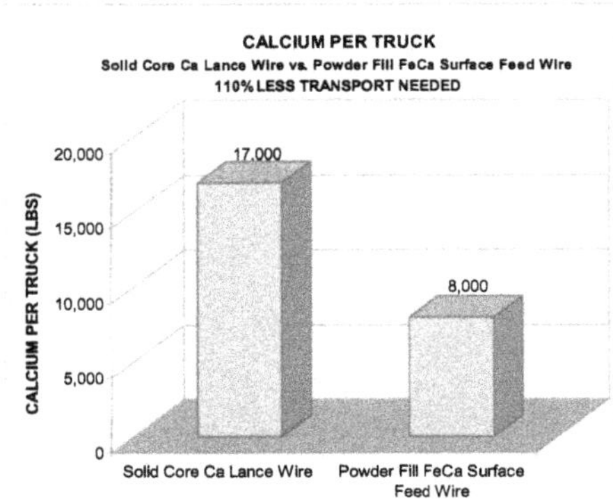

Figure 5a. Example showing over 2 times the Ca per truck, so fewer deliveries are needed and lower CO_2 emissions with solid calcium core deep lance injection.

The new solid core calcium deep lance technology results in even greater calcium yields, lower calcium consumption, reduced freight and lower emissions compared to powder fill wire. The following table shows typical solid core lance calcium versus surface feed powder fill wire results with over 3 times the calcium yield and over 70% lower calcium consumption.

Table II. Typical Solid Core Lance Calcium vs. Surface Feed Powder Fill Wire

Capacity - 1.8M Ton annual Si-free production		
Heat Size - 220 Ton		
	Solid Core Ca Lance Wire	Powder Fill FeCa Surface Feed Wire
Ca Yield in Ladle (%)	45-50%	15%
Ca Add (lbs./ton)	0.09	0.30
Ca Add (lbs./heat)	20	66
Ca Consumed (lbs./year)	160,000	540,000
Annual Deliveries (trucks/year)	10	85
Diesel Fuel Consumed (gals./year)	1,000	8,500
CO_2 Emissions (tons/year)*	11	95

The combination of improved packing efficiency and less calcium needed results in significantly less deliveries and overall emissions. In above example, there is over 88% reduction in annual deliveries, 10 per year with solid core calcium wire versus 85 deliveries per year with the powder fill wire. The 88% reduction in annual diesel fuel consumed (est. 1,000 gals versus 8,500 gals.) has a direct impact on reduced emissions. In this example, overall CO_2 emissions were reduced 88% to 11 tons per year with the solid calcium core deep lance technology versus 95 tons per year using the powder fill wire (Figure 5b).

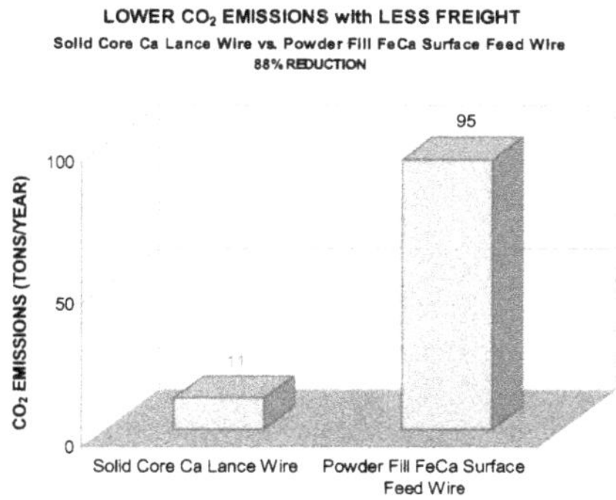

Figure 5b. Example showing 88% lower overall CO_2 emissions due to reduced freight and lower consumption with solid core calcium wire.

GREENER CALCIUM PRODUCTION PROCESS – IMPACT ON EMISSIONS

The aluminothermic calcium process generates far less greenhouse gases than the electrolytic process. Detail on production processes as follows:

Aluminothermic Ca Method:

$$CaCO_3 + \text{heat} \rightarrow CaO + CO_2 \text{ (fuel oil heating, gallons)}$$
$$2Al_2O_3 + 3C \rightarrow 4Al + 3CO_2 \text{ (electricity, kwh)}$$
$$3CaO + 2Al \rightarrow 3Ca + Al_2O_3 \text{ (propane gas, cubic feet)} \quad (1)$$

In the aluminothermic process (Figure 6), high calcium limestone $CaCO_3$, is quarried and calcined to form calcium oxide. The calcium oxide is ground to a small particle size and dry-blended with the desired amount of finely divided aluminum. This mixture is then compacted into briquettes to ensure good contact on reactants. The briquettes are placed in horizontal metal tubes (retorts) made of heat resistant steel and heated in a furnace to 1100-1200°C. The open ends of the retorts protrude from the furnace and are cooled by water jackets to condense the calcium vapor. The retorts are then sealed and evacuated to a pressure of less than 0.1 mm Hg. After the reaction has been allowed to proceed, the vacuum is broken with argon, and the condensed blocks of 99% pure Calcium metal (known as crowns) and calcium aluminate residue are removed.

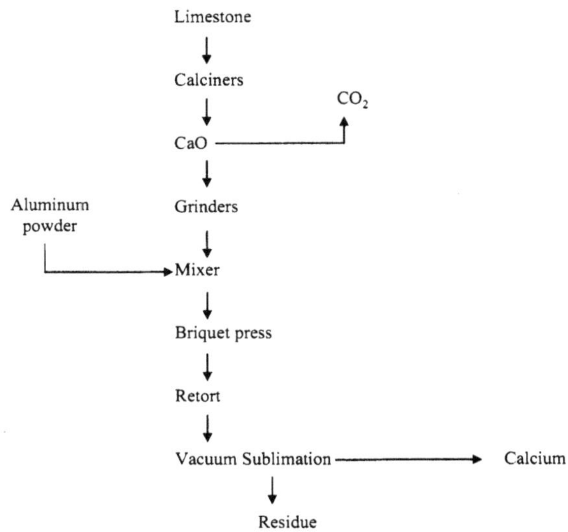

Figure 6. Flow sheet for aluminothermic process

Electrolytic Ca Method:

$$CaCO_3 + \text{heat} \rightarrow CaO + CO_2 \text{ (fuel oil heating, gallons)}$$
$$CaO + H_2O \rightarrow Ca(OH)_2$$
$$4Ca(OH)_2 + 4Cl_2\,(g) \rightarrow 4CaCl_2 + 2H_2O + O_2 \text{ (electricity, kwh)}$$
$$CaCl_2 + H_2O\,(l) + \text{heat} \rightarrow CaCl_2 + H_2O\,(g) \text{ (natural gas, cubic feet)}$$
$$CaCl_2 + Cu + \text{Electricity} \rightarrow CaCu + Cl_2\,(g) \text{ (electricity, kwh)}$$
$$CaCu + \text{heat} \rightarrow Ca\,(g) + Cu\,(l) \text{ (natural gas, cubic feet)} \qquad (2)$$

This process (Figure 7) produces calcium metal via the electrolytic dissociation of calcium chloride in a molten copper bath. The starting material is the same as for the aluminothermic process, high calcium limestone which is quarried and calcined to form calcium oxide. The calcium oxide is ground and mixed with water to form a calcium hydroxide solution through which chlorine gas (from the electrolysis process) is bubbled to form calcium chloride. The calcium chloride is then dried in preparation for addition to an electrolytic cell. An external voltage is applied to the electrodes which cause the negatively charged chlorine ions to flow to the anode and be released as a gas. The calcium collects at the cathode but remains as a liquid forming a calcium-copper alloy. This process is quite energy intensive, requiring approximately 10 kwh/pound of calcium produced. When the calcium content reaches approx. 40%, a portion of the liquid CaCu alloy is removed and transferred into a stainless steel distillation vessel (retort). Heat and vacuum are applied which causes the calcium to vaporize and condense in a water cooled section of the retort where it is removed as a solid. The residual copper is recycled back to the electrolytic cell.

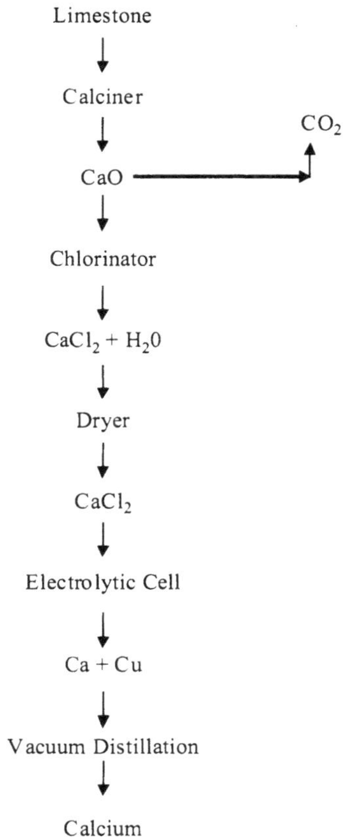

Figure 7. Flow sheet for electrolytic process

The electrolytic process is much more energy intensive than the aluminothermic process, resulting in the formation of significantly higher greenhouse gas emissions.

TYPICAL RESULTS AND LOWER CO_2 EMISSIONS BASED ON Ca PRODUCTION PROCESS – SURFACE FEED SOLID CORE ALUMINOTHERMIC Ca WIRE VS. POWDER FILL ELECTROLYTIC Ca WIRE

As discussed, calcium produced in the United States uses an aluminothermic process that produces significantly lower emissions compared with the alternative electrolytic process. In addition to improved calcium yield and less calcium needed with solid core Ca wire, the calcium production process is significant factor in reducing overall carbon footprint with solid calcium core wire. The following table and graph (Figure 8) show typical solid calcium core wire using aluminothermic produced Ca versus powder fill wire using electrolytic produced Ca results, with 75% lower CO_2 emissions for the aluminothermic Ca.

Table III.

Capacity - 1.8M Ton annual Si-free production		
Heat Size - 220 Ton		
	Solid core surface feed Aluminothermic Ca Wire	Powder fill surface feed Electrolytic Ca Wire
Ca Yield in Ladle (%)	31%	15%
Ca Add (lbs./ton)	0.16	0.3
Ca Add (lbs./heat)	36	66
Ca Consumed (lbs./year)	290,000	540,000
Fuel Oil (gallons/year)	15,000	28,500
Electricity (kwh/year)	880,000	5,850,000
Propane (cu.ft./year)	17,000	---
CO_2 Emissions (tons/year)*	1,300	5,300

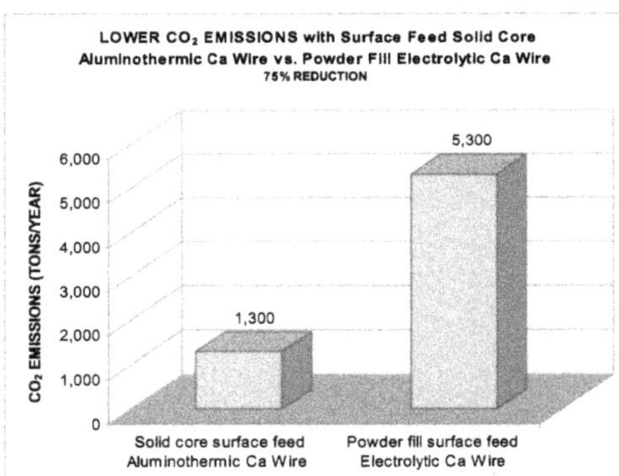

Figure 8. Example showing 75% lower CO_2 emissions with Aluminothermic Ca production method and solid core Ca wire.

TYPICAL RESULTS AND LOWER CO_2 EMISSIONS BASED ON Ca PRODUCTION PROCESS – SOLID CORE LANCE ALUMINOTHERMIC Ca WIRE VS. SURFACE FEED POWDER ELECTROLYTIC Ca WIRE

Again, calcium produced in the United States uses an aluminothermic process that produces significantly lower emissions compared with alternative electrolytic process. In addition to improved calcium yield and less calcium needed with solid core Ca wire lance, the calcium production process is a significant factor in reducing overall carbon footprint. The following table and graph (Figure 9) show typical solid core lance aluminothermic Ca wire versus powder fill electrolytic Ca wire, with 87% lower CO_2 emissions for the aluminothermic Ca.

Table IV.

Capacity - 1.8M Ton annual Si-free production		
Heat Size - 220 Ton		
	Solid core lance Aluminothermic Ca Wire	Powder fill surface feed Electrolytic Ca Wire
Ca Yield in Ladle (%)	45-50%	15%
Ca Add (lbs./ton)	0.09	0.3
Ca Add (lbs./heat)	20	66
Ca Consumed (lbs./year)	160,000	540,000
Fuel Oil (gallons/year)	8,300	28,500
Electricity (kwh/year)	490,000	5,850,000
Propane (cu.ft./year)	9,500	---
CO_2 Emissions (tons/year)*	700	5,300

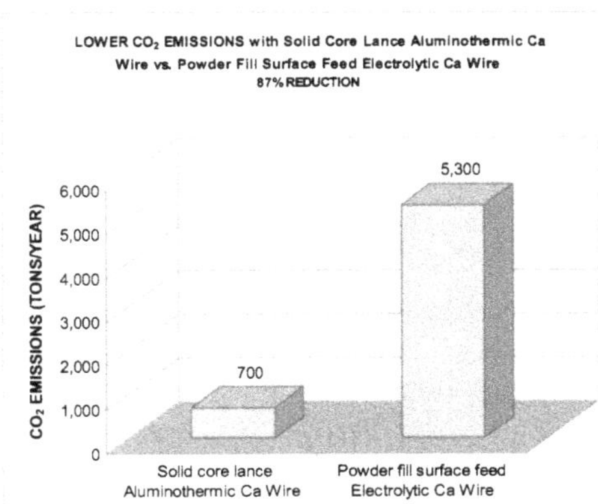

Figure 9. Example showing 87% lower CO_2 emissions with Aluminothermic Ca production method and solid core Ca wire deep lance injection

TOTAL EMISSIONS – TRANSPORT, CALCIUM PRODUCTION METHOD AND USAGE EFFICIENCIES

The following table and graph (Figure 10a) compare the total emissions for transport, calcium production method and usage efficiencies for typical surface feed solid core aluminothermic Ca versus powder fill electrolytic Ca, with 75% lower CO_2 emissions for the aluminothermic Ca.

Table V. Solid Core Surface Feed Aluminothermic Ca vs. Powder Fill Electrolytic Ca

Capacity - 1.8M Ton annual Si-free production		
Heat Size - 220 Ton	CO_2 Emissions (tons/year)	
	Ca solid core surface Aluminothermic Ca	FeCa powder fill surface Electrolytic Ca
Transport	30	95
Ca Metal Production	1,300	5,300
Total CO_2 Emissions (tons/year)*	1,330	5,395

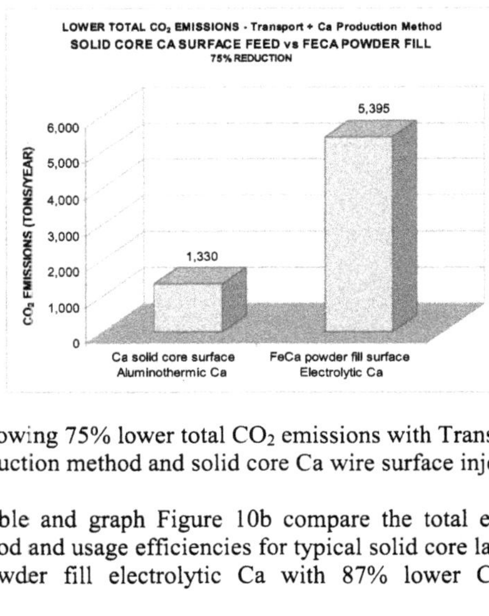

Figure 10a. Example showing 75% lower total CO_2 emissions with Transport, Aluminothermic Ca production method and solid core Ca wire surface injection.

The following table and graph Figure 10b compare the total emissions for transport, calcium production method and usage efficiencies for typical solid core lance aluminothermic Ca versus surface feed powder fill electrolytic Ca with 87% lower CO_2 emissions for the aluminothermic Ca.

Table VI. Solid Core Lance Aluminothermic Ca vs. Surface Feed Powder Fill Electrolytic Ca

Capacity - 1.8M Ton annual Si-free production		
Heat Size - 220 Ton	CO_2 Emissions (tons/year)	
	Ca solid core lance Aluminothermic Ca	FeCa powder fill Electrolytic Ca
Transport	10	95
Ca Metal Production	700	5,300
Total CO_2 Emissions (tons/year)*	710	5,395

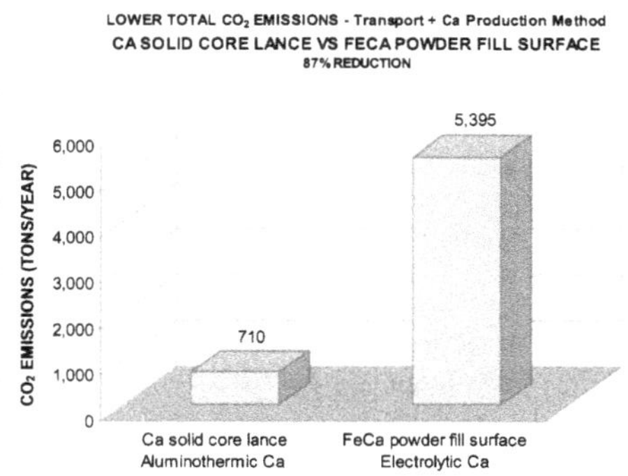

Figure 10b. Example showing 87% lower total CO_2 emissions with Transport, Aluminothermic Ca production method and solid core Ca wire deep lance injection.

CONCLUSION

(1) Significantly lower carbon emissions are shown using solid metallic calcium wires for steel treatment. By achieving higher calcium recovery, lower consumption and improved consistency compared to powdered calcium blends, solid core calcium wires require substantially less freight and energy. These advantages fit well with demand to reduce carbon emissions and move quickly to available, greener technology.

(2) Current practice in most of the new thin slab melt shops has been to take a further step – away from previous options which utilize the surface fed calcium wires and utilize deep lance injection of calcium wire. As discussed, this results in extremely high, consistent calcium recoveries, lowest calcium consumption, lowest freight and energy consumption, and lowest overall CO_2 emissions.

(3) Calcium produced in the United States uses an aluminothermic process that produces significantly lower emissions compared with alternative electrolytic processes used in imported calcium. In addition to improved calcium yield and less calcium needed with solid core Ca wire, the calcium production method is a significant factor in reducing overall carbon footprint in the calcium treatment of steel.

(4) The future will involve continuous change and improvement. With all the news about CO_2 emissions standards, greenhouse gasses, carbon footprint, etc., there is increasing demand on the steel industry to adopt new technologies that reduce carbon emissions. Producers and suppliers of calcium treatment technology can play a significant part in this environmental stewardship, being there to grow and improve along with their customers.

*All emissions in this paper were calculated using the carbon calculator at http://www.carbonify.com/carbon-calculator.htm.

REFERENCES

[1]M A Herrera Garcia and S Santos Padilla, Improvement in Surface Quality and Internal Cleanness of Thin Slab Casting at HYLSA, Vol. II., AISTech Conference (2004)

[2]L A Frank, Castability – From Alumina to Spinels New Orleans, LA: ISS 56th Electric Furnace Conference (1998)

[3]R Bischof, G J Hampton, and V Janacek, Application of Pure Metallic Calcium Wire at Trinecke Zelezarny a.s. When Making Steel Grades with Restricted Levels of Silicon, XIII Conference of Secondary Metallurgy and Steelworks Production, RODNOV pod RADHOSTEM, Czech Republic (1997)

[4]R Baum and J Yanker, Improved Calcium Treatment for Thin Slab Casting, International Symposium on Thin Slab Casting and Rolling, Nanjing, China (2009)

[5]G Cichy, K Ahlborg, and J Lewis, Process Improvements in HSLA Calcium-Treated Steels, Steelmaking Conference (1997)

[6]A Ploch, M Arns, Worner, Dieter, Installation and First Operating Results of a 380 t Ladle Furnace at the Bruckhausen Steel Plant of Thyssen Krupp Stahl AG, European Oxygen Steelmaking Conference, Birmingham, GB, (2000)

Nanotechnology for Energy

OPTICAL CHARACTERIZATION OF CHEMICALLY DEPOSITED SbCuS THIN FILMS

*[1]Chinedu EKUMA; [2]Mishack NNABUCHI, [2]Nwabueze AJA and [1]Israel OWATE

*[1]Department of Physics, University of Port Harcourt, Pmb 5323, Port Harcourt, Nigeria

[2]Department of Industrial physics, Ebonyi State University P. M. B. 053; Abakaliki Nigeria

ABSTRACT

Multi – component thin films of SbCuS have been developed on glass substrates using chemical bath deposition technique (CBD). The films were deposited for 1.5hr (sample A) and 2.0hr (sample B) both at room temperature. The optical properties of the films were characterized using spectrophotometer in the range of UV – VIS – NIR. The minimum transmittance for both samples occurred at 4.18% before increasing exponentially to a maximum of 71.85% for sample A and 60.76% for sample B within the same wavelength range. The maximum refractive index for both samples occurred at 2.29 at photon energy of 2.21eV before declining exponentially to almost an asymptotic value of 0.11. From the spectra, the band gap were found to be 3.08eV for sample A and 2.72eV for sample B giving band shift of 0.36eV within this proximity studied time interval. The extinction coefficient and optical dielectric constants (ε_r and ε_i) were also characterized. The band gap and optical properties indicate that the film has high transmittance and thus can be of high technological application for fabrication of solar cells, coating of poultry house and as window materials.

INTRODUCTION

Chemical bath deposition (CBD) technique is a simple, reproducible and cost effective technique of fabricating high quality semiconductor metal halide and large area metal chalcogenide thin films on both metallic and non-metallic substrates.[1,2,3] The method though simple produces films that have comparable structural and optoelectrical properties to those produced using various sophisticated thin film deposition techniques.[2,4,5,6,7] The CBD has been applied in fabricating emerging materials for solar thermal controls in buildings, solar cells, protective coating and other diverse applications in industries.[6,8]

The technology is based on slow but controlled precipitation of the desired compound from its ions (or atoms) in a reaction bath solution[2]; with ligand acting as a catalyst normally used to control the reaction in an appropriate medium as indicated by the pH to obtain crystal growth after initial nucleation[4]. The condition for crystal to be deposited from a solution bearing its ions is that its ionic product (I.P) should be at least greater than the solubility product (K_{sp}).[4,9]

This work reports for the first time (to the best of our knowledge) the successful deposition of ternary Antimony Copper Sulphide (SbCuS) thin films at room temperature (300K) and Dip time of 90min (sample A) and 120min (sample B) using chemical bath deposition (CBD) technique. Optical and solid state properties of the deposited films (which have not been verified by any method) were also studied. Possible applications of the films were suggested.

EXPERIMENTAL PROCEDURE

The Antimony Copper Sulphide (SbCuS) thin films which, were deposited on glass slides using the Chemical Bath Deposition (CBD) technique at room temperature (300K) were initially degreased in HCl for 24hr, cleaned in cold water with detergent, rinsed in distilled water and dried in open air. The reaction bath for the deposition were of analytical reagents containing 1M 2mL $SbCl_3$; 0.1M 5mL $CuCl_2$; 1.0M 25mL $Na_2S_2O_3$ as complexing agent; 5mL of C_3H_6O as solvent; 1.0M 5mL thiourea and 15mL of distilled water. The mixtures were carefully stirred to maintain homogeneous mixture.

The normal incidence transmittance and absorbance spectrum of the deposited films were characterized using UNICO UV-2102 PC spectrophotometer in the UV – VIS – NIR region and the optical properties deduced from the various theoretical relations already well established in literature.[4,7,10,11,12,13,14,15,16,17,18] The measurements were taken under the parametric conditions of normal incident, ambient temperature and uncoated microscopic glass slide used as blank.

RESULTS AND DISCUSSION

The results of the various optical properties of the thin films deposited at 90 and 120min respectively at room temperature are as shown in figures 1 – 6.

Fig. 1 shows the transmittance of the films as a function of the photon wavelength. It can be inferred that the films showed moderate transmittance in the UV region and high transmittance in the VIS – NIR regions. The minimum average transmittance for both samples is about 4.18% in the UV region before increasing to maximum values of 71.85% for sample A and 60.76% for sample B in the VIS – NIR region. The observed high transmittance makes the film good candidate for solar thermal application in flat plate collectors, fabrication of antireflective coating materials, material for coating poultry roofs and walls. The moderately lower transmittance in the UV region suggests that they can be harnessed for possible application in screening off UV portion of the electromagnetic spectrum which is known to be harmful to humans and domestic animals.

Figs. 2 and 3 show the plot of the refractive index and extinction coefficient as a function of photon energy respectively. The maximum refractive index for both samples (see fig. 2) occurred at 2.29 at photon energy of 2.21eV and declined exponentially to almost an asymptotic value of 0.11. The maximum extinction coefficient can be observed to be $12.1x10^{-2}$ for sample A and $6.78x10^{-2}$ for sample B at almost the same photon energy of 3.10eV.

Figs. 4 and 5 show the plot of the real and imaginary dielectric constant of the films as a function of photon energy respectively. From Fig. 4, both films exhibited maximum real dielectric constant of $5.23x10^{-3}$ corresponding to photon energy of 2.07eV before declining exponentially to an almost asymptotical value of $0.13\ x10^{-3}$ at photon energy of 2.9eV. The imaginary dielectric constant showed similar behaviour to that of the real dielectric constant but with a maximum value of $2.16\ x10^{-1}$ and $2.04\ x10^{-1}$ for samples A and B respectively at the same photon energy of about 2.24eV. This decrease may be attributed to the space charge polarization.[19]

From Fig. 6, it can be inferred that the synthesized samples show direct band gap with an optical band gap values of 3.08eV for sample A and 2.72eV for sample B and a band shift of 0.36eV. The important remark that can be made here is that the dip time significantly affected the band gap of the developed films as the band gap reduced with increasing dip time. The large band gap value is again an affirmation of large transmittance of the sample in the visible region than in the UV region.[20]

CONCLUSION

The synthesis and optical characterization of Antimony Copper Sulphide deposited by the Chemical bath deposition (CBD) technique has been investigated. The optical properties studied from the *T* and *R* values show that the film has interesting properties that can be harnessed for diverse technological applications such as fabrication of solar cells; as antireflective coatings; possible application in screening off UV portion of the electromagnetic spectrum; could be effective material as coatings for poultry houses and even as window layers for photocells etc.

REFERENCES

[1] K. Bindu, Jos´e Campos, M. T. S. Nair, A. S´anchez, P. K. Nair: *Semicond. Sci. Technol.*, **20**, 496 (2005).

[2]S. Chandra, R. K. Pandey, and R. C. Agrawal: J. Phys. D. Appl. Phys. **13**(9), 1957 (1980).

[3]K. L. Chopra, and S. R. Das: *Thin Film Solar Cells*, Plenum Press, New York, Chapters 2 and 5 (1983).

[4]G. Hodes: Chemical Solution Deposition of Semiconductor Films, Marcel Dekker Inc (2002).

[5]H. V. Campe, G. H. Hewig, W. Hoffman, H. Huschka, D. Schlurick, and J. Worner: 6th E.C. Photovoltaic Solar Energy Conf., 778 (1985).

[6]R. U. Osuji: *Nig. J. of Solar Energy* **14**, 90 (2003).

[7]J. Singh. (edn.): Optical Properties of Condensed Matter and Applications, John Wiley & Sons (2006).

[8]F. I. Ezema, and C. E. Okeke: *Nig. Jour. of Solar Energy* **14**, 66 (2003).

[9]J. A. Lange: *Lange's Hand Book of Chemistry*, 14th Edition, McGraw-Hill Book Co., New York, (1992).

[10]G. Furlan, D. Nobili, A. M. Sayigh, and B. O. Seraphin (eds.): *Workshop on Materials Science and the Physics of Non- Conventional Energy Sources*, World Scientific, Singapore, (1989).

[11]G. F. Cothian, *Absorption Spectrophotometry:* 2nd Edition, Hilger and Watts Ltd., London, (1958).

[12]D. E. Gray (ed.): *American Institute of Physics Handbook*, 3rd Edition, McGraw- Hill Book Co., New York, (1982).

[13]L. Eckertova: *Physics of Thin Films*, 2nd Edition, Plenum Press, New York, (1986).

[14]F. Abeles (ed.): *Optical Properties of Solids*, North Holland Pub. Co., Amsterdam, (1972).

[15]J. D. Meakin: *Workshop on Materials Science and Physics of Non Conventional Energy Sources*, World Scientific, Singapore, (1989).

[16]N. Suzuki, K. Sawai, and S. Adachi: *Jour. Appl. Phys.* **77**, 3 (1995).

[17]F. Wooten: *Optical Properties of Solid,* Academic Press, New York, (1972).

[18]C.U. Okujagu: Growth and Characterization of some semiconducting thin films. PhD Thesis, University of Nigeria, Nsukka (1991).

[19]C. Justin Raj; M. B. Lincoln, and S. Jerome Das: Cryst. Res. Technol. **43**, 8, 823 – 827 (2008).

[20]D. D. O. Eya; A. J. Ekpunobi, and C. E. Okeke: *Acad. Open Int. J.* 17 (2006).

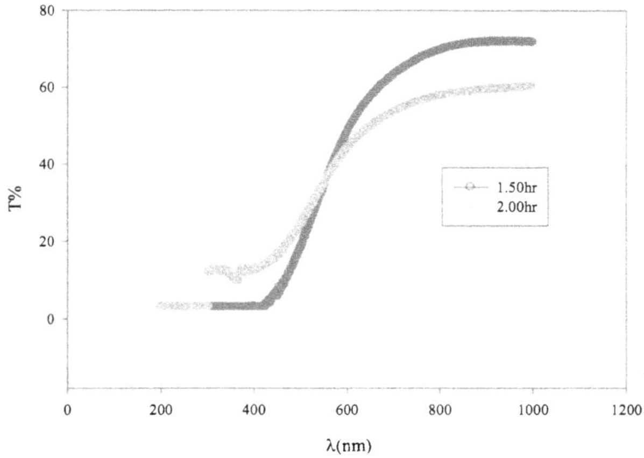

Fig. 1: Transmittance (T) as a function of wavelength (λ) for SbCuS thin films

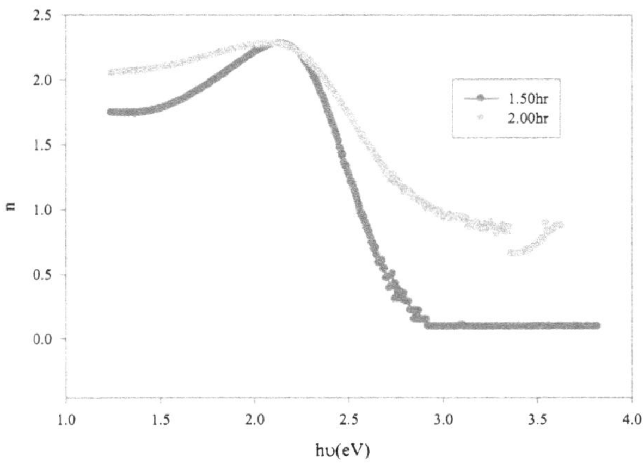

Fig. 2: Refractive index (n) as a function of photon energy (hv) for SbCuS thin films

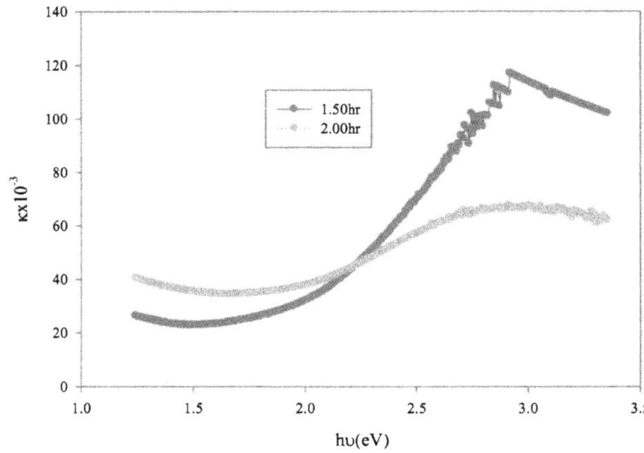

Fig. 3: Extinction coefficient (k) as a function of photon energy (hv) for SbCuS thin films

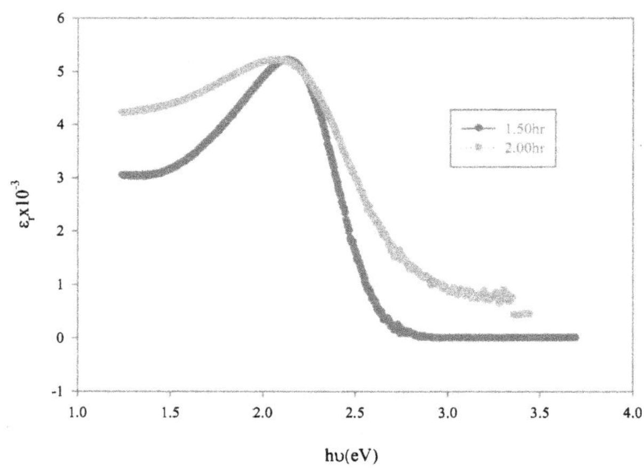

Fig. 4: Real dielectric constant as a function of photon energy (hv) for SbCuS thin films

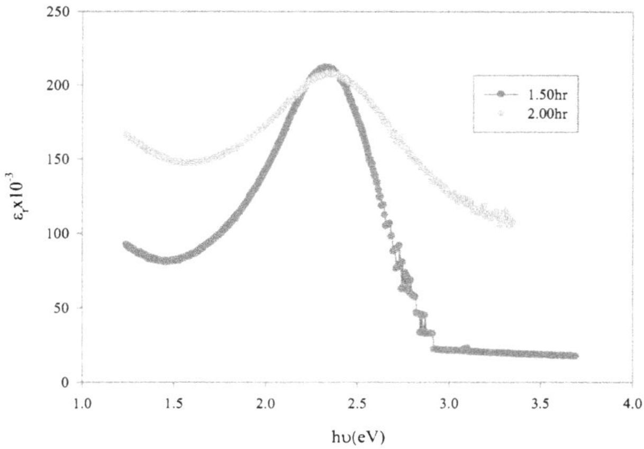

Fig. 5: Imaginary dielectric constant as a function of photon energy (hv) for SbCuS thin films

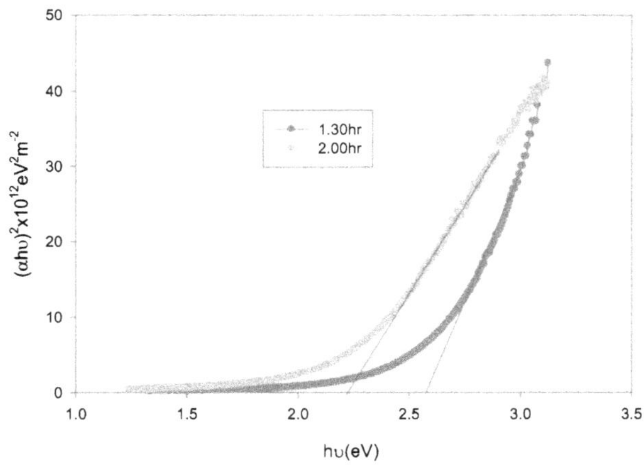

Fig. 6: Plots of $(\alpha hv)^2$ as a function of photon energy (hv) for SbCuS thin films

EXAMINING DEFECTS IN SOLID CORE 2-D PHOTONIC BAND-GAP FIBERS WITH HIGH INDEX INCLUSIONS

Neal T. Pfeiffenberger and Gary R. Pickrell
Virginia Tech, Blacksburg, VA, USA

ABSTRACT

Solid Core Photonic Band-Gap Fibers (SC-PBGF) are being closely examined today for use in the telecommunications industry because of their dispersive and spectral properties near the 1550nm wavelength. One of the benefits of these fibers, unlike the hollow core photonic band gap fibers, is the ease of fabrication and splicing. Research has shown that the optical properties of SC-PBG fibers depend on parameters including refractive index contrast, the wavelength of excitation, diameter of the high index cladding, and pitch. This research focuses on understanding the robustness of SC-PBG fibers and their ability to operate with defects in the periodic lattice. The effect of variations in size and position of these defects on the resulting optical properties will be examined.

INTRODUCTION

Design of the appropriate fiber structure is critical in today's sensor and data transmission markets. Many articles have been published on the use of optical fibers for sensors in a wide variety of applications including defense, civil infrastructure monitoring, and down-hole oil well sensing. Among the many advantages, optical fibers are immune to electromagnetic interference, which can greatly reduce the noise as compared to conventional electrical components. This is an important consideration in many applications.

The fibers under study in this work are referred to as Solid Core Photonic Band-Gap Fibers (SC-PBG). They are composed of a silica core surrounded by a periodic lattice of high index inclusions. Solid Core Photonic Band-Gap Fibers belong to a branch of micro-structured optical fibers (MOFs). These MOF fibers are divided into two main categories. The first group operates via total internal reflection. The second group operates under a photonic band gap and is the basis for this study.

The continual work on the characterization and fabrication of ordered hole fibers can be seen throughout the literature[i][ii][iii][iv][v] beginning with the work of Knight, Birks, Russell and Atkin. SC-PBG's are fabricated by the stack and draw method[vi] with a graded index that is doped with germanium to aid in production of a periodic structure. Solid Core Photonic Band-Gap Fibers are of interest because unlike hollow-core photonic band gap fibers, they are easily fabricated and spliced due to their "hole-free" nature. They are being investigated for use in the telecommunications industry because of their dispersive and spectral properties around the 1550nm wavelength[vii].

Light is guided in a Solid Core Photonic Band-Gap Fiber by a photonic band-gap effect in the cladding where the average cladding index is higher than that of the core's index. In Solid Core Photonic Band-Gap Fibers, confinement losses exist due to the fact that there are only a finite amount of high index inclusions in the cladding. Photonic band gap theory states that an infinite array of high index inclusions in the cladding would give a full 2D band-gap effect and therefore produce no loss[7].

The optical properties of SC-PBG fibers also depend on the diameter d of the Ge doped rods, the ratio of d/Λ, and the wavelength of excitation[7]. The pitch, Λ, is the distance between the centers of the high-index inclusions. Confinement loss, L_c, is directly related to the imaginary portion of the propagation constant[viii], γ through the equation:

$$L_c = 8.686*\alpha \quad (1)$$

where α is the attenuation constant. This equation gives the loss in units of dB/m[ix]. The propagation constant is defined as:

$$\gamma = \alpha + j\beta \quad (2)$$

where β is the phase constant. Maxwell's equations for a vector wave with an anisotropic PML is defined as:

$$\nabla \times ([s]^{-1}\nabla \times \boldsymbol{E}) - k_0^2 n^2[s]\boldsymbol{E} = 0 \quad (3)$$

This equation is obtained from the uniformity of the electric field, which gives us:

$$\boldsymbol{E}\,(x, y, z) = \boldsymbol{e}(x, y)\exp(-\gamma z) \quad (4)$$

This paper is focused on studying these structures through the use of Finite Element Modeling (FEM). The benefit of this type of FEM modeling is its accuracy and the efficiency of the results.

EXPERIMENTAL SECTION

Accurate numerical modeling is essential for optimal fiber design. Comsol Multiphysics uses high-order vectorial elements along with an automatic and iterative grid refinement calculator for optimum error estimation. This process is a necessity when accurate silica-germanium interfaces need to be sampled during the meshing stage. The nature of the silica-germanium interface puts a large burden on the computer's memory because of fast far-field variations and the fact that the electric field's normal component will become discontinuous. Comsol Multiphysics uses a direct linear system solver (UMFPACK) for its modal analysis. This linear solver is commonly used to compute the equations involved in the FEM discretization of the Maxwell eigenproblem[x].

The initial model of the SC-PBG fiber used in this work was replicated from a previously published paper[xi] to validate the working model. The fiber has 7 rings of Ge doped rods with a refractive index of 1.48. The doping process assumes a parabolic refractive index range of 3.0E-2 from the center of the Ge doped rod to the outer edge of the rod. This fiber maintains a d/Λ ratio equal to 0.683 with an outer PML diameter of 1.368E-4m and an inner diameter at 1.216E-4m. This gives a diameter d of 300μm and pitch Λ of 15.2μm. The core region is approximately 20μm in diameter. All models were examined at the free space wavelength λ_0=1550nm. When no defects are present in this fiber, Comsol Multiphysics gives a confinement loss L_c = 0.006968 dB/km. The loss with no defects can be seen over a wide range of free space wavelengths as seen in Figure 1 below.

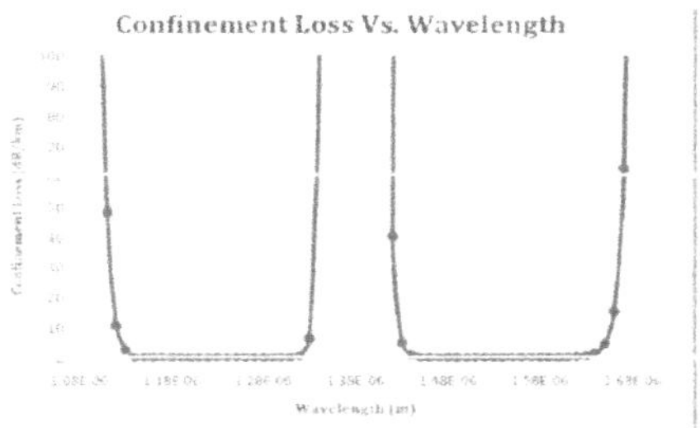

Figure 1. 2-D Comsol model of a Solid Core Photonic Band Gap Fiber

The objective of this paper was to see how robust Solid Core Photonic Band Gap Fiber's are when a defect is present. The results should follow the ARROW model[xii], which states that the closer a defect is to the core region, the greater the confinement loss will be. An air hole defect was thus added and its diameter and position was changed in order to observe how this would affect the confinement loss of the fundamental mode (LP_{01}). The air hole defect starts with a radius of 0.2595μm and increases by 0.2595μm each time, till it reaches a final value of 2.076μm. The x positioning stays constant at 0 while the y positioning changes starting from (0,0) and increasing by 1.86μm till the y position reaches a value of 18.6μm. An example of this can be seen below in Figure 2.

Figure 2. 2-D model of an air hole defect. The defect's diameter and y position was modified to examine its effect on confinement loss.

FEM ANALYSIS

The precision of the data directly relates to how refined the mesh can be. The only limitation of the mesh is the memory of the computer that will be solving the boundary equations. Figure 3 below shows a mesh consisting of 288892 elements run on a 4Gb Macintosh.

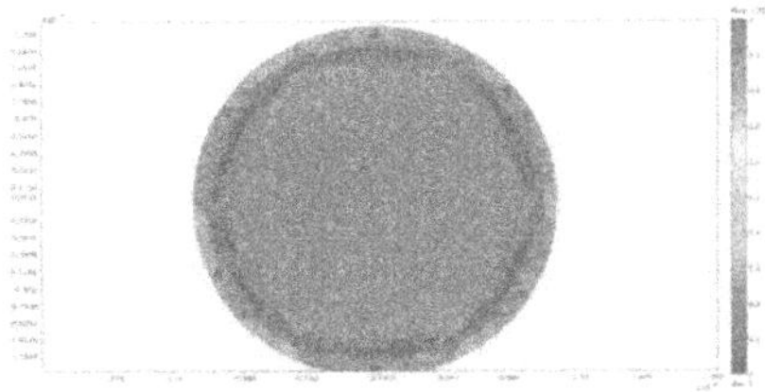

Figure 3. 2-D Comsol model of the meshing accuracy of the Solid Core Photonic Band Gap Fiber

The outer diameter of the fiber is set as a cylindrical perfectly matched layer (PML), which utilizes a reflectionless outer layer. This is used in our research in order to absorb all outgoing waves. The PML is inherently a lossy material that is reflectionless for any frequency, thus absorbing all outgoing radiation. The PML layer gives leaky eigenmodes, which radiate toward the outer silica layer[10].

When the mesh is complete, the boundary conditions must be met. Continuity was set inside the cylindrical PML while a perfect electrical conductor was set to the outer edge. The Ge doped rods were then set to n=1.48 with a parabolic refractive index range of 3.0E-2 and the silica glass was set to n=1.45. Comsol is then solved using a direct linear equation solver (UMFPACK) near the effective mode index of the silica glass. Figure 4, shown below is the resultant fundamental hybrid mode for the SC-PBG with no defect. This is as expected for the power flow to be concentrated directly in the center of the core.

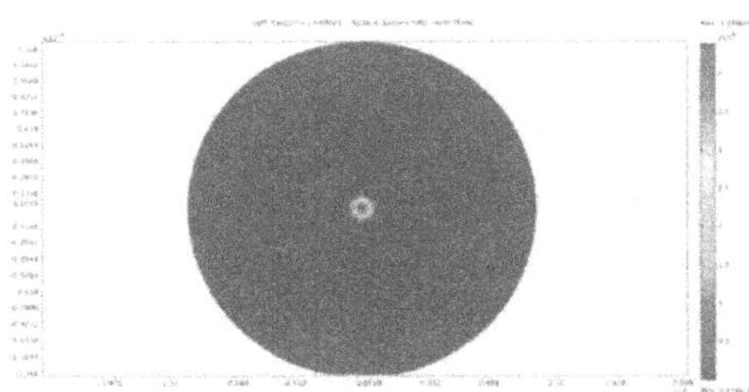

Figure 4. Transversal profile of the time-averaged power flow (W/m^2) for the LP_{01} mode in the z direction.

As stated in the Experimental Section above, an air hole defect was then added and its diameter and position was changed in order to observe how this would affect the confinement loss of the fundamental mode (LP_{01}). The results for these changes can be viewed below in Figure 5 and 6.

Figure 5 shows the confinement loss versus the radius of the air hole defect with the numbers in the legend indicating the distance from the core in (m). Figure 6 shows the confinement loss versus distance from the core in the y-axis with the numbers in the legend indicating the radius of the air hole defect. These results are also for the fundamental mode (LP_{01}) with a free space wavelength of 1.55μm.

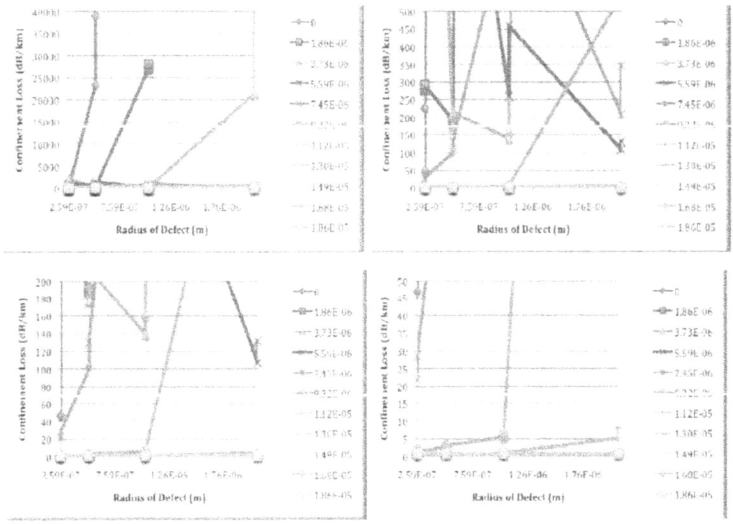

Figure 5. Confinement loss (dB/km) vs. Radius of Air Hole Defect (m)

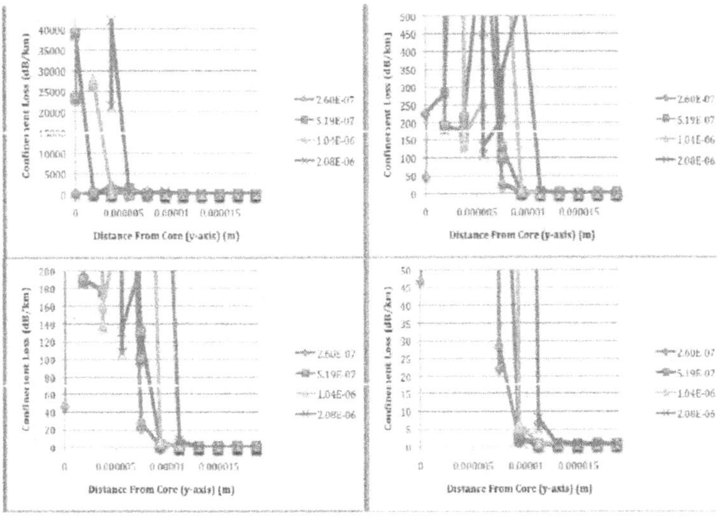

Figure 6. Confinement loss (dB/km) vs. Distance From Core (y-axis)(m)

CONCLUSIONS

This work presents the results of an addition of an air hole defect in a Solid Core Photonic Band-Gap Fiber (SC-PBG) using a multiphysics modeling program. The characteristics of the defect were modified in terms of spatial location in relation to the core, the size of the defect was changed, and the resulting changes in optical characteristics of the fiber were determined. This process allowed for easy modification of the current defect position/radius in order to further analyze what effect this will have on the confinement loss. The results follow the ARROW model and show that the closer a defect is to the core region, the greater the confinement loss will be. The results also show that the confinement loss is directly related to the size of the air hole defect.

REFERENCES

---

[i] J. C. Knight, T. A. Birks, P. St. J. Russell, and D. M. Atk in, Opt. Lett. 21, 1547 (1996)

[ii] T. A. Birks, J. C. Knight, and P. St. J. Russell, Opt. Lett. 22, 961 (1997).

[iii] J. C. Knight, T. A. Birks, P. St. J. Russell, and J. P. de Sandro, J. Opt. Soc. Am. A 15, 748 (1998).

[iv] T. M. Monro, P. J. Bennett, N. G. R. Broderick, and D. J. Richardson, Opt. Lett. 25, 206 (2000).

[v] B. J. Mangan, J. C. Knight, T. A. Birks, and P. St. J. Russell, in Digest of Conference on Lasers and Electro-Optics (CLEO), Optical Society of America, Washington, D.C., 559 – 560 (1999).
[vi] P.St.J. Russell, " Photonic crystal fibers," Science **299**, 358-362 (2003)
[vii] Vincent Pureur, Géraud Bouwmans, Mathias Perrin, Yves Quiquempois, and Marc Douay, "Impact of Transversal Defects on Confinement Loss of an All-Solid 2-D Photonic-Bandgap Fiber," J. Lightwave Technol. **25**, 3589-3596 (2007)
[viii] K. Saitoh and M. Koshiba, "Full-vectorial imaginary-distance beam propagation method based on finite element
scheme: Application to photonic crystal fibers," IEEE J. Quantum Electron. **38**, 927-933 (2002).
[ix] Kunimasa Saitoh and Masanori Koshiba, "Leakage loss and group velocity dispersion in air-core photonic bandgap fibers," Opt. Express **11**, 3100-3109 (2003)
[x] Jianming Jin, "The Finite Element Method in Electromagnetics," Second Edition, John Wiley & Sons, Inc,
New York 2002.
[xi] G. Bouwmans, L. Bigot, Y. Quiquempois, F. Lopez, L. Provino, and M. Douay, "Fabrication and characterization of an all-solid 2D photonic bandgap fiber with a low-loss region (< 20 dB/km) around 1550 nm," Opt. Express **13**, 8452-8459 (2005)
[xii] N. M. Litchinitser, S. C. Dunn, B. Usner, B. J. Eggleton, T. P. White, R. C. McPhedran, and C. M. de Sterke, "Resonances in microstructured optical waveguides," Opt. Express **11**, 1243-1251 (2003)

NANOPHASED MATERIALS IN SUPERCRITICAL CO_2: CERAMIC NANOPOWDER SYNTHESIS, ENCAPSULATION AND DEPOSITION

JC. Ruiz, F. Charton, S. Sarrade, A. Hertz, B. Fournel
French Atomic Energy Commission (CEA) – Marcoule Research Centre – DTCD/SPDE
BP 17171- F-30207 Bagnols sur Cèze cedex, France

A. Julbe
European Membrane Institute (UMR CNRS 5635) – UMII-CC047
2, place E. Bataillon – F – 34095 Montpellier cedex 5, France

C. Guizard
SAINT GOBAIN CREE, 550 Avenue A. Jauffret – BP 224 – F – 84306 Cavaillon cedex, France

ABSTRACT

The supercritical CO_2 (SC-CO_2) is a suitable and promising medium allowing the synthesis (above 7.4 MPa and 32°C) of high surface area nanostructured solids. This process makes it possible to perfectly control the solids' characteristics and morphologies. Therefore, this medium is appropriate to produce new materials or catalysts displaying high dispersion of the active phase and high reactivity.
The process' principle is based on a bottom-up of synthesis (chemical or thermal decomposition) in which metal precursors, metal-organics or salts, are hydrolyzed to form condensed species (clusters or colloids), leading to ultrafine oxide particles. In this way, it was possible to synthesize a great number of ceramic powders containing simple (SiO_2, TiO_2) or mixed oxides ($Ce_{1-x}Gd_xO_{2-\delta}$, $Zr_{1-x}Y_xO_{2-\delta}$). These powders have surface areas varying between 50 and 500 $m^2.g^{-1}$.
It is also possible to produce nanopowders composed of dispersed Pt or Pd nanoparticles in strong interaction with a substrate (oxide) by SC-CO_2 route. Impregnation of such composite nanostructured powder in direct SC-CO_2 synthesis on a macroporous substrate (monolith, foam) has also been developed.
The MATCOS facility is being developed in order to go beyond laboratory scale studies. MATCOS is a platform including different pilots up to a semi industrial scale. The facility is located in Marcoule (France) and is able to perform advanced material synthesis.

INTRODUCTION

Over the past decade, supercritical fluids have received considerable attention as solvent or reaction media for the synthesis of a number of ceramic and closely related oxide materials. Low viscosity of SC-CO_2 greatly enhances the mass transfer whereas its' high density enables the solubilization of solutes along with a high diffusivity value. This enhanced diffusion process is a condition for a faster reaction kinetic.
Synthesis of material in SC-CO_2 is thus leading to new particles' morphologies as well as high surface areas. It is then expected that the reactivity based on the use of these original materials will be greatly increased compared to classical routes for material synthesis (precipitation in a liquid, sol-gel) [1] [2] [3]. Moreover, a ceramic powder synthesized by this method needs no drying step, and their crystallization temperature is generally much lower compared to the traditional processes.

In a previous study we have improved TiO_2 synthesis in SC-CO_2 in order to increase its UV attenuation factor [4]. Experiments have been performed first by batch process and secondly in a semi-continuous reactor. The effects of operating conditions on the final product and their influence in the different steps of the process have been studied. Results have shown that product crystallinity, mean particle size and particle size distribution are related to the UV attenuation factor.
The second example concerns an original preparation to obtain by direct synthesis in SC-CO_2 medium, nanopowders composed of dispersed Pt or Pd nanoparticles in strong interaction with a TiO_2 substrate. These anatase mixed powders show surface areas varying between 200 and 410 $m^2.g^{-1}$, according to the amount of noble metal introduced. We also studied the impregnation of such composite nanostrutured powder in a direct SC-CO_2 synthesis on ceramic Al_2O_3/ZrO_2 foam.
Recently, we have investigated the preparation of nanosize Yttrium Stabilized Zirconia (YSZ) powders by a sol-gel process. The process was assisted by SC-CO_2 in order to increase the electrical properties of nanostrutured sintered YSZ ceramics for application in high temperature electrochemical devices. In this third example, the encapsulation of the YSZ particles with a polymer shows that nanophased powder can be successfully handled and aggregated by using a MMA polymerisation process in SC-CO_2.

APPARATUS AND PROCESSING PRINCIPLES

Ceramic powder synthesis

Basically, the process which has been developed by our research team starts from one or several organometallic precursors and combines a chemical conversion step with a crystallization step, using SC-CO_2 as solvent (or anti-solvent) and as chemical vector. Typical conversion steps are sol-gel reaction, thermal conversion and chemical reduction. In the case of sol-gel route (hydrolysis/condensation of an organometallic precursor), the hydrolysis reaction is initiated either by introduction of water into the reactor or by in situ dehydration of the precursor using a temperature increase.

Synthesis of ceramic nanoparticles: Batch process

Figure 1 shows the supercritical equipment used to synthesis ceramic nanoparticles by SC-CO_2 route, according to a process batch. The precursor is introduced into the vessel which is then controlled in temperature and pressure under agitation. Solutes or reactive species (water in the case of TiO_2) can be introduced in supercritical conditions via a HPLC pump. Once the reaction has started, a few seconds are enough for a complete reaction of the reactants. Under the effect of temperature, crystallization occurs in five minutes. The required single-crystal phase depends on the temperature of the process.

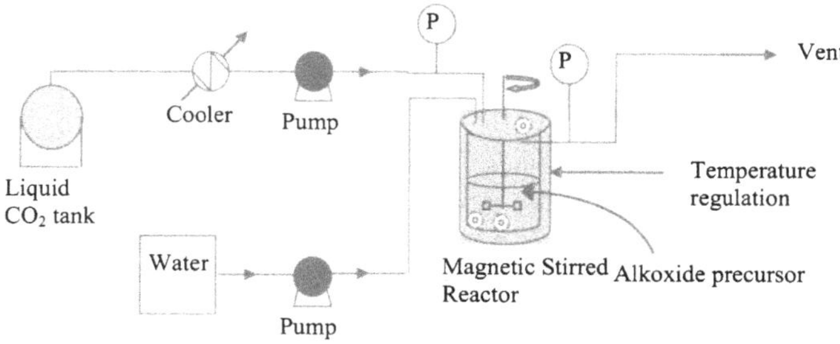

Figure 1: Schematic representation of the supercritical equipment used for the synthesis of nanoparticles: Batch process

<u>Synthesis of ceramic nanoparticles: Fast injection process</u>

In the case of the fast injection processing, solutes or reactive species are first introduced into the reactor. The reactor is loaded with pure CO_2 and temperature and pressure are increased. The organometallic precursor's solution is injected into the reactor between 30 and 40 MPa (figure 2). The pressure drop between the reactor and the injector chamber defines the flow rate of injection. An efficient mixing (turbine) allows to disperse the reactive solution in the medium, then to disperse the precursor solution during the injection. The fast injection process has been developed in order to avoid the problems linked to the kinetic and reactional ways of the batch process. This innovating process [5] consequently allows to control physical parameters and all characteristics of the powders.

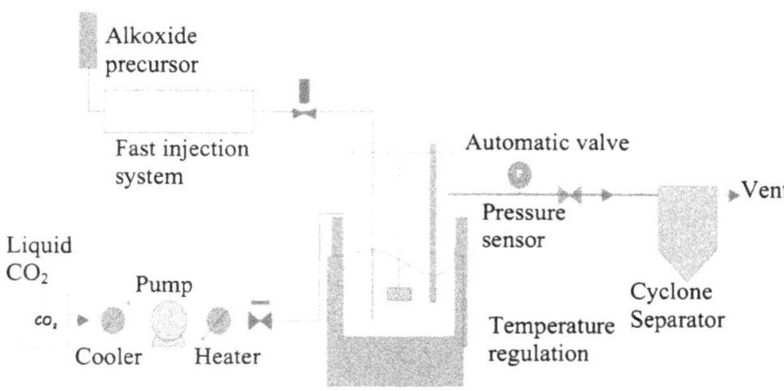

Figure 2: Schematic representation of the supercritical equipment used for the synthesis of nanoparticles: Fast injection process

<u>Characterization methods</u>

Morphology and particle sizes of the produced powders are studied using electron microscopy FESEM (Hitachi S4500) and transmission electron microscopy (TEM). The specific area of the powders is evaluated by means of nitrogen adsorption BET (Micromeritics-Asap 2010). The crystal structure of the TiO_2 powders is investigated by X-ray powder diffraction (XRD) using Cu Kα radiation (Brucker D800). The diameter of the nanoparticles is calculated from the DRX measurement using the Scherrer equation.

RESULTS

Nanosized TiO_2 powder synthesis: use of TiO_2 nanoparticles as inorganic UV filters

The powders synthesized in this study are elaborated using the reaction between titanium isopropoxide $Ti(OiPr)_4$ and water according to reactional mechanisms of classical sol-gel route [6].

The present study consisted in improving TiO_2 synthesis in supercritical carbon dioxide (SC-CO_2) in order to increase its UV attenuation factor. Experiments have been carried out first by batch process and then in a semi-continuous reactor (fast injection process). We studied the effects of the operating conditions on the final product and their influence in the different steps of the process (figure 3). The results show that product crystallinity, mean particle size and particle size dispersion influence the UV attenuation factor (figure 4). Thus, well crystallized TiO_2 anatase particles of about 15 nm in diameter have been obtained working at 250-300°C, 30 MPa with residence time of 5 min. Such particles show good radiation protection properties and specific surface area BET of around 100 $m^2.g^{-1}$.

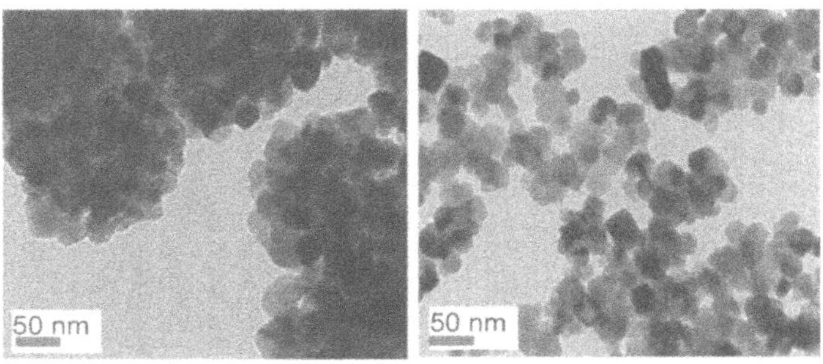

Figure 3: TEM pictures of TiO_2 particles obtained by fast injection process; a) by in situ dehydration route: b) by hydrolytic sol-gel route

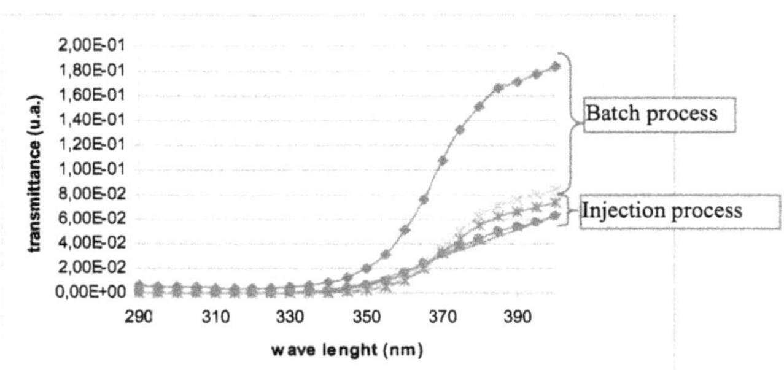

Figure 4: UV transmission of TiO_2 samples elaborated by batch and fast injection processes

Noble metals based TiO_2 nanopowders: use of crystalline Pd(Pt)-TiO_2 phases in pollution remediation

Catalytic nanoparticles synthesis of Pd(Pt)-TiO_2 was achieved in a batch process at 30 MPa and a temperature ranging from 100 to 300 °C [7]. The reaction is hydrolysis followed by condensation of a titanium alkoxide and a solution composed of a platinum or palladium

precursor. Precursors used for synthesis are titanium isopropoxide $Ti(OiPr)_4$, palladium nitrate $Pd(NO_3)_2$ and platinum acetyl-acetonate($Pt(acac))_2$.
$Pd\text{-}TiO_2$ powders are produced by the hydrolysis-condensation of $Ti(OiPr)_4$ and an aqueous solution of $Pd(NO_3)_2$ in $SC\text{-}CO_2$. $Pt\text{-}TiO_2$ powders are obtained by hydrolysis and condensation of an alcoholic solution of $Ti(OiPr)_4$ and $Pt(acac)_2$ in $SC\text{-}CO_2$.
The properties of the so obtained mixed nanomaterials were assessed in terms of catalytic decomposition of chlorinated volatile organic compounds, specifically in the oxidation of chlorobenzene, which is a standard reaction for studying the elimination of dioxins. The results obtained for the catalysts containing 0.1 wt% Pt indicate that the catalytic activity depends on the structure of the TiO_2 support.
SEM observations show agglomerated particles of size order 100 nm for all noble metal concentrations (Pd-Pt) (Figure 5). The particle size composing these aggregates is ranging from 15 to 30 nm. Powder composed of 0.5 % of noble metal seems more homogeneous than powders obtained with lower noble metal concentrations (0.01 and 0.1% weight).

Figure 5: $Pd\text{-}TiO_2$ 0,5% weight SEM image (200°C / 30 MPa)

$Pd\text{-}TiO_2$ deposits containing 0.1% of palladium have been realized on ceramic foams samples of Al_2O_3/ZrO_2. These foams have a macroporous structure in order to minimize gas pressure loss. To enhance surface exchanges, high and deep impregnation of the solid substrate are necessary. In addition, impregnation must be stable to high temperatures to avoid removal of the catalytic active layer. Deposition of the active layer has been achieved at 300 °C, 30 MPa during 15 minutes in the batch process. The rate of material deposition on the solid foam surface is about 7% (weight). SEM observation reveals a surface layer deposit as well as a deep deposition onto the pore surfaces (figure 6). The obtained active layer on Al_2O_3/ZrO_2 solid substrate is homogeneous and covers the entire support surface.

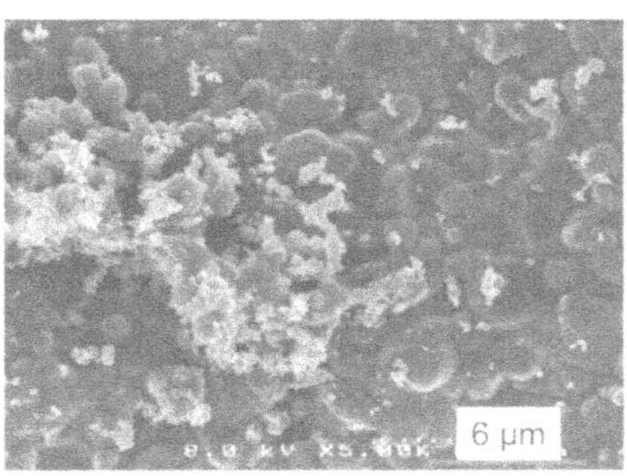

Figure 6: $Pd\text{-}TiO_2$ deposition (0.1%) on ceramic foam of Al_2O_3/ZrO_2

Nanosized YSZ powder synthesis: YSZ electrolyte membrane for SOFC

The synthesis of the nanosize YSZ powder has been carried out by the SC-CO_2-assisted sol-gel route. In this process, different mechanisms are involved in crystalline powder formation, based on hydrolysis and condensation reactions of the precursors as well as nucleation and particle growth due to the anti-solvent effect of SC-CO_2. Both, co-solvent selection and control of the heating rate, revealed to be decisive parameters for producing pure and fully crystallized YSZ phase from the two precursor mixtures in CO_2/pentane and CO_2/isopropanol. Using these results, we are able to prepare tetragonal stabilized, nanosize and monodisperse particles, 40-80 nm in size, and showing specific surface area up to 200 $m^2.g^{-1}$. Typical powder morphology is shown in figure 7.

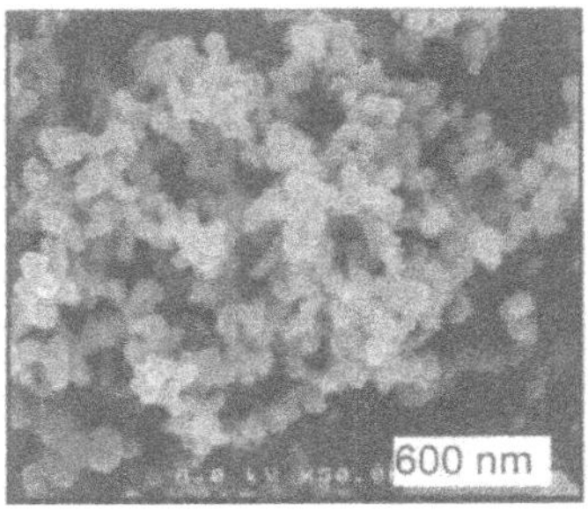

Figure 7: FESEM observation of YSZ powder with controlled characteristics.

In order to solve the problems of handling and aggregation, the YSZ particles have been successfully encapsulated with a polymer using a MMA polymerisation process in SC-CO_2 (figure 8).

In a first step, the CO_2-phobic part of a surfactant organizes itself at the surface of the particles, forming a CO_2-philic shell. Then steric interactions between surfactant shells help the dispersion of particles in SC-CO_2. Finally, the monomer adsorbs itself on the CO_2-philic part of the surfactant, and its polymerization is promoted by the presence of a free radical initiator. The powder's surface accessibility to MMA molecules dissolved in SC-CO_2 first determines the quantity of adsorbed monomer and then the quantity of formed polymer in the porous structure of the powder. This quantity can be controlled by the ratio of the monomer concentration to the total powder surface introduced in the reactor.

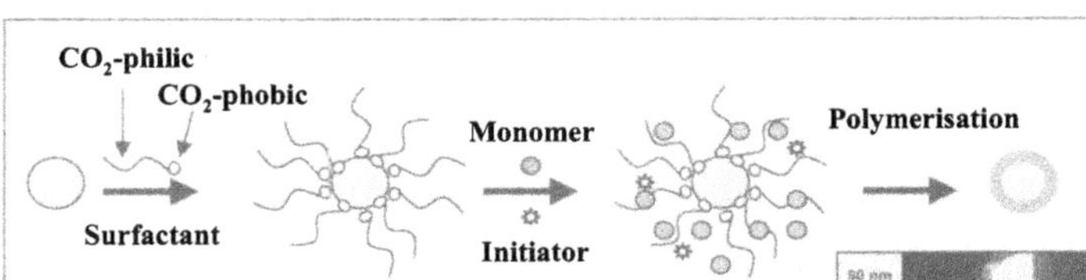

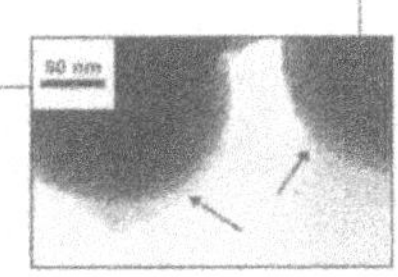

Figure 8: Schematic description of the mechanism for encapsulation of YSZ powder by polymerization in SC-CO_2 and HR-TEM image of the encapsulated aggregate.

CONCLUSION

The conditions of sol-gel synthesis adapted in SC-CO_2 medium allow interesting possibilities for the preparation of nanostrutured metallic oxide particles with controlled morphology. Fast diffusion of the species in this medium accelerates the reactions of hydrolysis and condensation of the precursors as well as nucleation and aggregate growth of nanocrystallites. The possibility of forming crystallized particles with hierarchical structure is one of the specialities of this process. Encapsulation of the particles by a polymer can also be carried out in the supercritical medium, directly after the synthesis of the powders. These powders have a great reactivity to sintering leading to remarkable functional properties.

A supercritical fluid is one of the suitable processes allowing production of high surface/volume ratio nanomaterials. Titanium dioxide (TiO_2) is a well known active semiconductor photocatalyst used in heterogeneous photocatalysis to destroy environmental pollutants. Fine particle of titanium dioxide allow to efficiently attenuate UVB (290-320 nm) and UVA (320-400 nm) sunlight radiation. We have shown that the size, the particle size distribution, the structure and the composition of TiO_2 nanopowders have a considerable influence on the absorption of UV light.

The SC-CO_2 medium is therefore appropriate to produce new catalyst displaying a high dispersion of the active phase and a high reactivity. Nanopowders composed of dispersed Pt or Pd nanoparticles in strong interaction with a TiO_2 substrate have been elaborated by direct synthesis in SC-CO_2 medium. The results obtained for the catalysts containing 0.1 wt% Pt indicate that the catalytic activity depends on the structure of the TiO_2 support.

First results on YSZ electrolyte membrane preparation lead to interesting densification rates (> 97% d_{th}) by Spark Plasma Sintering (SPS) with ceramic grain sizes close to 100 nm. These materials are now suitable for electrochemical characterization and ionic conductivity evaluation. Further work is in progress using these powders in other forming and sintering ceramic processes with the aim to produce YSZ electrolyte membranes with different shapes. Moreover, the YSZ particles have been encapsulated with a MMA polymerization process in SC-CO_2 in order to solve the problems of handling and aggregation.

REFERENCES

[1] F. Cansell, C. Aymonier and A. Loppinet-Serani, Review on materials science and supercritical fluids. Current opinion in Solid State and Materials Science, Vol 7, 2003, p 331-340

[2] S. Sarrade, L. Schrive, C. Guizard and A. Julbe, in PCT Int,. 1998, WO9851613, France

[3] A. Hertz, S. Sarrade, C. Guizard and A. Julbe, Journal of European Ceramic Society, Vol 26(7), 2006, p 1195 – 1203

[4] B.L.Diffey, J.Robson, J.Soc.Cosmet.Chem, Vol 40, 1989, p 127

[5] A. Hertz, B. Fournel, C. Guizard, A. Julbe, JC. Ruiz and S. Sarrade, in PCT Int,. 2007, WO2007131990, France

[6] JC. Ruiz, A. Michaux, V. Guyot Ferréol, JF. Tranchant and B. Fournel, Use of TiO_2 nanoparticle as inorganic UV filters: effect of operating conditions on the characteristics on the final product, in Proc. of 11th European Meeting on Supercritical Fluids, Barcelona, Spain, 2008, n°ISBN 2-905267-55-0

[7] JC. Ruiz, B. Fournel, Sabine Valange, L. Pinard and J. Barrault, Noble Metals Based Titania Nanopowders Elaboration in SC-CO_2 Medium and their Use for Gas Depollution, in Proc. of 8th International Symposium on Supercritical Fluids, Kyoto, Japan, 2006

INFLUENCE OF GAS FLOW RATE ON THE FORMATION OF ZnO NANORODS AND THEIR EFFECTS ON PHOTOELECTROCHEMICAL RESPONSE

Sudhakar Shet,[1,2] Kwang-Soon Ahn,[3] Ravindra Nuggehalli,[2] Yanfa Yan,[1] Todd Deutsch,[1] John Turner,[1] and Mowafak Al-Jassim[1]

[1] National Renewable Energy Laboratory, Golden, CO 80401 USA
[2] New Jersey Institute of Technology, Newark, NJ 07102 USA
[3] School of display and chemcial engineering, Yeungnam University, Dae-dong, Kyungsan, S. Korea 712-749

ABSTRACT

ZnO thin films are deposited in pure Ar and mixed Ar and N_2 gas ambient at substrate temperature of 500°C by rf sputtering ZnO targets. All the films were deposited on fluorine doped tin oxide coated glass. We find that the presence of optimum N_2 to Ar ratio in the deposition ambient promotes the formation of well aligned ZnO nanorods. ZnO thin films grown at 25 % N_2 gas flow rate promoted aligned nanorods along c-axis exhibit significantly enhanced photoelectrochemical response, as compared to ZnO thin films grown at other N_2 to Ar gas flow ratios. Our results suggest that chamber ambient is very important for the formation of aligned nanostructures, which offer potential advantages for improving the efficiency of photoelectrochemical water splitting for H_2 production.

INTRODUCTION

Transition-metal oxides are attractive candidates for photoelectrochemical (PEC) splitting of water for H_2 production due to their low cost and potential stability[1-4]. However, to date, only TiO_2 has received extensive attention[1,2,5-7]. ZnO has similar bandgap (~3.3 eV) and band-edge positions compared to TiO_2. Furthermore, ZnO has a direct bandgap and higher electron mobility than TiO_2[8]. Thus, ZnO could also be a potential candidate for PEC splitting of water[9].

The performance of thin-film electrodes are often affected by the morphological features of the thin films, such as grain size, grain shape, and surface areas. Thus, electrodes with nanostructures have been applied to improve PEC properties[10-12]. However, most of these nanostructures are not single crystals and are defective. It is known that defects typically act as recombination centers that can kill photon-generated electron-hole pairs before they can reach to surfaces for reactions. Therefore, single-crystal nanostructures are highly desirable. For ZnO, various forms of single-crystal nanostructures, such as nanobelts, nanorods, and nanowires, have been reported. The PEC properties of all these nanostructures deserve careful investigation.

Recently, we reported on the growth of aligned ZnO nanorods and their PEC performance[13-14]. Our results suggested that the deposition ambient can be used to control the morphology of ZnO thin films, and aligned single-crystal nanorods can be potentially beneficial for PEC performance. In this paper, we systematically examine the effects of growth conditions on the morphologies and PEC response of ZnO thin films sputtered in pure Ar and mixed Ar and N_2 ambient at different RF power. We find that deposition in pure Ar ambient produces polycrystalline ZnO films, whereas deposition in mixed Ar and N_2 ambient leads to the formation of aligned single-crystal ZnO nanorods along the *c*-axis when substrate temperature and RF power become appropriate.

EXPERIMENTAL

Three sets of samples are deposited. One set was deposited in mixed Ar and N_2 ambient and is referred to as ZnO(Ar/N_2). The second set of samples was deposited in pure Ar gas ambient. This set of samples is referred to as ZnO(Ar). The third set of samples was deposited in mixed Ar and N_2 ambient with different gas flow rate and is referred to as ZnO(N_2:Ar). ZnO targets were used in these depositions. Transparent conducting FTO (20–23 Ω/)-coated glass was used as the substrate to allow PEC measurements. Prior to deposition, the substrates were ultrasonically cleaned by an acetone-methanol-deionized (DI) water sequence. The distance between the ZnO target and substrate was about 10 cm, and the substrates were rotated 30 rpm to enhance deposition uniformity. The base pressure was below 1×10^{-6} torr and the working pressure was 5×10^{-3} torr. The chamber ambient was either pure Ar or the mixed Ar and N_2 with different gas flow rate. Prior to sputtering, a pre-sputtering cleaning was performed for 20 min to eliminate possible contaminants from the target. Sputtering was then conducted at RF power of 200W for ZnO(Ar), ZnO(Ar/N_2) and 300W for ZnO(N_2:Ar) samples. All the deposited samples were controlled to have similar film thickness of 1 ± 0.05 μm as measured by stylus profilometry.

The structural and crystallinity characterizations were performed by X-ray diffraction (XRD) measurements, using an X-ray diffractometer (XGEN-4000, SCINTAG Inc.), operated with a Cu Kα radiation source at 45 kV and 37 mA. The N concentration in the ZnO(Ar:N_2) films was evaluated by X-ray photoelectron spectroscopy (XPS). Monochromatic Al K_α radiation was used for all data sets, and the analyzer was set to a 59-eV pass energy. Argon ion sputtering (3 keV, 0.8 μAmm^{-2}, 120 s) was used to clean samples prior to analysis. The surface morphology was examined by atomic force microscopy (AFM) conducted in the tapping mode with a silicon tip, and field-emission scanning electron microscopy (FE-SEM). The ultraviolet-visible (UV-Vis) absorption spectra of the samples were measured by an n&k analyzer 1280 (n&k Technology, Inc.) to investigate the optical properties.

PEC measurements were performed in a three-electrode cell with a flat quartz-glass window to facilitate illumination to the photoelectrode surface[15-17]. The sputter-deposited films were used as the working electrodes. A Pt plate and Ag/AgCl electrode were used as counter and reference electrodes, respectively. A 0.5-M Na_2SO_4 mild aqueous solution was used as the electrolyte for the stability of the ZnO[15-17]. Photoelectrochemical response was measured using a fiber optic illuminator (150-W tungsten-halogen lamp) with a UV/infrared (IR) filter. Light intensity was measured by a photodiode power meter, in which total light intensity with the UV/IR filter was fixed to 125 mW/cm^2. The PEC response data under light on/off illumination were also measured to confirm the photoresponse of the films during the potential sweep (scan rate: 5 mV/s).

RESULTS AND DISCUSSION

We first see how the presence of N_2 in the ambient can promote the formation of aligned nanorods in ZnO thin films where the substrate temperature is higher than 300°C. Figures 1(a) and 1(b) show XRD curves for the ZnO(Ar/N_2) films and ZnO(Ar) films deposited at different substrate temperatures. The crystallinity of ZnO films increases gradually with the increase of substrate temperatures in both cases. Figure 1(c) shows the expanded intensity scale for the substrate and ZnO(Ar/N_2) films deposited at 100 and 200 °C. These two ZnO(Ar/N_2) films have random orientation. With the increase of substrate temperature to 500°C, the (0002) peak of the ZnO(Ar/N_2) film was enhanced greatly, as shown in Fig. 1(a). The measured full-width at half-maximum (FWHM) values of (0002) peaks are shown in Fig. 1(d). The N concentrations (at.%)

for the $ZnO(Ar/N_2)$ films measured by XPS are also given in Fig. 1(d). With the increase of substrate temperature, the N concentration decreased rapidly and disappeared at temperatures above 300°C. The FWHM of ZnO(Ar) decreased slightly with the increase of substrate temperature. $ZnO(Ar/N_2)$ films grown at temperatures below 200°C exhibited random orientation and larger FWHM values than the ZnO(Ar) films grown at the same temperatures. This is because the high concentration of N is incorporated in $ZnO(Ar/N_2)$ films at these temperatures. It is known from recent reports that incorporated N atoms can deteriorate the crystal structure and modify the growth mode[18-20].

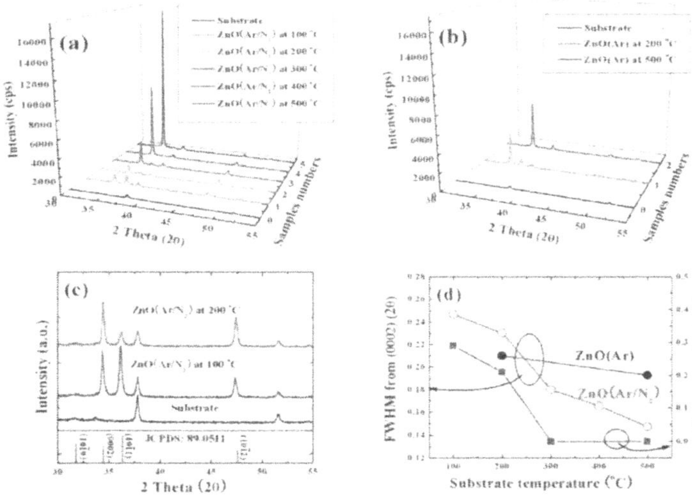

Figure 1. X-ray diffraction curves for (a) $ZnO(Ar/N_2)$ films and (b) ZnO(Ar) films. The sample numbers 1, 2, 3, 4, and 5 along the y-axis in (a) correspond to substrate temperature of 100°, 200°, 300°, 400°, and 500°C, and the sample numbers 1 and 2 in (b) correspond to substrate temperatures of 200° and 500°C. (c) Expanded intensity scale for the substrate and the $ZnO(Ar/N_2)$ films deposited at 100 and 200 °C. (d) FWHM values estimated from (0002) for the $ZnO(Ar/N_2)$ and ZnO(Ar) films and the N concentrations for the $ZnO(Ar/N_2)$ films as a function of the substrate temperature.

However, the FWHM values of $ZnO(Ar/N_2)$ films decrease rapidly when the substrate temperatures are above 300°C because no significant N can be incorporated at these temperatures. At substrate temperatures above 300°C, $ZnO(Ar/N_2)$ films exhibit much smaller FWHM values than the ZnO(Ar) films. The rapid growth of the FWHM values indicates either increased crystallinity or formation of nanorods or nanowires along the c-axis.

Figures 2(a) and 2(b) show FE-SEM top-views of the ZnO(Ar) and $ZnO(Ar/N_2)$ films, respectively, deposited at a substrate temperature of 500°C. FE-SEM images revealed that the significantly increased (0002) peak in the XRD curve obtained in $ZnO(Ar/N_2)$ at 500°C is largely due to the formation of aligned nanorods along the c-axis. It clearly shows that the nanorod structure was not present in the ZnO(Ar) film, whereas the ZnO(Ar/N_2) films at 500°C exhibited vertically aligned, single-crystal hexagonal-like nanorods with flat (0002) surfaces. However, the ZnO(Ar) film deposited at the same temperature is polycrystalline. It should be noted that at 500°C, the diameters of the nanorods are smaller than that of the grains in polycrystalline ZnO film. The smaller FWHM value for the ZnO(Ar/N_2) film is attributed to the nanorod feature[21-23].

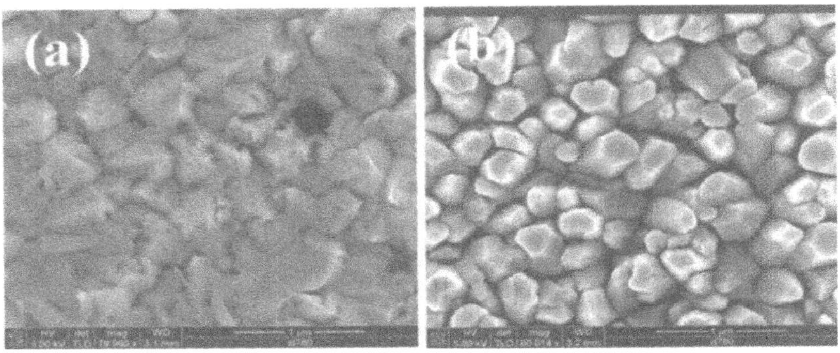

Figure 2. FE-SEM top-views of the (a) ZnO(Ar) film and (b) ZnO(Ar/N_2) nanorod film, respectively, deposited at 500°C.

No metal clusters were found at the end of the nanorods, indicating that the growth mechanism is not the catalyst-assisted vapor-liquid-solid (VLS) growth[22-24]. Recently, catalyst-free ZnO nanorods/nanowires have been synthesized by various chemical and physical techniques such as plasma-enhanced chemical vapor deposition, metal-organic vapor-phase epitaxy, and pulsed laser deposition[22-24]. The nanorod structures provide high surface areas and superior carrier transport (or conductivity) along the c-axis, which may lead to increased interfacial reaction sites and reduced recombination rate[10,25]. Therefore, the aligned nanorod films deposited at 500°C should lead to enhanced PEC response.

The PEC response for the ZnO films deposited in different ambient was also investigated. Figures 3(a) and 3(b) show photocurrent-voltage curves of the ZnO(Ar/N_2) and ZnO(Ar) films deposited at 500°C, respectively, under continuous illumination (red curve), dark condition (black curve), and light on/off illumination (blue curve) with an UV/IR filter. Both ZnO films show very small dark currents up to a potential of 1.4 V. The ZnO(Ar/N_2) nanorod film deposited at 500°C exhibited much higher photocurrents than the ZnO(Ar) film deposited at the same substrate temperature. The ZnO(Ar/N_2) film deposited at 500°C exhibits the best photoelectrochemical response—more than two times higher than the ZnO(Ar) films deposited at the same temperature.

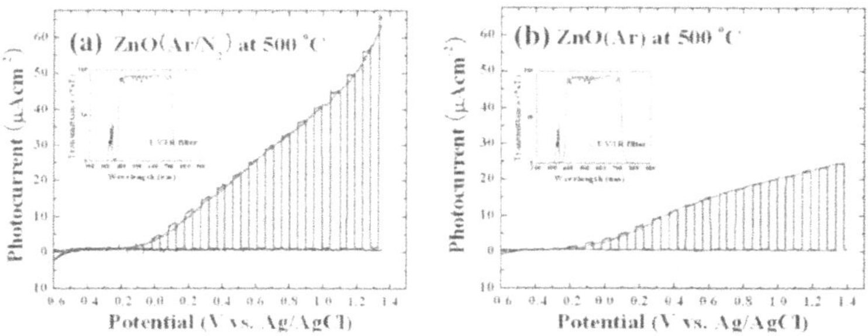

Figure 3. Photocurrent-voltage curves of (a) ZnO(Ar/N_2) nanorod and (b) ZnO(Ar) films, deposited at 500°C under (red curve) continuous illumination, (black curve) dark condition, and (blue curve) light on/off illumination with an UV/IR filter.

The enhancement can be attributed to the aligned nanorod structure along the c-axis and additional light absorption in the long-wavelength regions. The electron-hole pairs are generated by the absorption of photons with energies larger than the bandgap and separated by the electric field of the depletion region. In the case of n-type semiconductors, the excited electrons move through bulk region to the counter electrode where water reduction occurs. The generated holes move towards semiconductor/electrolyte interface where water oxidation takes place. The aligned nanorod structures along the c-axis provide high surface area and superior carrier transport along the c-axis, leading to the increased interfacial reaction sites and reduced recombination rate between the electrons and holes. As a result, the PEC performance of the nanorod structure is greatly enhanced.

Figures 4 shows X-ray diffraction curves for the ZnO thin films deposited at different nitrogen to argon ratio at substrate temperature of 500°C. From the above discussion, we reported that aligned ZnO nanorod can be formed in mixed N_2 to Ar chamber ambient at substrate temperature of 500°C. The gas flow rate of 25% nitrogen gas showed better crystallinity of ZnO film compared to the other ZnO thin films deposited at gas flow rate of 12.5% and 75% nitrogen. The ZnO thin film deposited at 25% nitrogen gas flow rate showed least FWHM value, whereas, 75% nitrogen ambient exhibited the maximum FWHM value. The much smaller FWHM value indicates either increased crystallinity or formation of nanorods or nanowires along the c-axis. No significant N concentrations (at.%) found for the ZnO thin films measured by X-ray photoelectron spectroscopy (XPS).

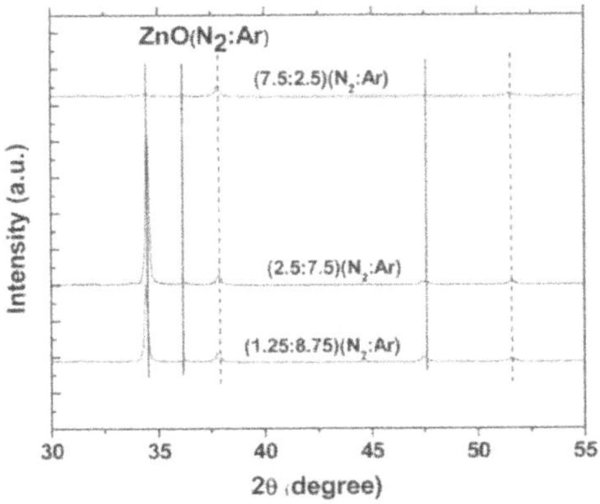

Figure 4. X-ray diffraction curves for $ZnO(N_2:Ar)$ thin films deposited at different nitrogen gas flow rate.

Figures 5 (a) and (b) shows FE-SEM top-views of the ZnO thin films deposited at 25% and 75% gas flow rate, respectively. It clearly shows that the nanorod structure was not present in the ZnO thin film grown at 75 % N_2 chamber ambient, whereas $ZnO(25\ \%N_2)$ film, exhibited vertically aligned, single crystal hexagonallike nanorods with flat (0002) surface.

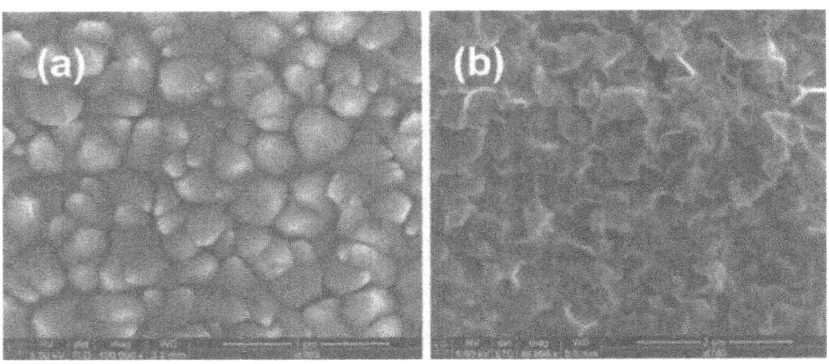

Figure 5. FE-SEM top-views of (a-c) the $ZnO(N_2:Ar)$ thin films deposited at the nitrogen gas flow rate of (a) 25% and (b) 75%, respectively.

The ZnO film grown at 12.5% nitrogen gas flow rate also not showed any nanorod structure (not shown here). ZnO thin film grown at 75 % nitrogen chamber ambient is polycrystalline.

Figure 6 (a) and (b) shows photocurrent-voltage curves of the ZnO (75 % N_2) and ZnO (25 % N_2) thin films deposited at 500°C respectively, under dark condition (black curve), and continuous illumination (red curve), with an UV/IR filter.

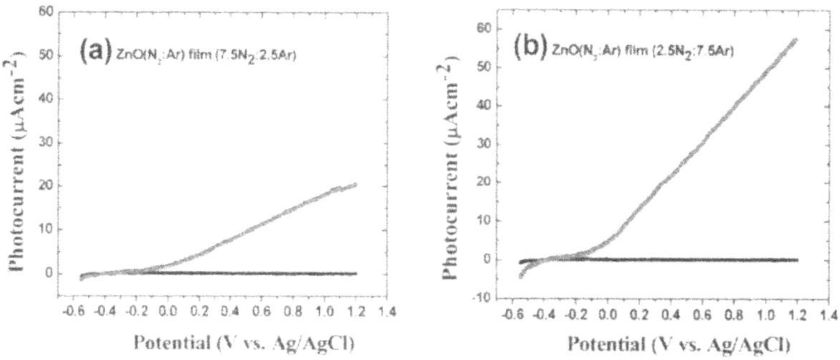

Figure 6. Photocurrent-voltage curves of (a) ZnO(0%N_2) and (b) ZnO(25%N_2) films, deposited at 500°C under (red curve) continuous illumination, (black curve) dark condition, with an UV/IR filter.

A very small dark current up to a potential of 1.2 V is shown by both the films. The ZnO (25 % N_2) thin film exhibited much higher photocurrent than the ZnO (75 % N_2) film due to the aligned nanorod formed along the c-axis. Our results shows that there is a optimum chamber ambient must be present for formation of nanorod along the c-axis in a sputter-deposited ZnO thin films, which is leading to greatly enhanced PEC response.

CONCLUSION

We have synthesized and characterized ZnO thin films deposited at various substrate temperatures and in mixed Ar and N_2 gas ambient. We find that the deposition in pure Ar ambient leads to polycrystalline ZnO thin films. However, the presence of N_2 in the deposition ambient promotes the formation of aligned nanorods at temperatures above 300°C. ZnO films with aligned nanorods deposited at 500°C exhibit significantly enhanced photoelectrochemical response, as compared to polycrystalline ZnO thin films grown at the same temperature. ZnO thin films grown at 25 % N_2 gas flow rate promoted aligned nanorods along c-axis exhibit significantly enhanced photoelectrochemical response, as compared to ZnO thin films grown at other N_2 to Ar gas flow ratios. Our results suggest that chamber ambient is very important for the formation of aligned nanostructures, which offer potential advantages for improving the efficiency of photoelectrochemical water splitting for H_2 production.

ACKNOWLEDGEMENT

This work was supported by the U.S. Department of Energy through the UNLV Foundation under Contract # DE-AC36-99-GO10337O.

REFERENCES

[1]A. Fujishima and K. Honda, *Nature* 238 (1972) 37.
[2]R. Asahi, T. Morikawa, T. Ohwaki, K. Aoki, and Y. Taga, *Science* 293 (2001) 269.
[3]O. Khaselev and J.A. Turner, *Science* 280 (1998) 425.
[4]V.M. Aroutiounian, V.M. Arakelyan, and G.E. Shahnazaryan, *Solar Energy* 78 (2005)

581.
[5]J. Yuan, M. Chen, J. Shi, and W. Shangguan, *Inter. J. Hydrogen Energy* 31 (2006) 1326.
[6]G.K. Mor, K. Shankar, M. Paulose, O.K. Varghese, and C.A. Grimes, *Nano Lett.* 5 (2005) 191.
[7]B. O'Regan and M. Grätzel, *Nature* 353 (1991) 737.
[8]K. Kakiuchi, E. Hosono, and S. Fujihara, *J. Photochem. & Photobiol. A: Chem.* 179, 81 (2006).
[9]T.F. Jaramillo, S.H. Baeck, A. Kleiman-Shwarsctein, and E.W. McFarland, *Macromol. Rapid Comm.* 25, 297 (2004).
[10]C.M. López and K.–S. Choi, *Chem. Commun.* (2005) 3328.
[11]J.B. Baxter and E.S. Aydil, *Sol. Energy Mater. Sol. Cells* 90 (2006) 607.
[12]M. Grätzel, *Nature* 414 (2001) 338.
[13]K.–S. Ahn, S. Shet, T. Deutsch, C. S. Jiang, Y. Yan, M. M. Al-Jassim, and J. Turner, *J. Power Sources* 176, (2008) 387-392.
[14]S. Shet, K.-S. Ahn, Y. Yan, N. M. Ravindra, T. Deutsch, J. Turner, M. Al-Jassim, "Effect of rf power and substrate temperature for the formation of aligned nanorods in ZnO thin films," Journal of Electronic Materials (JEM), submitted, (2009).
[15]S Shet, K.-S. Ahn, Y. Yan, T. Deutsch, K. M. Chrustowski, J. Turner, M. Al-Jassim, "Carrier concentration tuning of bandgap-reduced p-type ZnO films by co-doping of Cu and Ga for improving photoelectrochemical response," Journal of Applied Physics (JAP) 103, 073504, 2008.
[16]K.-S. Ahn, Y. Yan, S. Shet, T. Deutsch, J. Turner, M. Al-Jassim, "Enhanced photoelectrochemical responses of ZnO films through Ga and N codoping," Applied Physics Letters (APL), Vol. 91, issue 23, 231909, 2007.
[17]S. Shet, K.-S. Ahn, T. Deutsch, H. Wang, N. M. Ravindra, Y. Yan, J. Turner, and M. Al-Jassim, "Synthesis and characterization of bandgap-reduced ZnO:N and ZnO:(Al:N) films for photoelectrochemical water splitting," Journal of Materials research (JMR), accepted, 2009.
[18]X. Li, Y. Yan, T. A. Gessert, C. L. Perkins, D. Young, C. DeHart, M. Young, and T. J. Coutts, J. Vac. Sci. Technol. A, 21 (2003) 1342.
[19]C. Jagadish and S. J. Pearton, Zinc Oxide Bulk, Thin Films and Nanostructures, Elsevier, Amsterdam, 2006, chap. 3.
[20]D. Paluselli, B. Marsen, E. L. Miller, and R. E. Rocheleau, Electrochem. and Solid-State Lett., 8 (2005) G301.
[21]Y. Li, X. Li, C. Yang, and Y. Li, J. Mater. Chem. 13 (2003) 2641.
[22]X. Liu, X. Wu, H. Cao, and R. P. H. Chang, J. Appl. Phys. 95 (2004) 3141.
[23]W. I. Park, D. H. Kim, S. –W. Jung, and G. –C. Yi, Appl. Phys. Lett. 80 (2002) 4232.
[24]S. Choopun, H. Tabata, and T. Kawai, J. Cryst. Growth 274 (2005) 167.
[25]M. Law, L. E. Greene, J. C. Johnson, R. Saykally, and P. Yang, Nature Mater. 4 (2005) 455.

NANOCOATING ENHANCED OPTICAL FIBER SENSORS

M. Smietana [1,2], W.J. Bock [2], J. Szmidt [1], G.R. Pickrell [3]
[1] Institute of Microelectronics and Optoelectronics, Warsaw University of Technology, Warsaw, Poland
[2] Centre de recherche en photonique, Université du Québec en Outaouais, Gatineau, QC, Canada
[3] Materials Science and Engineering Department, Virginia Polytechnic Institute and State University, Blacksburg, VA, USA

ABSTRACT

This work presents an application of various nanometer thick coatings aiming to enhance the properties of optical fiber sensors. The coatings for modern sensors can create or improve sensitivity to some selected physical or chemical influences. Some of the most popular nanocoating deposition methods are presented and discussed in the context of their applicability to optical fibers sensors. The work focuses on plasma techniques that allow for deposition of good quality, high-refractive-index thin films as required for the sensors. The sensitivity of the long-period grating based structures to variations in the external medium's refractive index, humidity, hydrostatic pressure and temperature is modified by the deposition of nanocoating, the effect of which is discussed in this paper.

INTRODUCTION

Increasing concern over environmental pollution, demands for safe and accurate monitoring in industrial processes and new trends in medical diagnostics have created a need for small, precise, reliable and integrated sensing systems able to measure various environmental influences. Optical fiber sensing has been a continuously growing field during the last decade. The optical fiber sensors are small (from several to hundreds of microns in diameter), light weight, free from electromagnetic interference, can be multiplexed and have remote sensing capability. From the signal processing point of view, they exhibit large dynamic range, low attenuation and high signal to noise ratio. Moreover, this kind of sensing structure exhibits a high degree of flexibility and good thermal stability. From the point of view of medical devices, the optical fiber sensors can be sterilized and ensures lack of direct electrical connections to the patient. The sensors are able to measure mechanical forces and material deformation, pressure, variations in material composition or density, electric and magnetic field, electric current, temperature and number of other parameters.

There are many coatings for optical fibers that can improve or even initiate sensitivity to some selected physical or chemical influences [1-9]. The most important parameters of the coating from the point of view of optical sensing, are their optical properties and thickness, which can be changed by a specific measurand. In many realizations of chemical sensors, the overlay has an embedded indicator that responds to a chemical stimulus by changing its absorption, refractive index or luminescence. Some other sensors utilize nanostructured coatings, possessing nanometers-size pores in the structure, which allow the sorption effect to take place in the coating. That mechanism allows for the fabrication of highly sensitive and selective sensors [8, 9]. The chemical reactions, as well as in the film, can take place on the surface of the film.

Optical fiber sensors coated with a thin antibody film allows precise monitoring of bonding antibody-antigen reactions that cause refractive index changes [10]. There is also another group of coatings that, due to their high refractive index, that can tune the intrinsic sensitivity of the optical fiber devices to a certain range of the measurand. This kind of coating plays a passive role in the sensing mechanism, usually increasing the sensitivity to the refractive index [11].

NANOCOATINGS FOR OPTICAL FIBER SENSORS

During the last decade, significant work has been done in terms of developing novel coatings and their application in optical fiber sensing. The sensitivity to a number of influences is most often realized by variation of the refractive index of the coating subjected to the influence of external factors. In the past, some thick (over 100μm) coatings were applied in order to increase sensitivity to various parameters, with one such example being humidity [1]. Most of them were deposited with rather simple and unrepeatable dipping methods, where control of thickness and optical properties of the coating was limited. Moreover, these types of coatings (e.g. gelatin based) usually suffer from week mechanical properties and fast aging.

Deposition of high-refractive-index nanometric thin films (up to 300nm) can significantly modify the sensitivity of the structures to certain external influences. Advantages exist for thin film deposition techniques offering the ability to control the film optical properties and thickness on the nanometer scale. Such a coating allows for optimization of the interaction of the light guided in the fiber and coating [5].

Various nanocoating deposition techniques for optical fibers have been presented. In most of the cases already developed deposition techniques had to be modified or redesigned in order to let them be applied for other than flat substrates, for example, those of complex shapes like long and thin optical fibers. Several deposition methods were found to be inappropriate for this kind of application due to limited accuracy and repeatability of film properties, limitation to selected substrates, high temperature of the process (Hot Filament Chemical Vapour Deposition), high cost of the single deposition (Molecular Beam Epitaxy and Metalo-Organic Chemical Vapour Deposition) or the possibility of unilateral film deposition (Sputtering, Evaporation, Pulse Laser deposition). In order to overcome the last issue, some rotating mechanisms were applied, that might allow for all-around deposition [12]. This approach is the most often used in the case of metal deposition (gold, silver, palladium), but still the thickness of the film on the fiber is hard to define.

During the last few years the most popular methods of nanocoating deposition onto optical fibers were found to be those based on liquid precursors i.e. sol-gel, Langmuir-Blodgett (LB) and Electrostatic Self Assembly (ESA) [2, 6, 13]. However there are some other methods such as Radio Frequency Plasma Assisted Chemical Vapour Deposition (RF PACVD), where gas precursors are used, which seem to be very promising for future applications. Besides a number of advantages, all the methods have several disadvantages that limit their application. The mentioned methods will be discussed below.

SOL-GEL TECHNIQUE

The sol–gel technique relies on a process involving a solution or sol that undergoes a sol–gel transition [13]. The technique allows for immobilization of various indicators that show chemical sensitivity to certain measurands. This feature makes it very attractive from the point of view of application in chemical and biosensing. The mixture of sol material is introduced to the fiber by the dip coating technique with controlled upward pulling speed. The speed of pulling primarily determines the film thickness. The method shows several advantages like low-cost processing, precise control of the doping level and a porous structure desirable for gas sensor application [14]. Apart from simplicity, the films produced by the sol–gel process are well bonded to the fiber.

Besides the undeniable advantages of the method, it shows a number of limitations. The most important one is lack of film thickness control at the nanometric level. The average thickness of the film deposited using the method usually reaches 100-200nm [13-15], that for some application is might be considered as thick. Moreover, the deposited material determines the drying temperature (even 500°C [14]) and its duration. Optical properties of the films are also limited to available sol materials. To the best of our knowledge, the highest refractive index of sol-gel films already applied to optical fiber sensors was 1.76 [14].

LANGMUIR-BLODGETT TECHNIQUE (LB)

The method employs phenomena of that was reported in 18th century and further described and applied by Langmuir and Blodgett in the beginning of 20th century [16]. If a clean substrate is immersed in a liquid containing compressed monolayer of a material on its surface, a monolayer attaches to the substrate. During substrate withdrawn a second monolayer can attach [16]. When the surface pressure is maintained this process can be repeated and allows for building-up multilayers of any required thickness.

The method can be used at room temperature and normal pressure, giving nanometric precision of deposition. It offers resolution of 1-3nm depending on deposited material. For application in optical fiber sensors the resolution of about 6nm has been achieved [5].
The main feature of the method is that LB films may consist of a single chemical species [5]. This homogeneity is not available to the ESA method described below, which requires two types of materials. On the other hand there is a possibility of incorporating some sensitive indicators in the LB film to make the chemical sensor. That is why the method is often applied in the construction of biosensors.

Also in the case of the LB method, several disadvantages can be found. The technique is rather limited with respect to the set of molecular components suitable for LB deposition [17]. The molecules are often firmly trapped on the surface, and can rearrange themselves during or after the deposition. Major problems with the use of LB films include their extreme fragility and poor thermal stability. Moreover, the LB technique is relatively slow and requires specialized and expensive equipment to precisely control the pressure on the liquid surface. These drawbacks of the LB films favor the ESA techniques described below as a more suitable process for practical devices, due to its enhanced reliability, tunability and film quality [18].

ELECTROSTATIC SELF ASSEMBLY MONOLAYERS (ESA)

The technique was applied for the first time in the 1950s and the growth of the film is based on the electrostatic attraction between the opposite charges of molecules [5]. Besides the Electrostatic Self Assembly Monolayers (ESAM or ESA)[19], the method is also often called Ionic Self-Assembly Monolayers (ISAM or SAM) [6] or Layer-by-Layer Deposition [17]. As a first step of the deposition procedure, a substrate is cleaned and treated to create a charged surface [11]. Then, the substrate is exposed to a solution of a polyion of opposite charge for a short time (minutes), causing adsorption of a monolayer of polyions on the surface. In this way, the substrate is alternately dipped into solutions of cationic and anionic polymers to create a multilayer thin film that consists of a polyanion–polycation bilayer. After each monolayer is formed, it is necessary to rinse the sample in pure water to remove the excess molecules that are not bounded. The last step of each monolayer deposition is substrate drying. This process is repeated and increases, layer by layer, the thickness of the material deposited. The method allows for precise control of thickness of deposited films and maintains their properties during the process. Using ESA technique, it is possible to provide low-cost deposition of polymer materials at room temperature and normal humidity [11]. The reported thickness of the polymer bilayer deposited onto optical fiber was about 12nm [11]. The ESA deposition is possible on substrates of various sizes, shapes and materials, including metals, plastics, ceramics and semiconductors [5]. The technique can be used for deposition of variety of multilayered optical materials of magnetic, electrically conductive, and nonlinear properties. The ESA films show good reliability and thermal stability. The refractive index of ESA deposited films can be controlled (typically in range of 1.55 to 1.8 [21]) using different film materials or altering the pH of the polyion solutions.

The technique has several advantages that make it especially suitable for biosensors [13]. The method allows for mixing specific indicators or reactants (e.g. clay platelets, proteins, virus particles, etc.) with polyions resulting in trapping them in polymeric film and selective chemical sensing [19, 20]. There is a possibility of fabrication of several layer structure where each of the layers shows different sensing properties [21]. That feature makes the method more universal than LB.

There are also a number of disadvantages of ESA method [17, 21]. During the deposition, a number of process parameters must be precisely controlled in order to achieve repeatable films. The repeatability of the process is usually maintained using continuous monitoring of the deposited films and by robotizing the ESA process. Moreover, the procedure of multilayer deposition is time-consuming that stands as limitation of the method to laboratory use.

PLASMA ASSISTED CHEMICAL VAPOR DEPOSITION (RF PACVD)

The Radio Frequency Plasma Assisted Chemical Vapor Deposition (RF PACVD) technique has been already widely used in industrial mass production during the last decades, due to the possibility of low cost fabrication and high efficiency. The growth of the film is due to activation of the gas phase precursor in a glow discharge (plasma) environment. The chemical reactions activated by plasma take place over the substrate as well as at the substrate. The RF plasma deposition is a relatively low-temperature process (from room temperature to 250°C [22, 23]) and allows for high deposition rate [24].

The method gives the possibility of deposition of a wide range of optically transparent materials, whose properties can be tuned by process parameters. The well controlled parameters of the process e.g. RF power, substrate temperature or gas flow, give repeatable results in nanometric scale. The plasma deposited films, besides the attractive optical properties i.e. low absorption or refractive index from 1.4 to 2.6 at λ=550 nm, show chemical and thermal stability, high hardness, wear resistance and good adhesion to many substrates [23]. Using the RF PACVD system, one can deposit uniform films even on three-dimensional objects [25]. The films can be deposited as graded-refractive-index or stack of nanofilms showing different properties.

The films find typical applications as antireflective coatings, optical filters, waveguides and for decorative purposes. RF plasma deposited dense films were applied for optical fiber sensing structures for the first time by authors of this paper [3, 4, 26]. We introduced to the sensing optical fibers one of the hardest plasma deposited materials - diamond-like carbon (DLC). The developed modified RF PACVD method gives deposition of DLC films of well defined thickness and refractive index on optical fibers. The DLC films have already found their applications as protective coatings for metals, semiconductors, dielectrics and ceramics [27, 28]. Successful deposition of DLC onto various polymers was also demonstrated e.g. silicone rubber [29], PMMA [30], PET [24] or polyurethane [31]. In the case of coatings for medical implants, where DLC is assumed to be one of the best materials, biocompatibility and resistance to bacterial colonization on the coating's surface are essential [28, 31]. Besides the hardness, the DLC films show a desirable refractive index (typically 1.7÷2.1 at λ=1550 nm [32]), very low absorption in the infrared, low roughness and good adhesion to fused silica. All the features make the films perfect not only as a sensitivity modifying nanocoating for optical fibers but also as films to protect the structures against mechanical damages or chemical corrosion. It has been shown that DLC films can also be modified by addition of dopants [33] or by attaching molecules to the surface [34, 35] in order to change their properties for specific applications.

The plasma method, like all the others mentioned above, also shows some disadvantages. Some materials that are sensitive to temperature or UV radiation cannot be coated using the method. Moreover, there is a limited number of materials that can be deposited. The repeatability of the process depends on maintaining several process parameters at the appropriate levels, as well as controlling the conditions of the substrate material used. The films are usually very dense and cannot be penetrated by gases or chemical reactants, so the chemical reactions are limited to the surface of the film.

LONG-PERIOD GRATING (LPG) BASED OPTICAL FIBER SENSORS

There have been many optical fiber sensing concepts presented, where nanocotings can be successfully applied. The most intensively investigated ones are those based on Fabry-Perot cavity, fiber tapers, surface plasmon resonance and fiber gratings, especially Long-Period Gratings (LPGs). The LPGs are considered as one of the most sensitive to several external influences compared to all optical devices so far developed [36].

LPGs have been known for over a decade [37]. The LPGs are a periodic modulation of the refractive index along the length of the fiber. Under special phase-matching conditions the grating couples the fundamental core mode to discrete cladding modes that are rapidly attenuated due to absorption and scattering. The coupling from the guided mode to cladding modes is wavelength dependent, so one can obtain a spectrally selective loss.

Originally the LPG structures were applied as band-rejection filters for erbium doped fiber amplifiers [37], but also a number of sensors based on these structures have been presented for temperature, strain, bending and refractive index sensing. Several improvements have been reported for the LPG structures used for sensing applications. It was found that the sensitivity increases with the decrease of the period of the grating (typically below 300μm) [38]. Higher order cladding modes coupled by the LPG structure with shorter period intensively penetrate the surrounding medium giving higher sensitivity to variations of its refractive index, but unfortunately this also results in a higher sensitivity to temperature [36]. There were also reported attempts of using various types of optical fibers in order to improve the sensitivity of the

LPG structures for certain measurands and simultaneously to obtain structures with low cross-sensitivity to other measureands [39].

NANOCOATINGS EMPLOYED FOR MODIFICATION OF LPGS' EXTERNAL REFRACTIVE INDEX SENSITIVITY

Due to coupling of the cladding modes by the LPG structure, there is a significant dependence between properties of the cladding, or the cladding/coating system and its surface, and the spectral response of the LPG. The highest sensitivity to the refractive index of the uncoated LPG can be observed when the external medium's refractive index values are close to the refractive index of the cladding, which typically is made of fused silica. In this particular case, the core of the fiber is surrounded by an infinite medium, and the cladding modes cannot be produced, so the coupling cannot take place [40]. The sensitivity of such a structure, defined as the ratio between the resonance wavelength shift and the external refractive index variation, can reach $d\lambda/dn= -2000$ nm per refractive index unit (RIU), which for a typical spectrum analyzers gives a measuring resolution of $5\cdot10^{-5}$ RIU [41].

It was shown that high-refractive-index thin coatings (<300nm) can be used to produce a high sensitivity of the LPG [2, 9, 14, 42]. The coating significantly modifies the conditions of propagation for the guided cladding modes. The sensitivity in the specified range of external refractive index can be precisely adjusted by the thickness and optical properties of the coatings. The coatings of higher refractive index imply higher variations of the effective refractive index of the cladding modes as a function of external refractive index and a higher shift in the attenuation bands [42, 43]. Moreover, according to the work of Dong [44], materials with higher refractive index can be deposited as thinner layers in order to obtain comparable increases in sensitivity as those obtained with lower refractive-index materials. Using high RI materials, it is possible to reduce deposition time and decrease material usage.

A number of authors reported deposition of nanocoatings of various refractive index. James et al. [5] employed both LB and ESA techniques for increasing the refractive index sensitivity of LPG. So far, the highest refractive index that has been achieved reached between 1.71 to 1.8 based on measurements of the ESA films, and this was used for tuning the sensitivity of the LPG to a certain range of refractive index [18]. Deposition of higher-refractive-index materials seems to be problematic using most of the methods employed so far, which are mainly suitable for deposition of various polymers. The way to overcome that limitation can be the idea suggested by Del Villar et al. [43], who fund that deposition of a two-cladding structure (low and then high refractive index) avoids the necessity of materials of higher refractive indices than those typical of polymers. We also believe that plasma deposited nanocoatings can solve the problem as well. Moreover, it is known that ESA and LB nanodeposition techniques present the problem of a complex refractive index [21]. Consequently, due to the non-zero extinction coefficient, there is an additional phenomenon of vanishing of the attenuation bands in the transmission spectrum. As a good alternative, plasma deposited films can be applied. In Fig. 1 is presented dispersion of optical properties for plasma deposited thin films. Both presented films exhibit high refractive index and negligible absorption in infrared spectral range.

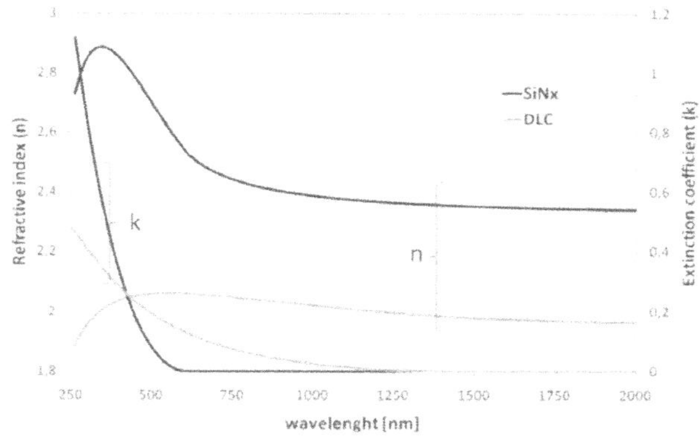

Figure 1 Dispersion relations of refractive index (n) and extinction coefficient (k) for RF PACVD deposited diamond-like carbon (DLC) and silicon nitride (SiN_x) thin films.

The advantage of depositing materials of a high-refractive-index contrast with the cladding of the LPG is that the sensitivity to variations in the external refractive index is highly increased [43]. Tuning the maximum sensitivity to variations of external refractive index using thin overlays was shown by a number of authors [e.g. 45]. Response of uncoated LPG and those tuned using 60nm-thick DLC films with the refractive index of 1.83 and 1.94 (at λ=1550nm) is shown in Fig. 2. Increasing thickness or refractive index of the coating one can shift maximum sensitivity to lower external refractive index. Using only 60nm film in thickness, that deposition took only 3 minutes, we could shift the maximum sensitivity from external refractive index of 1.46 to 1.44 and 1.38 for coating refractive index of 1.83 and 1.94, respectively. Moreover, the sensitivity in range where water pollutants can be detected (n=1.33) was increased over 2 times using the coating with n=1.94.

There is another sensing concept, where LPGs are used for dynamic humidity detection. The issue is important from the point of view of some breath related disease. The exhaled vapors condense as a thin water film on the surface of the fiber causing sudden change of its external refractive index. Simultaneously, thickness of the water film increases with the amount of water vapors delivered to the sensor surface. The process is hardly noticeable using uncoated LPGs, due to their low sensitivity in range of external refractive index from 1 to 1.33. The nanocoatings can be used for tuning the LPGs in order to let them be very sensitive in that specific range of external refractive index. In the case of using the film of 60nm in thickness and refractive index of 1.94, the sensitivity can be 3 times higher when compared to an uncoated LPG sample (Fig. 2b). Moreover, the DLC films due to their density and extremely smooth surface, cannot be penetrated by water molecules. This feature, together with the hydrophobic character of the DLC films makes them very promising for LPG based humidity sensors. Utilization of just the surface of the coating material increases the dynamics of the sensor response. Besides, for chemical sensing, depending on the deposited film, some molecules of the target chemical can be trapped at the surface of the coating. A similar mechanism will take place there, detected by LPG as slight variations of the external refractive index. In both cases, application of high-refractive-index nanocoatings is beneficial.

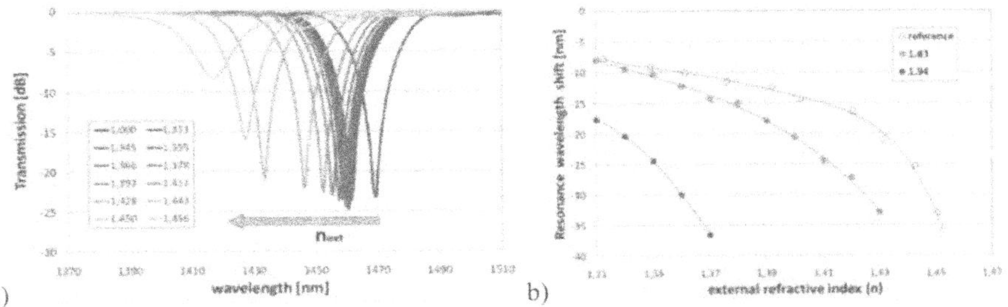

a) b)

Figure 2 Response of the 7th cladding mode of the LPG to variations of external refractive index in range form 1 to 1.456, where: (a) transmission spectra and (b) resonance wavelength shifts referred to n=1 of for bare (reference) and coated LPGs with 60nm thin films of 1.83 and 1.94 in refractive index.

The high sensitivity of the LPGs to external refractive index changes can disturb measurements of other influences e.g. temperature or pressure. For this group of LPG sensors, the refractive index sensitivity might be significantly reduced using nanocatings. When the measurements are performed in water, the resonances of uncoated grating are deep and well visible. The problem appears when the refractive index of the surrounding medium is close to that of the cladding e.g. when measurements need to be done in oils (typically n~1.47). At this point, the maximum sensitivity to variations of external refractive index occurs and the vanishing of the resonances take place. The phenomena significantly disturbs measurements of other influences and in some cases even makes them impossible. Using a high-refractive-index nanocoating, one can shift the maximum refractive index sensitivity out of the range of the working liquid. The effect of deposition of the nanocoating for such a purpose is shown in Fig. 3a. The nanocoating of the adjusted thickness and refractive index makes the resonance visible in high-refractive-index oil and simultaneously makes it invisible in water. The influence of pressure (in range of 1 to 240 bars) on the transmission spectra for measurements preformed in oil is shown in Fig. 3b. It can be seen that plasma deposited nanocoating makes this kind of measurement possible. Moreover, dense and highly durable plasma deposited coatings can stand harsh environmental conditions. These features make them suitable for sensors working at high pressure and temperature or in the presence of corrosive chemicals.

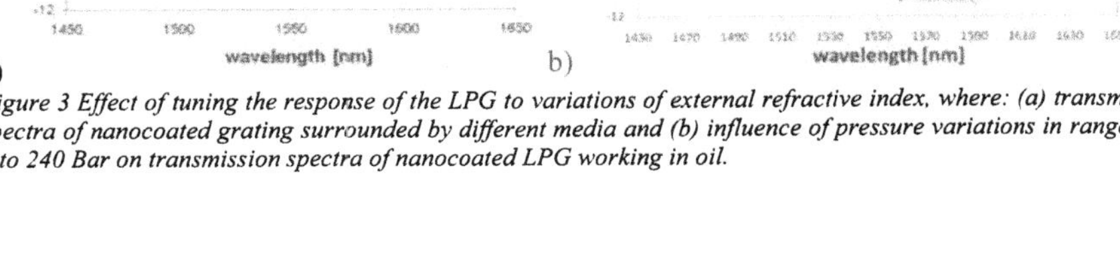

a) b)

Figure 3 Effect of tuning the response of the LPG to variations of external refractive index, where: (a) transmission spectra of nanocoated grating surrounded by different media and (b) influence of pressure variations in range from 1 to 240 Bar on transmission spectra of nanocoated LPG working in oil.

CONCLUSIONS

In this paper we discussed the nanocoating deposition methods that are the most often used for optical fiber sensors. From a point of view of highly selective chemical sensing, the most promising seems to be the ESA technique, that allows for embedding chemical indicators in the nanofilm and simultaneously gives precise control of optical properties and thickness of the coating. On the other hand for the applications where the nanocoating is used only to tune the sensitivity to variations of external refractive index, the favorable method seems to be RF PACVD. This technique allows for fast and controllable deposition of very high refractive index coatings onto various optical fiber sensors. Besides the modification of the LPG response to refractive index variations, the use of hermetic plasma deposited coatings for silica glass fibers improves the fiber reliability and adds protection against moisture and corrosive chemicals penetrating into the structure. The coatings show very promising applications in LPG based external refractive index, humidity, pressure and temperature sensing.

ACKNOWLEDGMENTS

The authors gratefully acknowledge support for this work from the Natural Sciences and Engineering Research Council of Canada, the Canada Research Chairs Program and the Ministère de l'Éducation, du Loisir et du Sport du Québec. Support from the development grant R02 022 01 from founds of Polish Ministry of Science and Higher Education in years 2006-2009 is also acknowledged.

REFERENCES

[1] J. Sun, Y. Li, D. Liu, Widely tunable long-period fiber gratings with interpolymers as sensitivity-enhanced materials, *Opt. Comm.* Vol 249, 2005, p 193–200

[2] I.M. Ishaq, A. Quintela, S.W. James, G.J. Ashwell, J.M. Lopez-Higuera, R.P. Tatam, Modification of the refractive index response of long period gratings using thin film overlays, *Sensors and Actuators B,* Vol 107, 2005, p 738-741

[3] M. Smietana, J. Szmidt, M. Dudek, P. Niedzielski, Optical properties of diamond-like cladding for optical fibres, *Diamond and Related Materials* Vol 13, 2004, p 954-957

[4] M. Smietana, J. Szmidt, M.L. Korwin-Pawlowski, W.J. Bock, J. Grabarczyk, Application of diamond-like carbon films in optical fibre sensors based on long-period gratings, *Diamond and Related Mat.* Vol 16, 2007, p 1374-1377

[5] S.W. James, R.P. Tatam, Fibre Optic Sensors with Nano-Structured Coatings, *J. Opt. A: Pure Appl. Opt.,* Vol 8, 2006, p S430-S444

[6] Z. Wang, J.R. Heflin, R.H. Stolen, S. Ramachandran, Highly sensitive optical response of optical fiber long period gratings to nanometer-thick ionic self-assembled multilayers, *Applied Physics Letters,* Vol 86, 2005, 223104

[7] S. W. James, R. P. Tatam, Optical fibre long period grating sensors: Characteristics and application, *Meas. Sci. Technol.* Vol 14, 2003, p R49-R61

[8] J. Zhang, X. Tang, J. Dong, T. Wei, H. Xiao, Zeolite Thin Film-Coated Long Period Fiber Grating Sensor for Measuring Trace Chemical, *Opt. Express* Vol. 16, 2008, p 8317-8323

[9] A. Cusano, P. Pilla, L. Contessa, A. Iadicicco, S. Campopiano, A. Cutolo, M. Giordano, G. Guerra, High-sensitivity optical chemosensor based on coated long-period gratings for sub-ppm chemical detection in water, *Appl. Phys. Lett.* Vol 87, 2005, p 234105

[10] M.P. DeLisa, Z. Zhang, M. Shiloach, S. Pilevar, Ch.C. Davis, Evanescent Wave Long-Period Fibre Bragg Grating as an Immobilized Antibody Biosensor, *Anal. Chem.* Vol 72, 2000, p 2895-2900

[11] I.R. Matias, F.J. Arregui, J.M. Corres, J. Bravo, Evanescent Field Fiber-Optic Sensors for Humidity Monitoring Based on Nanocoatings, *IEEE Sensors*, Vol 7, No 1, 2007, p 89-95

[12] T. Vo-Dinh, P. Kasili, Fiber-optic nanosensors for single-cell monitoring, *Anal Bioanal Chem* Vol 382, 2005, p 918–925

[13] S.T. Lee, J. Gin, V.P.N. Nampoori, C.P.G. Vallabhan, N.V. Unnikrishnan, P. Radhakrishnan, A sensitive fibre optic pH sensor using multiple sol–gel coatings, *J. Opt. A: Pure Appl. Opt.* Vol 3, 2001, p 355–359

[14] Z. Gu, Y. Xu, Design optimization of a long-period fiber grating with sol–gel coating for a gas sensor, *Meas. Sci. Technol.* Vol 18, 2007, p 3530–3536

[15] V. Musat, A.M. Rego, R. Monteiro, E. Fortunato, Microstructure and gas-sensing properties of sol–gel ZnO thin films, *Thin Solid Films* Vol 516, 2008, p 1512–1515

[16] F. Davis, S.P.J. Higson, Structured thin films as functional components within biosensors, *Biosensors and Bioelectronics* Vol 21, 2005, p 1–20

[17] G. Decher, J.B. Schlenoff, Multilayer Thin Films – Sequential Assembly of Nanocomposite Materials, *Wiley-VCH*, 2003, p. 1-8

[18] Z. Wang J. R. H. Rogers, H. Stolen S. Ramachandran, Analysis of optical response of long period fiber gratings to nm-thick thin-film coatings, *Optics Express* Vol 13, 2005, p 2808-2813

[19] I. Del Villar, I.R. Matias, F.J. Arregui, R.O. Claus, Fiber-Optic Hydrogen Peroxide Nanosensor, *IEEE Sensors* Vol 5, 2005, p 365-371

[20] D. Galbarra, F.J. Arregui, I.R. Matias, R.O. Claus, Ammonia optical fiber sensor based on self-assembled zirconia thin films, *Smart Mater. Struct.* Vol 14, 2005, p 739–744

[21] J.M. Corres, I. del Villar, I.R. Matias, F.J. Arregui, Two-Layer Nanocoatings in Long-Period Fiber Gratings for Improved Sensitivity of Humidity Sensors, *IEEE Trans. Nanotech.* Vol 7, 2008, p 394-400

[22] A. Grill, Plasma-deposited diamondlike carbon and related materials, *IBM Journal of Research and Development, Plasma processing*, Vol 43, 1999, 1/2

[23] L. Martinu, D. Poitras, Plasma deposition of optical films: A review, *J. Vac. Sci. Technol. A* 18, 2000, p 2619-2645

[24] A. Shirakura, M. Nakaya, Y. Koga, H. Kodama,T. Hasebe, T. Suzuki, Diamond-like carbon films for PET bottles and medical applications, *Thin Solid Films* Vol 494, 2006, p 84 – 91

[25] H. Anma, J. Toki, T. Ikeda, Y. Hatanaka, Uniform deposition of SiC thin films on plastics surfaces, *Vacuum* Vol 59, 2000, p 665-671

[26] M. Smietana, M.L. Korwin-Pawlowski, W.J. Bock, G.R. Pickrell, J. Szmidt, Refractive index sensing of fiber optic long-period grating structures coated with a plasma deposited diamond-like carbon thin film, *Meas. Sci. Technol.* Vol 19, 2008, 085301

[27] I. Ahmad, P.D. Maguire, P. Lemoine, S.S. Roy, J.A. McLaughlin, Deposition of carbon films onto metal and silicon substrates by filtered cathodic vacuum arc, plasma enhanced CVD and unbalanced magnetron sputtering, *Diamond and Related Materials* Vol 13, 2004, p 1346–1349

[28] E. Salgueiredo, M. Vila, M.A. Silva, M.A. Lopes, J.D. Santos, F.M. Costa, R.F. Silva, P.S. Gomes, M.H. Fernandes, Biocompatibility evaluation of DLC-coated Si3N4 substrates for biomedical applications, *Diamond & Related Materials* Vol 17, 2008, p 878–881

[29] K. Donnelly, D.P. Dowling, M.L. McConnell, M. Mooney, The adhesion of hydrogenated amorphous carbon films on silicone, *Thin Solid Films* Vol 394, 2001, p 102-108

[30] M. Ban, T. Yuhara, Chemical resistance of DLC thin film deposited PMMA substrates, *Surface & Coatings Technology* Vol 203, 2009, p 2587–2590

[31] D.S. Jones, C. P. Garvin, D. Dowling, K. Donnelly, S.P. Gorman, Examination of Surface Properties and In Vitro Biological Performance of Amorphous Diamond-Like Carbon-Coated Polyurethane, *J Biomed Mater Res Part B: Appl Biomater*, Vol 78B, 2006, p 230–236

[32] R. McCann, S.S. Roy, P. Papakonstantinou, G. Abbas, J.A. McLaughlin, The effect of thickness and arc current on the structural properties of FCVA synthesised ta-C and ta-C:N films, *Diamond & Related Materials* Vol 14, 2005, p 983–988

[33] R. Hauert, A review of modified DLC coatings for biological applications, *Diamond and Related Materials* Vol 12, 2003, p 583–589

[34] T. Nakamura, T. Ohana, M. Suzuki, M. Ishihara, A. Tanaka, Y. Koga, Chemical modification of DLC films with perfluorooctyl functionality, *Diamond & Related Materials* Vol 14, 2005, p 1019–1022

[35] S. Lousinian, S. Logothetidis, A. Laskarakis, M. Gioti, Haemocompatibility of amorphous hydrogenated carbon thin films, optical properties and adsorption mechanisms of blood plasma proteins, *Biomolecular Engineering* Vol 24, 2007, p 107–112

[36] H.J. Patrick, A.D. Kersey, F. Bucholtz, Analisis of the Response of Long Period Fiber Gratings to External Index Refraction, *J.Lightwave Tech.* Vol 16, 1998, p 1606-1612

[37] A. M. Vengsarkar, P. J. Lemaire, J. B. Judkins, V. Bhatia, T. Erdogan, J. E. Sipe, Long-Period Fiber Gratings as Band-Rejection Filters, *J.Lightwave Tech.* Vol 14, 1996, p 58-65

[38] X. Shu, L. Zhang, Sensitivity Characteristics of Long-Period Fiber Gratings, *J. Lightwave Technol.* Vol 20, 2002, p 255-266

[39] T. Allsop, D.J. Webb, I. Bennion, A comparision of the sensing characteristics of long period gratings written in three different types of fiber, *Optical Fiber Technology* Vol 9, 2003, p 210-223

[40] T. Erdogan, Fiber Grating Spectra, *J. Lightwave Technology*, Vol 15, 1997, p 1277-1294

[41] R. Falate, R.C. Kamikawachi, M. Muller, H.J. Kalinowski, J.L. Fabris, Fiber optic sensors for hydrocarbon detection, *Sensors and Actuators B* Vol 105, 2005, p 430–436

[42] I. Del Villar, I. Matias, F. Arregui, P. Lalanne, Optimization of sensitivity in Long Period Fiber Gratings with overlay deposition, *Opt. Express* Vol 13, 2005, p 56-69

[43] I. Del Villar, I.R. Matias, F.J. Arregui, Deposition of coatings on long-period fiber gratings: tunnel effect analogy, *Optical and Quantum Electronics* Vol 38, 2006, p 655–665

[44] X. Dong, L. Pei, S. Jian, Widely tunable long-period fiber grating with nm-thick higher refractive index film overlay, *Optik* Vol 117, 2006, p 462–467

[45] A. Cusano, A. Iadicicco, P. Pilla, A. Cutolo, M. Giordano, S. Campopiano, Sensitivity characteristics in nanosized coated long period gratings, *Appl. Phys. Lett.* Vol 89, 2006, 201116

SURFACE PLASMON RESONANT ENHANCED OPTICAL TRANSMISSION THROUGH ZNO/AG/ZNO MULTILAYERED FILMS

Pangpang Wang (1, 2), Dongyan Zhang (1, 2), Chang Huang (1), Xiaoping Song (1), and Ri-ichi Murakami (2)
[1]Department of Materials Physics, School of Science, Xi'an Jiaotong University
Xi'an, Shaanxi 710049, People's Republic of China
[2]Department of Mechanical Engineering, The University of Tokushima
Tokushima, Tokushima 770-8506, Japan

ABSTRACT

The optical transmission enhancement in the visible range through a composite structure, zinc oxide/silver/zinc oxide (ZnO/Ag/ZnO), has been investigated in this work. The ZnO/Ag/ZnO multilayered films were deposited on glass substrates using dc magnetron sputtering method. The structure of the silver mid-layer was considered to play an important role in the enhanced optical transmission. The silver mid-layer was composed of randomly distributed silver particles (islands) which were connected by a thin continuous silver layer. When light impinged normal to the multilayered films, the layer-plus-islands silver could excite surface plasmons (SPs). The resonance between the incident light and the SPs could confine more light on the silver surface and enhanced the near field strength. The light could be reemitted on the exit side by decoupling the confined SPs, and the optical transmission could be enhanced. The surrounding ZnO layers also affected the optical transmission. The optical transmittances of multilayered films with 30 and 50 nm thick ZnO layers were higher those of multilayered films with 100 nm thick ZnO layers. The mechanism for enhanced optical transmission was discussed by considering the metal/dielectric interface model. The electrical properties were also studied in this work. The ZnO/Ag/ZnO multilayered films with high optical transmittance and low sheet resistance have the potential applications in solar cells and flat plate displays as transparent electrodes.

INTRODUCTION

It is well known that surface plasmons (SPs) are considered as electromagnetic waves that propagate along the surface of a conductor, usually a metal like gold or silver [1]. Surface plasmons were widely recognized in the field of surface science following the work of Ritchie in the 1950s [2]. The interaction between the incident light and the free electrons of a metal thin film or metallic nanoscale particles can give rise to collective oscillations, known as surface plasmon resonance (SPR). The properties of SPs in gold or silver nanoparticles can be tailored by tuning the structure, shape or size. This offers the potential for developing new types of photonic devices such as subwavelength optics, magneto-optic data storage, light generation, optical microscopy, solar cells and bio-photonics [3-8].

Localized surface plasmons (LSPs) of metallic nanoparticals have also been studied widely. Gold and silver nanoparticles with different shapes have been fabricated by chemical and electrochemical reduction methods, and the SPR peaks showed shift from visible range to near-infrared range by controlling the particle structures. Normally, the localized SPR absorption peaks appear in the extinction spectra, where extinction equals scattering (or diffraction) plus absorption. This effect was found for absorption of gold or silver colloid particles in solution [9. 10].

In contrast, nanometer scaled holes array in silver film could enhance the optical transmittance in a special wavelength range due to the SPR on the corrugated surface of silver film [11-13]. Either the absorption or the transmission could be attributed to SPR. However, the absorption phenomenon was due to the resonance between light and separated metallic particles, but the enhanced optical transmission was due to the resonance between light and continuous corrugated metallic films.

High quality transparent electrode was necessary to improve both electrical and optical properties in light emitters, solar cells and panel displays. Some research groups use a thin silver island film to enhance the optical properties by SPR [14-18]. However most of these reports focused on SPR absorption with opaque substrates, or were concerned only with the good properties. Mechanism has not been discussed clearly to explain the improvement of both optical and electrical properties simultaneously.

In this work, we will discuss the optical and electrical properties of ZnO/Ag/ZnO multilayered films on glass substrates. The optical transmission can be enhanced by considering SPR between the incident light and a composite silver layer. A layer-plus-islands structure is used to explain the improvement of electrical and optical properties simultaneously. The layer-plus-islands structure is expected to be useful for not only experimental designs of optical devices but also the fundamental research in surface plasmons.

EXPERIMETAL

The ZnO/Ag/ZnO multilayered films were deposited layer-by-layer on glass substrate s (Matsunami, S-2215, 76×26×0.9~1.2 mm) by a dc magnetron sputtering system (NACHI, SP-1530-1) at room temperature. The sputtering was performed using pure Zn and Ag sputter targets (Furuuchi Chemical, 99.99%, 300×62×5 mm). The base pressure before deposition was approximately 3.0×10^{-3} Pa, and the sputtering was carried out at a pressure of 1.0×10^{-1} Pa under argon or argon/oxygen atmosphere. For ZnO sputtering, the Ar and O_2 mixture atmosphere with Ar and O_2 flows were 50 and 3 sccm, respectively. For Ag sputtering, only Ar was used, and the flow of Ar was 50 sccm. In order to estimate the sputtering rate, a thick film was prepared before practical sputtering. The thickness of the thick film was determined using a surface profile measuring system (Veeco DEKTAK^3). The thickness of the ZnO layers (bottom and top) were set as 30, 50 and 100 nm with sputtering times of 60, 100 and 200 minutes (the sputtering rate was about 0.5 nm per minute). The Ag layers were varied as 4, 8, 12 and 16 nm with sputtering times of 10, 20, 30 and 40 seconds (the sputtering rate was about 0.4 nm per second). For comparison and morphologic observation, Ag/ZnO multilayered films were also prepared on glass substrate.

The crystalline structures of Ag/ZnO multilayered films were investigated by x-ray diffraction analysis (XRD, Rigaku-RINT2500) using a Cu-K_α radiation source (λ=0.154056 nm), and the optical transmittance spectrums of films were measured using a UV-Vis-NIR spectrometer (Jasco-V-570) in the wavelength range from 300 to 800 nm in air. The surface morphology of the silver layer was observed using field-emission scanning electron microscopy (FESEM, Hitachi-S4700). Finally, a Hall Effect measurement system (EGK-HEM-2000) operating with a magnetic field of 0.37 T (perpendicular to the sample surface), was used to demonstrate the electrical properties by means of the van der Pauw method [19].

RESULTS AND DISCUSSION

Figure 1 shows the top view SEM images of Ag layers of different thicknesses on 30 nm thick ZnO film. It is clear that the Ag layers are composed of Ag particles (islands) distributed randomly. The electrical property, sheet resistance, is shown in Figure 2. The sheet resistances for ZnO/Ag/ZnO films with same ZnO layer thicknesses decrease as the Ag layer increases. The low sheet resistance means that the Ag layer should have a continuous part expect the independent Ag islands. According to the SEM results and electrical measurements, the Ag layer should have a composite layer-plus-islands structure, and the thickness of the continuous thin Ag layer increased with the sputtering time. The sheet resistance of the ZnO/Ag/ZnO multilayered films decreased due to the increase of the continuous Ag layer.

The ZnO (002) and Ag (111) diffraction peaks can be observed in the XRD patterns shown in Figure 3, which mean that both ZnO and Ag layers are crystalline structures. It is known that metal film growing on a crystal surface or interface always follows a layer-plus-island growth mechanism, known as Stranski-Krastanov (SK) mechanism [20].

Figure 1. Top view SEM images of Ag layers with different thicknesses a) 4 nm, b) 8 nm, c) 12 nm and d) 16 nm on 30 nm thick ZnO films.

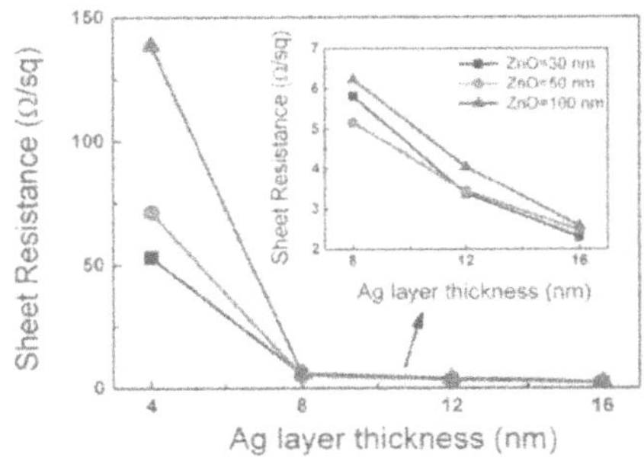

Figure 2. Sheet resistances of ZnO/Ag/ZnO multilayered films with different Ag mid-layer thicknesses of 4, 8, 12 and 16 nm.

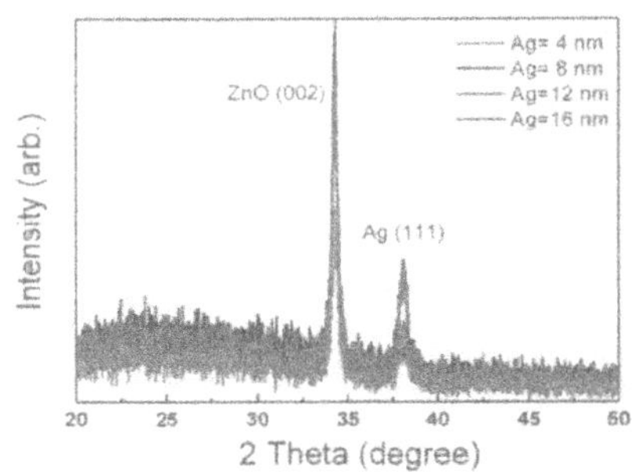

Figure 3. XRD patterns for Ag/ZnO films with different Ag layer thicknesses of 4, 8, 12 and 16 nm on glass substrate.

The optical transmission spectrums for ZnO/Ag/ZnO films with different ZnO layer thicknesses are shown in Figure 4, 5 and 6. In Figure 4 and 5, the enhanced optical transmission range can be observed obviously. However, in Figure 6, the enhanced optical transmission range cannot be observed. The enhanced optical transmission through the ZnO/Ag/ZnO multilayered films can be attributed to the interaction between incident light and surface plasmons on the layer-plus-islands Ag layer. The interaction, known as surface plasmon resonance, confines the incident light on the surface and enhanced the near field strength. Then the enhanced near field can excite the surface plasmons on the corrugated Ag surface. The exciting surface plasmons can be reemitted to light at the exit side, which enhance the optical transmission. The transmittance decreased in the samples with 100 nm thick ZnO layer due to the strong absorption by the ZnO layers. There are no surface plasmon resonance occurred.

For comparison, three structures for Ag layer are shown in Figure 7, the continuous layer structure, the islands structure and the layer-plus-islands structure. Figure 8 shows the optical transmitting mechanism for the three structures. In the case of continuous structure, the incident light will be reflected back, and has only a 20 nm thick transmission depth into the Ag layer [1, 21]. For the Ag layer thicknesses in this work, there will be not enhanced transmission phenomenon for the continuous structure. In the case of independent Ag islands, the incident light will be strongly diffracted and absorbed to localized SPs, and the optical transmission will be reduced. In the case of layer-plus-islands structure, the excitation of SPs is possible, which could lead to a resonance with the incident light. The resonance mechanism could be considered as a metal/dielectric grating interface model [1, 22]. The relation between light wave vector k_{in} and grating constant G_x and G_y (for two-dimensional grating) are shown in Equation (1) and Equation (2).

$$\vec{k}_{SP} = \vec{k} \pm i\vec{G}_x \pm j\vec{G}_y, \ (\vec{G}_x = \vec{G}_y = \frac{2\pi}{a_0}) \quad (1)$$

$$k_{SP} = k_{in}\sqrt{\frac{\varepsilon_d \varepsilon_m}{\varepsilon_d + \varepsilon_m}}, \quad (2)$$

Where k_{in} is the light wave-vector in free space, $k_{//}$ is the in-plane component of light wave-vector, k_{SP} is the SPs momentum along the corrugated layer-plus-islands Ag structure, a_0 is the approximate grating constant of the Ag islands grating (mean distance between Ag islands), ε_d and ε_m are the real parts of dielectric functions for surrounding ZnO dielectrics and Ag layer, and i and j are integers for the grating constant. For a dielectric/metal interface, the SPs resonance wave-vector k_{SP} can be also written by solving the Maxwell equations as shown in Equation (2).

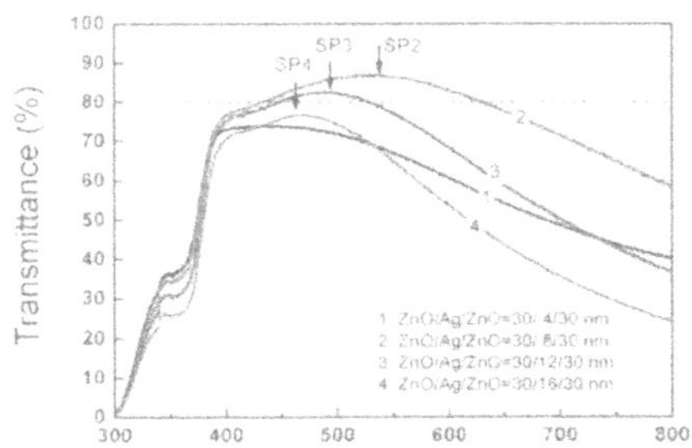

Figure 4. Optical transmission spectrums for ZnO/Ag/ZnO films with ZnO thickness of 30 nm and Ag layer thicknesses of 4, 8, 12 and 16 nm.

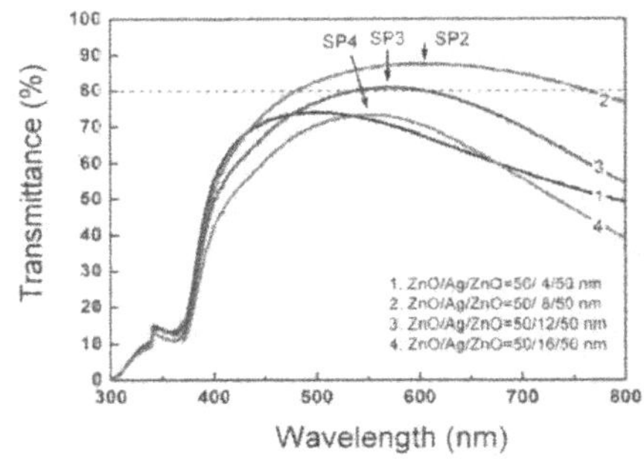

Figure 5. Optical transmission spectrums for ZnO/Ag/ZnO films with ZnO thickness of 50 nm and Ag layer thicknesses of 4, 8, 12 and 16 nm.

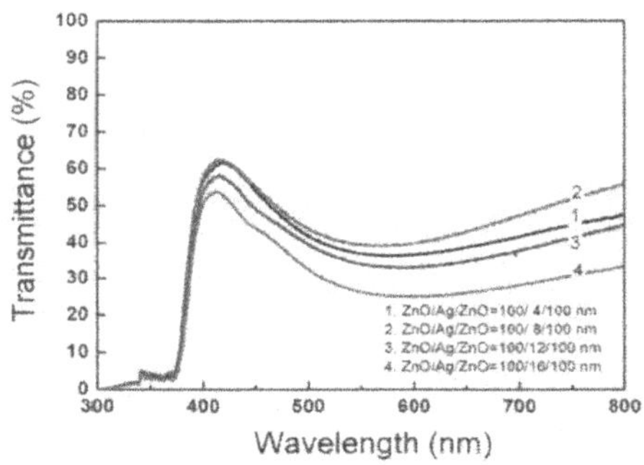

Figure 6. Optical transmission spectrums for ZnO/Ag/ZnO films with ZnO thickness of 100 nm and Ag layer thicknesses of 4, 8, 12 and 16 nm.

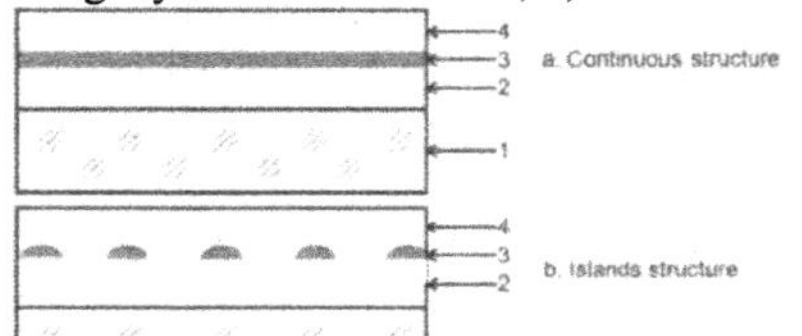

Figure 7. Schematic diagrams for different growth structures for Ag mid-layer, a) continuous, b) islands, and c) layer-plus-islands.

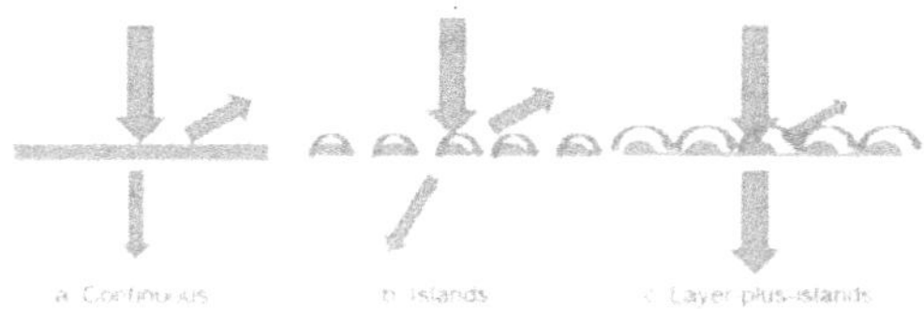

Figure 8. Optical transmission through Ag layer with different growth structures.

Generally, k_{SP} is always larger than k_{in} according to the Equation (2). In our case, the mismatched part between k_{SP} and k_{in} could be fulfilled using the Ag islands grating coupler as shown in Equation (1). The mean grating constant a_0 for the Ag islands decreased with the increasing Ag layers as shown in the SEM images (Figure 1). By reducing the Equations (1) and (2), we can get Equation (3).

$$\lambda_{SP} = \frac{a_0}{\sqrt{i^2 + j^2}} \sqrt{\frac{\varepsilon_d \varepsilon_m}{\varepsilon_d + \varepsilon_m}} . \tag{3}$$

Where λ_{SP} means the resonant SPs wavelength, which is proportional to the mean grating constant a_0. Therefore, with the increase of the Ag mid-layer thickness, the resonance wavelength should shift from red to blue side in the visible range, which was consistent with the observation in optical transmission spectrums. The layer-plus-island structure for silver mid-layer in ZnO/Ag/ZnO multilayered systems can explain improvement in optical and electrical properties, simultaneously. So this model is expected to be utilized in photo-electronic applications and fundamental study in surface plasmon resonance.

CONCLUSION

In summary, a thin Ag film enhanced the transmittance of ZnO/Ag/ZnO in a special range of light wavelengths by interaction with incident light due to the surface plasmon resonance. A layer-plus-island structure has been considered to explain the mechanism of surface plasmon resonance. Improvements of both optical and electrical properties of ZnO/Ag/ZnO multilayered films are explained by this layer-plus-island structure. The ZnO/Ag/ZnO multilayered films with high transparency and low sheet resistance were expected using as transparent electrode in solar cells and flat plate displays.

REFERENCES

[1]H. Raether, Surface Plasmons on Smooth and Rough Surfaces and on Gratings, Springer, (1988).
[2]R. H. Ritchie, Plasma Losses by Fast Electrons in Thin Films, *Phys. Rev.*, 106, 874 (1957).
[3]D. K. Gramotnev and D. F. P. Pile, Single-mode subwavelength waveguide with channel plasmon-ploartons in triangular grooves on a metal surface, *Appl. Phys. Lett.*, 85, 6323 (2004).
[4]W. Chuang, J. Wang, C. C. Yang, and Y. Kiang, Transient behaviors of surface plasmon coupling with a light emitter, *Appl. Phys. Lett.*, 93, 153104 (2008).
[5]S. Chang, S. K. Gray, and G. C. Schatz, Surface plasmon generation and light transmission by

isolated nanoholes and arrays of nanoholes in thin metal films, *Opt. Exp.*, 13, 3510 (2005).
[6]Z. Fang, X. Zhang, D. Liu, and X.Zhu, Excitation of dielectric-loaded surface plasmon polariton observed by using near-field optical microscopy, *Appl. Phys. Lett.*, 93, 073306 (2008).
[7]S. Kim, S. Na, J. Jo, D. Kim, and Y. Nah, Plasmon enhanced performance of organic solar cells using electrodeposited Ag nanoparticles, *Appl. Phys. Lett.*, 93, 073307 (2008).
[8]P. Berini, R. Charbonneau, and N. Lahoud, Long-range surface plasmons on ultrathin membranes, *Nano Lett.*, 7, 1376 (2007).
[9]Y. Xiong, J. Chen, B. Wiley, Y. Xia, Y. Yin, and Z. Li, Size-dependence of surface plasmon resonance and oxidation for Pd nanocubes synthesized via a seed etching process, *Nano Lett.*, 5, 1237 (2005).
[10]Y. Xia and Naomi J. Halas, Shape-Controlled Synthesis and Surface Plasmonic Properties of Metallic Nanostructures, *MRS Bulletin*, 30, 338 (2005).
[11]T. Park, N, Mirin, J. Britt Lassiter, C. L. Nehl, Naomi J. Halas, and P. Nordlander, Optical Properties of a Nanosized Hole in a Thin Metallic Film, *ACS Nano*, 2, 25 (2008).
[12]K. Baba, T. Okuno, and M. Miyagi, Silver-gold compound metal island films prepared by using a two step evaporation method, *Appl. Phys. Lett.*, 62, 437 (1993)
[13]A. P. Abiyasa, S. F. Lau, Eunice S. P. Leong, and H. Y. Yang, Enhancement of ultraviolet lasing from Ag-coated highly disordered ZnO films by surface plasomn resonance, *Appl. Phys. Lett.*, 90, 231106 (2007).
[14]S. Pillai, K. R. Catchpole, T. Trupke, and M. A. Green, Surface plasmon enhanced silicon solar cells, *J. Appl. Phys.*, 101, 093105 (2007).
[15]J. B. You, X. W. Zhang, Y. M. Fan, S. Qu, and N. F. Chen, Surface plasmon enhanced ultraviolet emission from ZnO films deposited on Ag Si 001 by Magnetron Sputtering, *Appl. Phys. Lett.*, 91, 239107 (2007).
[16]K. Okamoto, I. Niki, A. Shvartser, Y. Narukawa, T. Mukai, and A. Scherer, Surface-plasmon-enhanced light emitters based on InGaN quantum wells, *Nat. Mater.*, 3, 601 (2005).
[17]H. Han, N. D. Theodore, and T. L. Alford, Improved conductivity and mechanism of carrier transport in zinc oxide with embedded silver layer, *J. Appl. Phys.*, 103, 013708 (2008).
[18]P. Cheng, D. Li, Z. Yuan, P. Chen, and D. Yang, Enhancement of ZnO light emission via coupling with localized surface plasmon of Ag island film, *Appl. Phys. Lett.*, 92, 041119 (2008).
[19]L. J. van der Pauw, A method of measuring the resistivity and hall coefficient on lamellae of arbitrary shape, *Philips Tech. Rev.*, 20, 220 (1958).
[20]T. Muller and H. Nienhaus, Ultrathin Ag films on H:Si(111)-1×1 surfaces deposited at low temperatures, *J. App. Phys.*, 93, 924 (2003).
[21]V. M. Shalaev and S. Kawata, Nanophotonics with surface plasmons. Amsterdam, Boston, MA, Elsevier, (2007).
[22]W. L. Barnes, A. Dereus, and T. W. Ebbesen, Surface plasmon subwavelength optics, *Nature*, 424, 824 (2003).

CONTROLLED SHAPE SYNTHESIS OF $BaTiO_3$-$(Mn_{0.5}Zn_{0.5})Fe_2O_4$ NANOCOMPOSITES

Yaodong Yang*, Wenwei Ge, Shashank Priya, Yu U. Wang, Jie-Fang Li, and D. Viehland

Department of Materials Science and Engineering, Virginia Tech,
306 Holden Hall, Blacksburg, VA 24061 USA
*E-mail: yaodongy@vt.edu

ABSTRACT

We have developed method for synthesis of single-grain two-phase $BaTiO_3$-$(Mn_{0.5}Zn_{0.5})Fe_2O_4$ (BTO-MZF) nanomaterials, via solid state reaction. The morphologies of synthesized nanocomposites were found to be dependent on external substrate orientation and the concentration of starting materials. After confirming with the XRD and TEM we found both BTO and MZF coexisted in one grain. Such a synthesis of two-phase perovskite/spinel single crystal grains with different grain geometries offers the potential to build new types of multi-ferroic materials.

Multi-functional devices require the bringing together of two, or even more, functional materials with dissimilar structures[1]. In order to achieve the optimization of two or more independent properties (for example, magnetization and polarization in magnetostrictive/ piezoelectric composites), we need to enhance coupling between the subsystems. Composites of dissimilar functionalities can have unique product tensor properties, which neither phase possesses individually: for example, magnetoelectricity (ME).[2,3]

One good example of magnetoelectric material to achieve this goal is perovskite-spinel self-assembled epitaxial thin films. Using pulsed laser deposition, two dimensional epitaxially grown self-assembled nano-composite films have been deposited onto various substrates.[4] A number of studies on the synthetic means to improve their piezoelectric properties can be found in literature.[5,6] However, there remains an inherent disadvantage to such thin films in that the substrate limits the distortion[7]. If one could transfer 2D BTO thin film structures onto other nano-structures without substrates, such as simple self-assembly synthesis of single grain particles, then novel performance might be further enabled.[8,9]

To fabricate nano-composite grains, barium acetate, TiO_2, MnO, ZnO, Fe_2O_3, NaCl and NP-30 (nonylphenyl ether) were mixed with different molar ratios. After mixing uniformly, the mixture was annealed at 850°C for 5 hrs. Here, we choose two representative final compositions for our two phase materials which correspond to: (i) $(BaTiO_3)_2$-$Mn_{0.5}Zn_{0.5}Fe_2O_4$, designated as B2M1 or BTO_2-MZF_1; and (ii) $BaTiO_3$-$(Mn_{0.5}Zn_{0.5}Fe_2O_4)_2$, designated as B1M2 or BTO_1-MZF_2.

Evidence of a change in the shape of particles for these different compounds was obtained by SEM. In Figure 1, we show grain morphologies for (a, b) pure MZF; (c, d) B1M2; (e, f) B2M1; and (g, h) pure BTO. The grains of BTO-rich samples had a rod-like morphology with high aspect ratio and were approximately 200nm in diameter, whereas the grains of MZF-rich samples were octahedral-like and also approximately 200nm in size. Both of them are similar to pure BTO and pure MZF, respectively.

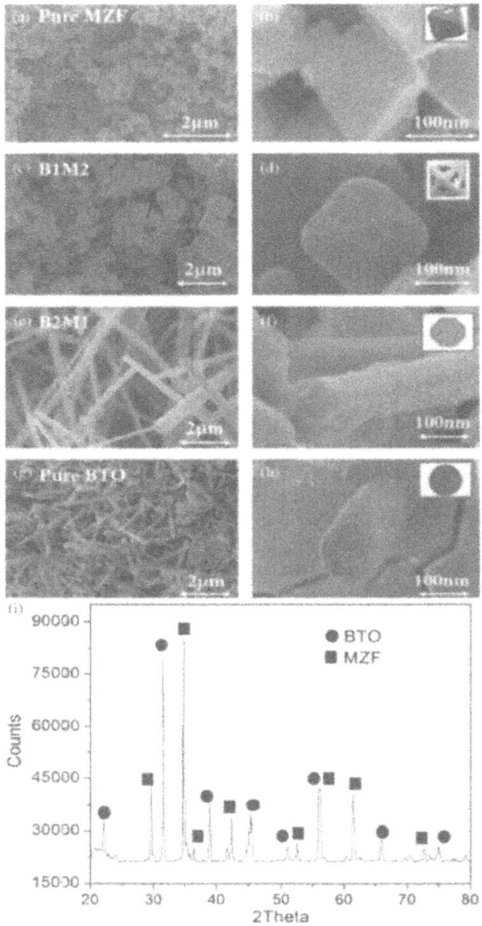

Figure 1. Scanning electron microscopy (SEM) images of (a, b) pure MZF nanoparticles, (c, d) BTO_1-MZF_2 nanoparticles, (e, f) BTO_2-MZF_1 nanorods and (g, h) pure BTO nanorods, (i) XRD pattern measured from a sample shown in (c). The insets show schematics of the particle structures (b & d) and the cross sectional geometries of the rods in images (f & h). [10]

Here, the images in left column of Fig. 1 were taken at low magnification and illustrate the ability to make such nano-particles in larger quantities by solid state reaction, which is an advantage if compared with PLD. The images in the right column were obtained under much higher magnification and better illustrate how the grain geometries varied with change in composition. An XRD pattern (Fig.1i) shows that the two phases coexisted. Clearly, introduction of MZF into BTO changes the cross-sectional geometry of the particles, and alters the nanoparticles and nanorods' facets.

From high-resolution TEM (HRTEM) images of the nanorods, as shown in Figure 2(a), we obtained a power spectrum from the area boxed-off in this lattice image, as shown in Figure 2(b). The power spectrum clearly reveals the presence of two sets of reflections. A HRTEM image was then taken at a higher magnification, as shown in Figure 2(c). We can see two regions, separated by a buffer zone that was approximately 1nm in thickness. A BTO phase region is identified whose inter-planar spacing was smaller than that of a MZF one. The lattice planes of two phases were coherent, but yet in part elastically relaxed by the buffer zone. We obtained power spectrums from both the BTO and MZF phase regions of the lattice image, as shown in the insets of Fig.2c. Unlike that in Figure 2b, these spectra contained a single set of reflections, demonstrating that they were single phase BTO and MZF regions.

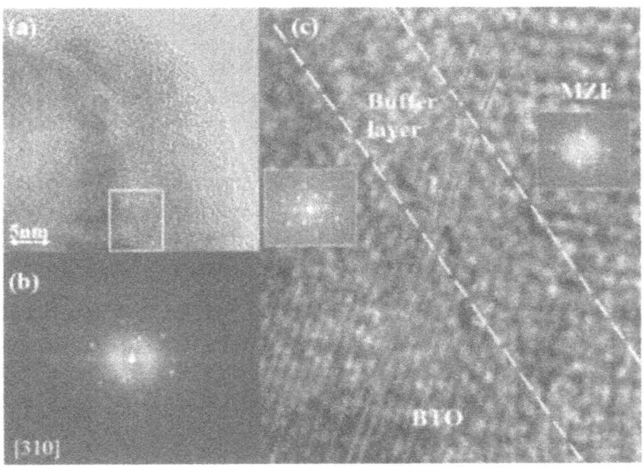

Figure 2. (a) High resolution TEM image of our B2M1 nanorod; (b) power spectrum taken from a selected area of lattice image given in (a); and (c) lattice image of higher resolution, demonstrating a buffer zone between phases, where the insets show power spectrums taken from the BTO and MZF areas, respectively.[10]

The magnetic hysteresis measured from these rod-like (B2M1) and particle (B1M2) samples were quite different. Comparisons of the M-H curves of B2M1, B1M2 sample revealed both higher residual magnetism and larger coercive fields, simply because there is more MZF phase in it than B2M1.

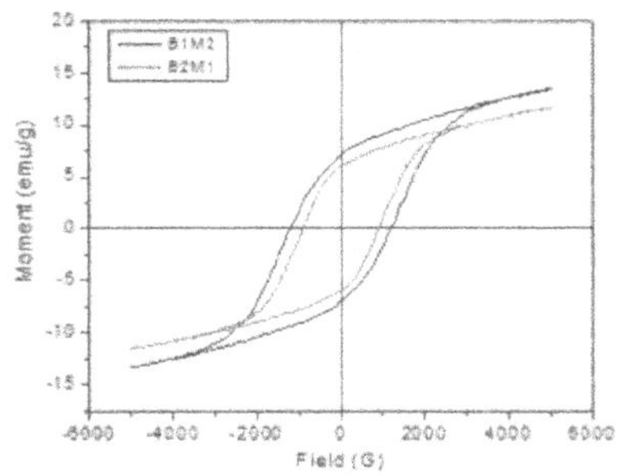

Figure 3. Magnetic hysteresis loops measured from B1M2 and B2M1 samples, respectively.

A shape and grain morphology controllable nanocomposite will also be easy to fit different needs in design of structures. Besides controlling the nano-scale geometries by changing the chemical prescription, we also controlled it via introducing different oriented substrates. Here, we only choose the BTO-rich prescription and $SrTiO_3$ (STO) single crystal substrates for different orientations as a demonstration. The experimental process was quite similar as previously described: after mixing the starting materials uniformly, they were annealed at 850°C for 2 hrs on STO single crystal substrates. (001), (011) and (111) oriented substrates were used here. A short time ultrasonic clean process was used to remove the reaction leftover.

The control mechanism was expected to occur by heteroepitaxy, and develop into facets through atomic rearrangements on a single crystal substrate with different orientations. [11] We expected that the surface energy would change with orientation of the substrates, due to variation in atomic arrangements. The facets and the final shape of nano-composites can be defined by the anisotropy of the surface-energy and can be predicted by Wulff construction[12]. Thus, the final product minimized the surface and interfacial energies between the individual components and substrates, and between the components[13].

Different geometries of nanoparticles were found for three differently oriented substrates, as clearly shown in Figure 4. On (111) substrates, the particle morphology was triangular, similar to BTO perovskite when viewed along [111] with (100), (001) and (010) facets. However, in our B2M1 sample, three extra facets appear on the particles that were not present on BTO grains. These are (101), (110) and (011) (see part a & b). A similar situation was found on (011) substrates, as illustrated in Figs. 4c-d. On (001) substrates, B2M1 particles were again found to have three additional facets: these are ($\bar{1}$12), (112) and (001) compared with pure MZF, as shown in Figs. 4e-f. Note that these additional facets are similar to those of MZF grains, which had a pyramidal morphology.

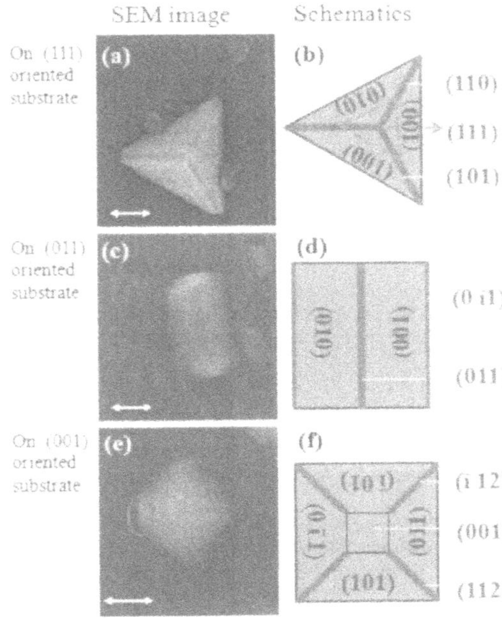

Figure 4. SEM images of B2M1 nanoparticle and its schematics on different substrates. Same row is on same oriented STO substrates from (111) to (001) orientation. Following is the detailed information of each image: (a, c and e) are B2M1 samples on (111), (011) and (001) substrates, respectively. All the scales are 200nm. (b, d and f) are schematics of marked facets of corresponding orientations in each row.

Based on the above discussion, we can identify two different approaches to control the nano-composite particle shape effectively. First adjust the concentration in the starting materials. In this case, if the composite has higher BTO phase concentration, then the shape will simply be similar to pure BTO nanorods; whereas if there is higher MZF phase concentration, then the final shape will be similar to pure MZF. Second, introduce differently orientated single crystal substrates. Using a Winterbottom and Wulff construction of the perovskite and spinel phases, one can verify the growth and morphology of our composite grains on these differently oriented substrates.[13, 14] A strong ME coupling is expected from this two phase coexisted nanocomposite.

ACKNOWLEDGEMENTS

Support of this work was provided by National Science Foundation under DMR-0757502. We also thank NCFL in VT for SEM, EDX and HRTEM works and useful discussion.

REFERENCES

1. Nan, C. W.; Bichurin, M. I.; Dong, S. X.; Viehland, D.; Srinivasan, G., Multiferroic magnetoelectric composites: Historical perspective, status, and future directions. *J. Appl. Phys.* **2008,** 103, 031101-35.
2. Priya, S.; Islam, R.; Dong, S. X.; Viehland, D., Recent advancements in magnetoelectric particulate and laminate composites. *J. Electroceram.* **2007,** 19, 147-164.
3. Zhai, J.; Xing, Z.; Dong, S.; Li, J.; Viehland, D., Magnetoelectric laminate composites: An overview. *J. Am. Ceram. Soc.* **2008,** 91, 351-358.
4. Yan, L.; Li, J. F.; Viehland, D., Structure of (001)-, (110)-, and (111)-oriented Pb(Fe1/2Nb1/2)O-3 epitaxial thin films on SrRuO3-buffered SrTiO3 substrates. *J. Mater. Res.* **2008,** 23, 663-670.
5. Qiao, L.; Bi, X. F., Domain configuration and phase transition for BaTiO3 thin films on tensile Si substrates. *J. Cryst. Growth* **2008,** 310, 5327-5330.
6. Yan, L.; Li, J. F.; Viehland, D., Deposition conditions and electrical properties of relaxor ferroelectric Pb(Fe1/2Nb1/2)O-3 thin films prepared by pulsed laser deposition. *J. Appl. Phys.* **2007,** 101, 104107-5.
7. Zheng, H.; Wang, J.; Lofland, S. E.; Ma, Z.; Mohaddes-Ardabili, L.; Zhao, T.; Salamanca-Riba, L.; Shinde, S. R.; Ogale, S. B.; Bai, F.; Viehland, D.; Jia, Y.; Schlom, D. G.; Wuttig, M.; Roytburd, A.; Ramesh, R., Multiferroic BaTiO3-CoFe2O4 nanostructures. *Science* **2004,** 303, 661-663.
8. Yang, Y. D.; Qu, L. T.; Dai, L. M.; Kang, T. S.; Durstock, M., Electrophoresis coating of titanium dioxide on aligned carbon nanotubes for controlled syntheses of photoelectronic nanomaterials. *Adv. Mater.* **2007,** 19, 1239-1243.
9. Guo, Z.; Lee, S. E.; Kim, H.; Park, S.; Hahn, H. T.; Karki, A. B.; Young, D. P., Fabrication, characterization and microwave properties of polyurethane nanocomposites reinforced with iron oxide and barium titanate nanoparticles. *Acta Mater.* **2009,** 51, 267-277.
10. Yang, Y. D.; Priya, S.; Li, J. F.; Viehland, D., Two-Phase Coexistence in Single-Grain BaTiO3-(Mn0.5Zn0.5)Fe2O4 Composites, Via Solid-State Reaction. *J. Am. Ceram. Soc.* **2009,** 92, 1552-1555.
11. Chen, H.; He, Y.; Shiflet, G. J.; Poon, S. J., Deformation-Induced Nanocrystal Formation in Shear Bands of Amorphous-Alloys. *Nature* **1994,** 367, 541-543.
12. Bodineau, T.; Ioffe, D.; Velenik, Y., Rigorous probabilistic analysis of equilibrium crystal shapes. *Journal of Mathematical Physics* **2000,** 41, 1033-1098.
13. Zheng, H. M.; Straub, F.; Zhan, Q.; Yang, P. L.; Hsieh, W. K.; Zavaliche, F.; Chu, Y. H.; Dahmen, U.; Ramesh, R., Self-assembled growth of BiFeO3-CoFe2O4 nanostructures. *Adv. Mater.* **2006,** 18, 2747-2752.
14. Yang, Y. D.; Priya, S.; Wang, Y. U.; Li, J. F.; Viehland, D., Solid-state synthesis of perovskite-spinel nanocomposites. *J. Mater. Chem.* **2009,** 19, 4998-5002.

Author Index

www.ingramcontent.com/pod-product-compliance
Ingram Content Group UK Ltd.
Pitfield, Milton Keynes, MK11 3LW, UK
UKHW020242290926
15470UKWH00002B/684

9 780470 927298